——— SECOND EDITION ———

PROBABILITY METHODS FOR COST UNCERTAINTY ANALYSIS

A Systems Engineering Perspective

SECOND EDITION

PROBABILITY METHODS FOR COST UNCERTAINTY ANALYSIS

A Systems Engineering Perspective

Paul R. Garvey
Stephen A. Book
Raymond P. Covert

CRC Press
Taylor & Francis Group
Boca Raton   London   New York

CRC Press is an imprint of the
Taylor & Francis Group, an **informa** business

A CHAPMAN & HALL BOOK

CRC Press
Taylor & Francis Group
6000 Broken Sound Parkway NW, Suite 300
Boca Raton, FL 33487-2742

© 2016 by Taylor & Francis Group, LLC
CRC Press is an imprint of Taylor & Francis Group, an Informa business

No claim to original U.S. Government works

Printed on acid-free paper
Version Date: 20151012

International Standard Book Number-13: 978-1-4822-1975-3 (Hardback)

Visit the Taylor & Francis Web site at
http://www.taylorandfrancis.com

and the CRC Press Web site at
http://www.crcpress.com

To my wife, Maura, and my daughters, Alana and Kirsten. I also dedicate this book to the memory of my daughter, Katrina, and my parents, Eva and Ralph.

Ad Majorem Dei Gloriam
Paul R. Garvey

To Dr. Stephen and Ruth Book.
Raymond P. Covert

The family of Stephen Book is extremely indebted to Paul R. Garvey and Raymond P. Covert for enabling Stephen's research to be published in Section II of this book. Apart from spending time with his family, working on research was Stephen's favorite and consuming activity. His mind was always engaged in solving the next problem, debunking the latest myth, and saving taxpayers from inefficiencies in the next program.

We hope the incorporation of Stephen's major papers throughout this book will continue the conversation and elicit new results and progress in cost risk analysis. It was a subject that challenged him, extended him, and brought out the best of his far-ranging intelligence. We know that Stephen was personally very proud and honored to be associated with all his colleagues in the systems engineering cost analysis community. It is our sincere hope the professional legacy he leaves will be as inspiring to you as his personal legacy is to us.

Ruth L. Book
Robert & Mandy Book
Lewis & Layla Book
Alexander & Elizabeth Book Kratz
Richard & Victoria Book Lupia
Lev & Jacqueline Book Novikov

Contents

Section II   Practical Considerations and Applications

Preface to the Second Edition

Acquiring today's systems is more sophisticated and complex than ever before. Increasingly, systems are engineered by bringing together many separate systems, which, as a whole, provide a capability otherwise not possible. Systems are now richly connected. They involve and evolve webs of users, technologies, and systems-of-systems through environments that offer cross-boundary access to a wide variety of resources and information repositories. Today's systems create value by delivering capabilities over time that meet user needs for increased agility, robustness, and scalability. System architectures must be open to allow the insertion of innovation that advances the efficacies of capabilities and services to users.

Many systems no longer physically exist within well-defined boundaries. They are increasingly ubiquitous and operate as an enterprise of technologies and cooperating entities in a dynamic that can behave in unpredictable ways. Pervasive with these challenges are economic and budgetary realities that necessitate greater accuracy in the estimated life cycle costs and cost risks of acquiring these systems.

Today, systems engineering is more than designing, developing, and bringing together technologies to work as a whole. Designs must be adaptable to change, flexible to meet user needs, and resource-managed. They must be balanced with respect to performance and affordability goals while being continuously risk-managed throughout a system's life cycle. Systems engineers must also understand the social, political, and economic environments within which a system operates. These factors can significantly influence risk, affordability, design options, and investment decisions.

In systems engineering, costs are estimated to reveal the economic significance of technical and programmatic choices that guide procuring a system that is affordable, cost-effective, and risk-managed. Identifying risks enables decision-makers to develop, execute, and monitor management actions based on the knowledge of potential cost consequences of inactions. Together, cost and cost uncertainty analyses are undertaken to address the paramount considerations of affordability, cost-effectiveness, and risk.

The mathematics of cost uncertainty analysis can be advanced utilizing the concepts of correlation, probability distributions, and means and variances. Today's cost analysts must be grounded in the underlying theory of this subject and convey their analysis findings clearly and concisely to audiences with broad backgrounds. Recognizing this, the second edition of this book is presented in two sections. Section I is unchanged from the first edition. It contains the original seven chapters on the underlying theory of cost uncertainty analysis. Section II is focused on the application of theory to problems encountered in practice. Section II presents the following chapters:

Chapter 8 provides a review of elementary concepts and key terms from Chapters 1 through 7. This includes a review on the scope of cost uncertainty analysis (what is captured, what is not captured), what it means to present and interpret cost as a probability distribution, and insights the analysis brings to decision-makers.

Chapter 9 discusses the importance of correlation as a critical consideration in cost uncertainty analysis. Shown throughout the preceding chapters, correlation can have a significant effect on the measure of a program's cost risk. This chapter presents several approaches to capture and incorporate correlation in cost uncertainty analyses. Guidelines on when one approach is preferred over another are offered.

Advances in computing technologies and mathematical methods have made regression a desirable approach for building statistical cost estimating models. These models contain features that readily incorporate into cost uncertainty analyses. Chapter 10 focuses on cost estimating models and the use of statistical regression methods to develop them. Chapter 10 describes classical statistical regression techniques and presents the general error regression method (GERM)—a major advance in practice and technique.

Chapter 11 introduces a phenomenon associated with producing an item over and over again. The phenomenon is called cost improvement. It refers to a lessening in the cost of an item produced in large quantities in the same way over a period of time. This chapter presents two main topics associated with the cost improvement in an item's recurring production cost. The first describes the phenomenon of cost improvement and methods to measure and mitigate its effects on the uncertainty in production cost estimates. The second illustrates how GERM can be applied to build cost estimating relationships of an item's recurring production costs in the presence of cost improvement effects.

Chapter 12 presents the last formal method for cost uncertainty analysis discussed in this book. Called the enhanced scenario-based method (eSBM), it was developed from a need in the cost analysis community to simplify the aspects of probability-based approaches. This chapter describes eSBM, identifies key features that distinguish it from other methods, and provides illustrative examples.

Chapter 13 provides recommended practices and considerations when performing cost uncertainty analyses. They reflect the authors' insights and experiences in developing, refining, and applying many of the techniques presented in this book.

Chapter 14 lists the major technical works of Dr. Stephen A. Book that advanced cost risk analysis theory and practice. The chapter is organized into works that were formally published in professional journals and those that were delivered as briefings in various conferences and technical gatherings.

This second edition is a memorial to the works of Dr. Stephen A. Book (1941–2012). Dr. Book was a mathematician and a world-renowned cost analyst whose innumerable contributions to this subject made it reachable and practical to cost analysts with a variety of academic and professional backgrounds. Dr. Book was beginning to work on this edition when he unexpectedly passed away. Despite this loss, the authors have incorporated Dr. Book's major works throughout the Section II chapters and in Appendix E. In keeping with his great witticisms, we close this preface with the following quote:

> It's not what you don't know that hurts you—it's what you *do* know that isn't true.
>
> **Dr. Stephen A. Book (1995)**

Paul R. Garvey
Raymond P. Covert

Preface to the First Edition

Cost is a driving consideration in decisions that determine how systems are developed, produced, and sustained. Critical to these decisions is understanding how uncertainty affects a system's cost. The process of identifying, measuring, and interpreting these effects is known as *cost uncertainty analysis*. Used early, cost uncertainty analysis can expose potentially crippling areas of risk in systems. This provides managers time to define and implement corrective strategies. Moreover, the analysis brings realism to technical and managerial decisions that define a system's overall engineering strategy. In *Juan De Mairena* (1943), Antonio Machado wrote "All uncertainty is fruitful ... so long as it is accompanied by the wish to understand." In the same way are insights gleaned from cost uncertainty analysis fruitful— provided they, too, are accompanied by the wish to understand and the will to take action.

Since the 1950s a substantial body of scholarship on this subject has evolved. Published material appears in numerous industry and government technical reports, symposia proceedings, and professional journals. Despite this, there is a need in the systems engineering community to synthesize prior scholarship and relate it to advances in technique and problem sophistication. This book addresses that need. It is a reference for systems engineers, cost engineers, management scientists, and operations research analysts. It is also a text for students of these disciplines.

As a text, this book is appropriate for an upper-level undergraduate (or graduate-level) course on the application of probability methods to cost engineering and analysis problems. It is assumed readers have a solid foundation in differential and integral calculus. An introductory background in probability theory, as well as systems and cost engineering, is helpful; however, the important concepts are developed as needed. A rich set of theoretical and applied exercises accompany each chapter.

Throughout the book, detailed discussions on issues associated with cost uncertainty analysis are given. This includes the treatment of correlation between the cost of various system elements, how to present the analysis to decision-makers, and the use of bivariate probability distributions to capture the joint interactions between cost and schedule. Analytical techniques from probability theory are stressed, along with the Monte Carlo simulation method. Numerous examples and case discussions are provided to illustrate the practical application of theoretical concepts. The numerical precision shown in some of the book's examples and case discussions is intended only for pedagogical purposes. In practice, analysts and engineers must always choose the level of precision appropriate to the nature of the problem being addressed.

Chapter 1 presents a general discussion of uncertainty and the role of probability in cost engineering and analysis problems. A perspective on the rich history of cost uncertainty analysis is provided. Readers are introduced to the importance of presenting the cost of a future system as a probability distribution.

Chapter 2 is an introduction to probability theory. Topics include the fundamental axioms and properties of probability. These topics are essential to understanding the terminology, technical development, and application of cost uncertainty analysis methods.

Chapter 3 presents the theory of expectation, moments of random variables, and probability inequalities. Examples derived from systems engineering projects illustrate key concepts.

Chapter 4 discusses modeling cost uncertainty by the probability formalism. A family of continuous univariate probability distributions, used frequently in cost uncertainty analysis, is fully described. A context for applying each distribution is also presented.

Chapter 5 introduces joint probability distributions, functions of random variables, and the central limit theorem. The application of these concepts to cost uncertainty analysis problems is emphasized. In addition, distributions are developed for a general form of the software cost-schedule model. The chapter concludes with a discussion of the Mellin transform, a useful (but little applied) method for working with cost functions that are products, or quotients, of two or more random variables.

Chapter 6 presents specific techniques for quantifying uncertainty in the cost of a future system. The reader is shown how methods from the preceding chapters combine to produce a probability distribution of a system's total cost. This is done from a work breakdown structure perspective. Case studies derived from systems engineering projects provide the application context.

Chapter 7 extends the discussion in Chapter 6 by presenting a family of joint probability distributions for cost and schedule. This family consists of the classical bivariate normal, the bivariate normal-lognormal, and the bivariate lognormal distributions; the latter two distributions are rarely discussed in the traditional literature. Examples are given to show the use of these distributions in a cost analysis context.

The book concludes with a summary of recommended practices and modeling techniques. They come from the author's experience and many years of collaboration with colleagues in industry, government, and academia.

The author gratefully acknowledges a number of distinguished engineers, scientists, and professors who contributed to this book. Their encouragement, enthusiasm, and insights have been instrumental in bringing about this work.

- *Stephen A. Book*: Distinguished engineer, The Aerospace Corporation, Los Angeles, California. A long-time professional colleague, Dr. Book peer-reviewed the author's major technical papers, some of which became chapters in this book. In addition, he independently reviewed and commented on many of the book's chapters as they evolved over the writing period.

- *Philip H. Young*: Director of research, Lori Associates, Los Angeles, California, and formerly of The Aerospace Corporation, conducted a detailed review of selected areas in this book. Also a long-time professional colleague, Philip Young shared with the author his formulas for the moments of the trapezoidal distribution (presented in Chapters 4 and 5), as well as a derivation of the correlation function of the bivariate normal-lognormal distribution. This derivation is provided as Theorem B.1 in Appendix B.

- *Nancy E. Rallis*: Associate professor of mathematics, Boston College, led the book's academic review. For two years, Professor Rallis studied the entire text from a theoretical and computational perspective. Her years of experience as a statistical consultant and cost analyst at the NASA Goddard Spaceflight Center, TRW Inc., and the Jet Propulsion Laboratory (California Institute of Technology) brought a wealth of insights that greatly enhanced this book. *Sarah E. Quebec*, a graduate mathematics student at Boston College, assisted Professor Rallis' review. I am grateful for her diligence in checking the many examples and case discussions.

- *Wendell P. Simpson III* (Major, USAF-Ret) and *Stephen A. Giuliano* (Lieutenant Colonel, USAF-Ret): Assistant professors, United States Air Force Institute of Technology. Professors Simpson and Giuliano developed and taught the school's first graduate course on cost risk analysis. The course used early drafts of the manuscript as required reading. Their comments on the manuscript, as well as those from their students, contributed significantly to the book's content and presentation style.

- Colleagues at The MITRE Corporation...

     Chien-Ching Cho: Principal staff, Economic and Decision Analysis Center. A long-time professional colleague, I am grateful to Dr. Cho for many years of technical discussions on theoretical aspects of this subject. I particularly appreciate his independent review of Case Discussion 6.2 and his commentary on Monte Carlo simulation, presented in Chapter 6.

     Barbara E. Wolfinger: While a group leader in the Economic and Decision Analysis Center, Ms. Wolfinger reviewed original drafts of Chapters 1 and 2. A creative practitioner of cost uncertainty analysis, her experiences and analytical insights were highly valued, particularly in the early stages of this work.

     Neal D. Hulkower: While a department head in the Economic and Decision Analysis Center, Dr. Hulkower reviewed a number of the author's technical papers when they were early drafts. A veteran cost analyst, his leadership on the necessity of presenting a system's future cost as a probability distribution fostered the award-winning research contained in this book.

William P. Hutzler: While director of the Economic and Decision Analysis Center, Dr. Hutzler provided the senior managerial review and leadership needed to bring the manuscript into the public domain. His enthusiasm and encouragement for this work will always be gratefully appreciated.

Francis M. Dello Russo and *John A. Vitkevich, Jr.*: Francis Dello Russo (department head, Economic and Decision Analysis Center) and John Vitkevich (lead staff, Economic and Decision Analysis Center) reviewed the book's first case discussion (Chapter 3). From an engineering economics perspective, they provided valuable commentary on issues associated with cost-volume-profit analyses.

Hank A. Neimeier: Principal staff, Economic and Decision Analysis Center. Hank Neimeier provided a careful review of the Mellin transform method (Chapter 5) and independently checked the associated examples. His expertise in mathematical modeling provided a valuable context for the application of this method to cost engineering and analysis problems.

Albert R. Paradis: Lead staff, Airspace Management and Navigation. Dr. Paradis reviewed an early version of the manuscript. His comments are highly valued. They helped fine-tune the explanation of a number of important and subtle concepts in probability theory.

Raymond L. Fales: A long-time professional colleague, Dr. Fales introduced the author to cost uncertainty analysis. He was among the early practitioners of analytical methods at MITRE and a mentor to many technical staff in the Economic and Decision Analysis Center.

Ralph C. Graves: A gifted and insightful systems engineer, Ralph Graves and the author worked jointly on numerous cost studies for the United States Air Force. During these studies, he introduced the author to Monte Carlo simulation (Chapter 6) as a practical approach for quantifying cost uncertainty.

The author also appreciates the staff at Marcel Dekker, Inc. for their diligence, professionalism, and enthusiasm for this work. Many thanks to Graham Garratt (executive vice president), Maria Allegra (acquisitions editor and manager), Joseph Stubenrauch (production editor), and Regina Efimchik (marketing and promotions).

Paul R. Garvey (2000)

Reserved Notation

Reserved notation used in this book:

FY	Fiscal year
$K	Dollars Thousand
$M	Dollars Million
SM	Staff Months
Eff	Effort for an activity (SM)
Eff_{SysEng}	Systems engineering effort (SM)
$\mathit{Eff}_{SysTest}$	System test effort (SM)
Eff_{SW}	Software development effort (SM)
I	Number of delivered source instructions (DSI)
P_r	Software development productivity rate (DSI/SM)
T_{SW}	Software development schedule (months)
$Cost_{Pgm},$	Total cost of a program or system as represented by its
$Cost_{Sys}, Cost_{WBS}$	work breakdown structure (WBS)

Section I

Theory and Foundations

1

Uncertainty and the Role of Probability in Cost Analysis

1.1 Introduction and Historical Perspective

This book presents methods for quantifying the cost impacts of uncertainty in the engineering of systems. The term "systems" is used in this book to include physical systems (those that occupy physical space [Blanchard and Fabrycky 1990]) and today's globally networked systems that enable communication and information exchanges to uncountably many users.

Systems engineering is a process that encompasses the scientific and engineering efforts needed to develop, produce, and sustain systems. Systems engineering is a highly complex technical and management undertaking. Integrating custom equipment with commercial products, designing external system interfaces, achieving user requirements, and meeting aggressive schedules while keeping within cost are among the many challenges faced in managing a systems engineering project.

When the cost of a future system* is considered, decision-makers often ask, "What is the chance its cost will exceed a particular amount?" "How much could cost overrun?" "What are the uncertainties and how do they drive cost?" Cost uncertainty analysis provides decision-makers insight into these and related questions. In general, *cost uncertainty analysis* is a process of quantifying the cost impacts of uncertainties associated with a system's technical definition and cost estimation methodologies.

Throughout a system's life-cycle, cost uncertainty analysis provides motivation and structure for the vigorous management of risk. When appropriately communicated to decision-makers, the insights produced by the analysis directs management's attention to critical program risk drivers. This enables risk mitigation strategies to be defined and implemented in a timely and cost-effective manner.

---

* This includes existing systems planned for modernization, consolidation, or re-engineering.

Cost uncertainty analysis has its genesis in the field known as military systems analysis (Hitch 1955), founded in the 1950s at the RAND Corporation. Shortly after World War II, military systems analysis evolved as a way to aid defense planners with long-range decisions on force structure, force composition, and future theaters of operation. Cost became a critical consideration in military systems analysis models and decision criteria. However, cost estimates of future military systems, particularly in the early planning phases, were often significantly lower than the actual cost or an estimate developed at a later phase. In the book *Cost Considerations in Systems Analysis*, Fisher (1971) attributes this difference to the presence of uncertainty—specifically, cost estimation uncertainty and requirements uncertainty.

Cost estimation uncertainty can originate from inaccuracies in cost-schedule estimation models, from the misuse (or misinterpretation) of cost-schedule data, or from misapplied cost-schedule estimation methodologies. The economic uncertainties that influence the cost of technology, the labor force, or geopolitical policies further contribute to cost estimation uncertainty (Garvey 1996).

Requirements uncertainty can originate from changes in the system's mission objectives, from changes in performance requirements necessary to meet mission objectives, or from changes in the business or political landscapes that affect the need for the system. Requirements uncertainty most often results in changes to the system's specified hardware-software configuration, which is also known as the system's architecture.

Uncertainty is also present in elements that define a system's configuration (or architecture). This is referred to as *system definition uncertainty*. Examples include uncertainties in the amount of software to develop, the extent code from another system can be reused, the number of workstations to procure, or the delivered weight of an end-item (e.g., a satellite) (Garvey 1996).

The early literature on cost uncertainty analysis concentrated on defining the sources, scope, and types of uncertainties that impacted the cost of future systems. Technical papers published in the period between 1955 and 1962 were not explicitly focused on establishing and applying formal methods to quantify cost uncertainty. However, by the mid-1960s a body of techniques began to emerge. An objective of this book is to discuss these techniques, present advances in methodology, and illustrate how these methods apply from a systems engineering perspective.

1.2 Problem Space

In systems engineering, three types of uncertainties must be considered. Described in the preceding section they are cost estimation uncertainty, requirements uncertainty, and system definition uncertainty. Figure 1.1 (Garvey 1996) illustrates how these uncertainties are related.

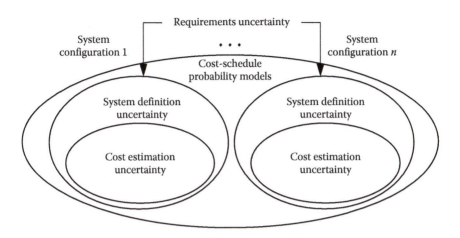

FIGURE 1.1
Types of uncertainty captured by cost-schedule probability models.

The n-system configurations shown are in response to requirements uncertainty. For a given system configuration, cost-schedule probability models (as described in this book) capture *only* system definition and cost estimation uncertainties. They provide probability-based assessments of a system's cost and schedule for that system configuration. When requirements uncertainty necessitates defining an entirely new configuration, a new cost-schedule probability model is likely to be needed. The new model must be tuned to capture the system definition and cost estimation uncertainties specific to the new configuration.

1.3 Presenting Cost as a Probability Distribution

Cost is an uncertain quantity. It is highly sensitive to many conditions and assumptions that change frequently across a system's life-cycle. Examining the change in cost subject to varying certain conditions (while holding others constant) is known as *sensitivity analysis*. In a series of lectures to the United States Air Force, Fisher (1962) emphasized the importance of sensitivity analysis as a way to isolate cost drivers. He considered sensitivity analysis to be "...a prime characteristic or objective in the cost analysis of advanced systems and/or force structure proposals." Although sensitivity analysis can isolate elements of a system that drive its cost, it is a deterministic procedure defined by a postulated set of "what-if" scenarios. Sensitivity analysis alone does not offer decision-makers insight into the question, "What is the chance of exceeding a particular cost in the range of possible costs?" A probability distribution is a way to address this question. Simply stated, a *probability distribution* is a mathematical rule associating a probability α to each possible outcome, or event of interest.

There are two ways to present a probability distribution. It can be shown as a probability density or as a cumulative probability distribution. Figure 1.2 presents an illustration of this concept from a cost perspective.

In Figure 1.2, the range of possible values for *Cost* is given by the interval $a \leq x \leq b$. The probability *Cost* will not exceed a value $x = c$ given by α_c. In Figure 1.2a, this probability is the area under $f(x)$ between $x = a$ and $x = c$. In Figure 1.2b, this probability is given by $F(c)$.

To develop a cost probability distribution, methods from probability theory were needed. Some of the earliest applications of probability theory to model cost uncertainty took place in the mid-1960s at the MITRE and RAND Corporations. In 1965, Steven Sobel [MITRE] published *A Computerized Technique to Express Uncertainty in Advanced System Cost Estimates* (Sobel 1965). It was among the earliest works on modeling cost uncertainty by the probability formalism. Sobel pioneered using the method of moments technique to develop a probability distribution of a system's total cost.

Complementary to Sobel's analytical approach, in 1966 Paul F. Dienemann [RAND] published *Estimating Cost Uncertainty Using Monte Carlo Techniques* (Dinemann 1966). The methodology applied Monte Carlo simulation, developed by operations analysts during World War II, to quantify the impacts of uncertainty on total system cost. With the advent of high-speed computers, Monte Carlo simulation grew in popularity and remains a primary approach for generating cost probability distributions. A discussion of Monte Carlo simulation is presented in Chapter 6.

An overview of the cost uncertainty analysis process is shown in Figure 1.3. The variables $X_1, X_2, X_3, \ldots, X_n$ are the costs of the n work breakdown

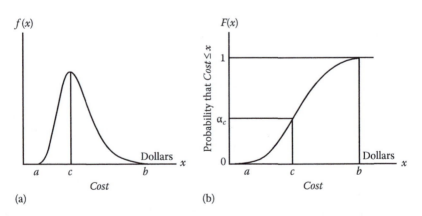

(a) (b)

FIGURE 1.2
Illustrative probability distributions. (a) Probability density. (b) Cumulative probability distribution.

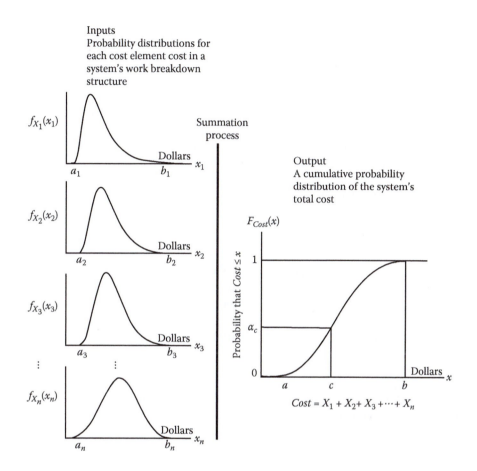

FIGURE 1.3
Cost uncertainty analysis process.

structure (WBS)* cost elements that comprise the system. For instance, X_1 might represent the cost of the system's prime mission hardware and software; X_2 might represent the cost of the system's systems engineering and program management (SEPM); X_3 might represent the cost of the system's test and evaluation. When specific values for these variables are uncertain, we can treat them as *random variables*. Probability distributions are developed for $X_1, X_2, X_3, \ldots, X_n$, which associate probabilities to their possible values. Such distributions are illustrated on the left-side of Figure 1.3. The random variables $X_1, X_2, X_3, \ldots, X_n$ are summed to produce an overall probability distribution of the system's total cost, shown on the right-side of Figure 1.3.

---

* A full discussion of the work breakdown structure is presented in Chapter 6.

The "input" part of this process has many subjective aspects. Probability distributions for $X_1, X_2, X_3, \ldots, X_n$ are either specified directly or they are generated. Direct specification relies on expert judgment to characterize a distribution's shape. The probability density is the usual way to make this characterization.

Generated distributions have shapes that are produced from a mathematical process. This is illustrated in the following discussion.

Suppose X_2 represents the cost of a system's systems engineering and program management (SEPM). Furthermore, suppose the cost of SEPM is derived as a function of three random variables* *Staff*, *PrgmSched*, and *LaborRate* as follows:

$$X_2 = Staff \cdot PrgmSched \cdot LaborRate \tag{1.1}$$

Suppose the engineering team assessed ranges of possible (feasible) values for these variables and directly specified the shapes of their probability distributions, as shown in Figure 1.4. Combining their distributions according to the rules of probability theory generates an overall distribution for X_2, which is the cost of SEPM. In this case, we say the probability distribution of X_2 has been generated by a mathematical process. Figure 1.4 illustrates this discussion.

As shown in Figure 1.4, it is good practice to reserve the direct specification of distributions to their lowest level variables in a cost equation (e.g., Equation 1.1). Often, expert judgment about the shapes and ranges of distributions are best at this level. Furthermore, this "specification" approach

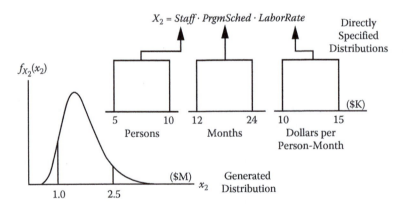

FIGURE 1.4
The specification of probability distributions.

---

* *Staff* (Persons), *PrgmSched* (Months), *LaborRate* ($/Person-Month).

structures the overall analysis in a way that specific "cost-risk-driving" variables can be revealed. Identifying these variables and quantifying how they affect a system's cost are critical findings to communicate to decision-makers.

A term conventional to cost engineering and analysis is point estimate. The *point estimate* of a variable whose value is uncertain is a single value for the variable in its range of possible values. From a mathematical perspective, the point estimate is simply one value among those that are feasible. In practice, a point estimate is established by an analyst (using appropriate cost analysis methods) prior to an assessment of other possible values. It provides an "anchor" (i.e., a reference point) around which other possible values are assessed or generated. This is illustrated in Figure 1.5.

In cost uncertainty analysis it is common to see more probability density to the right of a point estimate than to its left; this is seen in the generated distribution in Figure 1.5. Although this is a common occurrence, the point estimate *can* fall anywhere along the variable's probability distribution; it is just one value among those that are feasible.

Suppose a system's total cost is given by

$$Cost = X_1 + X_2 + X_3 + \cdots + X_n$$

where the random variables $X_1, X_2, X_3, \ldots, X_n$ are the costs of the system's n WBS cost elements. Suppose point estimates are developed for each X_i $(i = 1, \ldots, n)$. Their sum is the *point estimate of the cost of the system*. Let this sum be denoted by $x_{PE_{Cost}}$, where

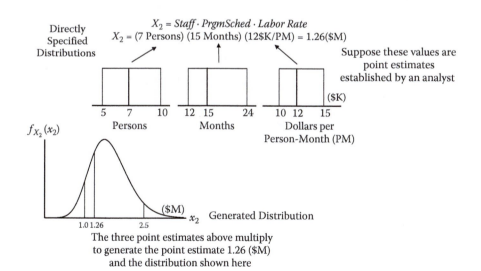

FIGURE 1.5
Point estimates: an illustration.

$$x_{PE_{Cost}} = x_{1PE_{X_1}} + x_{2PE_{X_2}} + x_{3PE_{X_3}} + \cdots + x_{nPE_{X_n}} \qquad (1.2)$$

and $x_{iPE_{X_i}}$ $(i = 1, \ldots, n)$ is the point estimate of X_i. Computing $x_{PE_{Cost}}$ according to Equation 1.2 is known among practitioners as the "roll-up" procedure.*

In cost engineering and analysis, it is traditional to consider $x_{PE_{Cost}}$ a value for *Cost* that contains *no reserve dollars*. As a point estimate, $x_{PE_{Cost}}$ provides the "anchor" from which to choose a value for *Cost* that *contains reserve dollars*. Decision-makers trade-off between $x_{PE_{Cost}}$ and the amount of reserve dollars to add to $x_{PE_{Cost}}$, such that the value of *Cost* determined by the expression $[x_{PE_{Cost}} + (\text{reserve dollars})]$ has an acceptable probability of *not being exceeded*. Figure 1.6 illustrates this discussion.

In Figure 1.6, suppose the point estimate of a system's cost is 100 ($M); that is, $x_{PE_{Cost}} = 100$. This value of *Cost* has just over a 30% probability of not being exceeded. A reserve of 20 ($M) added to $x_{PE_{Cost}}$ is associated with a value of *Cost* that has a 67% probability of not being exceeded. A reserve of 40 ($M) added to $x_{PE_{Cost}}$ is associated with a value of *Cost* that has just over a 90% probability of not being exceeded.

It is possible for $x_{PE_{Cost}}$ to fall at a high confidence level on its associated distribution function. Such a circumstance may warrant the addition of no reserve dollars; it suggests there is a good chance for *Cost* to actually be lower than perhaps anticipated. However, it may also flag a situation where cost

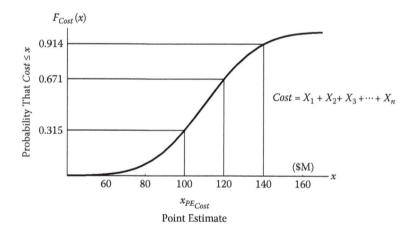

FIGURE 1.6
A cumulative probability distribution of system cost.

---

* From a probability perspective there are important subtleties associated with the roll-up procedure. These subtleties are illustrated in Case Discussion 5.1.

reserve was built, a priori, into the point estimate of each WBS cost element. These reserve dollars would be included in the "roll-up" of the individual point estimates. This result can make $x_{PE_{Cost}}$ hard to interpret, particularly if trade-off studies are needed. In practice, it is recommended keeping $x_{PE_{Cost}}$ "clean" from reserve dollars. This provides analysts and decision-makers an anchor point that is "cost reserve-neutral"—one where the trade-off between cost reserve and a desired level of confidence can be readily understood for various alternatives (or options) under consideration.

1.4 Benefits of Cost Uncertainty Analysis

Cost uncertainty analysis provides decision-makers many benefits and important insights. These, including the following:

> *Establishing a cost and schedule risk baseline*: Baseline probability distributions of a system's cost and schedule can be developed for a given system configuration, acquisition strategy, and cost-schedule estimation approach. This baseline provides decision-makers visibility into potentially high-payoff areas for risk reduction initiatives. Baseline distributions assist in determining a system's cost and schedule that simultaneously have a specified probability of not being exceeded (Chapter 7). They can also provide decision-makers an assessment of the likelihood of achieving a budgeted (or proposed) cost and schedule, or cost for a given feasible schedule (Garvey 1996).
>
> *Determining cost reserve*: Cost uncertainty analysis provides a basis for determining cost reserve as a function of the uncertainties specific to a system. The analysis provides the direct link between the amount of cost reserve to recommend and the probability that a system's cost will not exceed a prescribed (or desired) magnitude. An analysis should be conducted to verify the recommended cost reserve covers fortuitous events (e.g., unplanned code growth, unplanned schedule delays) deemed possible by the system's engineering team (Garvey 1996).
>
> *Conducting risk reduction trade-off analyses*: Cost uncertainty analyses can be conducted to study the payoff of implementing risk reduction initiatives (e.g., rapid prototyping) on lessening a system's cost and schedule risks. Furthermore, families of probability distribution functions can be generated to compare the cost and cost risk impacts of alternative system requirements, schedule uncertainties, and competing system configurations or acquisition strategies (Garvey 1996).

The validity and meaningfulness of a cost uncertainty analysis relies on the engineering team's experience, judgment, and knowledge of the system's uncertainties. Formulating and documenting a supporting rationale, that summarizes the team's collective insights into these uncertainties, is *the critical part of the process*. Without a well-documented rationale, the credibility of the analysis can be easily questioned.

The details of the analysis methodology are important and should also be documented. The methodology *must be technically sound* and offer value-added problem structure, analyses, and insights otherwise not visible. Decisions that successfully eliminate uncertainty, or reduce it to acceptable levels, are ultimately driven by human judgment. This at best is aided by, not directed by, the methods presented in this book.

Exercises

1.1 State and define the three types of uncertainties that affect the cost of a systems engineering project. Give specific examples for each type.

1.2 Define, from a cost perspective, the term point estimate. How is the point estimate of a variable used to establish a range of other possible values? Explain what is meant by the "roll-up" procedure.

1.3 In the following figure, suppose the *point estimate* of a system's cost is 23.5 dollars million ($M) and assume the three values shown along the vertical axis are paired with the three values shown along the horizontal axis. How many reserve dollars are needed such that the value of *Cost* associated with that reserve has a 70% chance of *not being* exceeded? Similarly, what is the reserve needed such that the value of *Cost* has only a 5% chance of *being* exceeded?

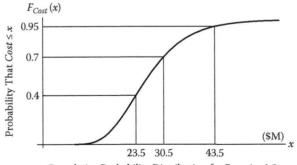

Cumulative Probability Distribution for Exercise 1.3

References

Blanchard, B. S. and W. J. Fabrycky. 1990. *Systems Engineering and Analysis*, 2nd edn. Englewood Cliffs, NJ: Prentice-Hall, Inc.

Dienemann, P. F. 1966. *Estimating Cost Uncertainty Using Monte Carlo Techniques*, RM-4854-PR. Santa Monica, CA: The RAND Corporation.

Fisher, G. H. 1962. *A Discussion of Uncertainty in Cost Analysis*, RM- 3071-PR. Santa Monica, CA: The RAND Corporation.

Fisher, G. H. 1971. *Cost Considerations in Systems Analysis*. New York: American Elsevier Publishing Company, Inc.

Garvey, P. R. 1996 (Spring). Modeling cost and schedule uncertainties—A work breakdown structure perspective. *Military Operations Research*, V2, 2(1), 37–43.

Hitch, C. J. 1955. *An Appreciation of Systems Analysis*, p.699. Santa Monica, CA: The RAND Corporation.

Sobel, S. 1965. *A Computerized Technique to Express Uncertainty in Advanced System Cost Estimates*, ESD-TR-65-79. Bedford, MA: The MITRE Corporation.

2

Concepts of Probability Theory

2.1 Introduction

Whether it is a storm's intensity, an arrival time, or the success of a financial decision, the words "probable" or "likely" have long been part of our language. Most people have a practical appreciation for the impact of chance on the occurrence of an event. In the last 300 years, probability theory has evolved to explain the nature of chance and how it may be studied.

Probability theory is the formal study of random events and random processes. Its origins trace back to seventeenth-century gambling problems. Games that involved playing cards, roulette wheels, and dice provided mathematicians a host of interesting problems. The solutions to many of these problems yielded the first principles of modern probability theory. Today, probability theory is of fundamental importance in science, engineering, and business.

2.2 Sample Spaces and Events

If a six-sided die* is tossed there are six possible outcomes for the number that appears on the upturned face. These outcomes can be listed as elements in the set $\{1, 2, 3, 4, 5, 6\}$. The set of all possible outcomes of an experiment is called the *sample space*, which we will denote by Ω. The individual outcomes of Ω are called *sample points*, which we will denote by ω.

A sample space can be finite, countably infinite, or uncountable. A *finite sample space* is a set that consists of a finite number of outcomes. The sample space for the toss of a die is finite. A *countably infinite sample space* is a set whose outcomes can be arranged in a one-to-one correspondence with the set of positive integers. An *uncountable sample space* is one that is infinite but not countable. For instance, suppose the sample space for the duration t (in hours) of an electronic device is $\Omega = \{t : 0 \le t < 2500\}$, then Ω is an

---

* Unless otherwise noted, dice are assumed in this book to be six-sided.

uncountable sample space; there are an infinite but not countable number of possible outcomes for t. Finite and countably infinite sample spaces are also known as *discrete sample spaces*. Uncountable sample spaces are known as *continuous sample spaces*.

An *event* is any subset of the sample space. An event is *simple* if it consists of exactly one outcome.* Simple events are also referred to as *elementary events* or *elementary outcomes*. An event is *compound* if it consists of more than one outcome. For instance, let B be the event an odd number appears and C be the event an even number appears in a single toss of a die. These are compound events, which may be expressed by the sets $B = \{1, 3, 5\}$ and $C = \{2, 4, 6\}$. Event B occurs if and only if one of the outcomes in B occurs; the same is true for event C.

As discussed, events can be represented by sets. New events can be constructed from given events according to the rules of set theory. The following presents a brief review of set theory concepts.

> *Union*: For any two events A and B of a sample space Ω, the new event $A \cup B$ (which reads A *union* B) consists of all outcomes either in A or in B or in both A and B. The event $A \cup B$ occurs if *either A or B occurs*. To illustrate the union of two events, consider the following: If A is the event an odd number appears in the toss of a die and B is the event an even number appears, then the event $A \cup B$ is the set $\{1, 2, 3, 4, 5, 6\}$, which is the sample space for this experiment.
>
> *Intersection*: For any two events A and B of a sample space Ω, the new event $A \cap B$ (which reads A *intersection* B) consists of all outcomes that are *both* in A and in B. The event $A \cap B$ occurs *only if both A and B occur*. To illustrate the intersection of two events, consider the following: If A is the event a six appears, in the toss of a die, B is the event an odd number appears, and C is the event an even number appears, then the event $A \cap C$ is the simple event $\{6\}$; on the other hand, the event $A \cap B$ contains no outcomes. Such an event is called a *null event*. The null event is traditionally denoted by $\emptyset$. In general, if $A \cap B = \emptyset$, we say events A and B are *mutually exclusive (disjoint)*. The intersection of two events A and B is sometimes written as AB, instead of $A \cap B$.
>
> *Complement*: The *complement* of event A, denoted by A^c, consists of all outcomes in the sample space Ω that are *not* in A. The event A^c occurs if and only if A does not occur. The following illustrates the complement of an event. If C is the event an even number appears in the toss of a die, then C^c is the event an odd number appears.
>
> *Subset*: Event A is said to be a *subset* of event B if all the outcomes in A are also contained in B. This is written as $A \subset B$.

---

* As we shall see, probabilities associated with simple events are sensitive to the nature of the sample space. If Ω is a *discrete sample space*, the probability of an event is completely determined by the probabilities of the simple events in Ω; however, if Ω is a *continuous sample space*, the probability associated with each simple event in Ω is zero. This will be discussed further in Chapter 3.

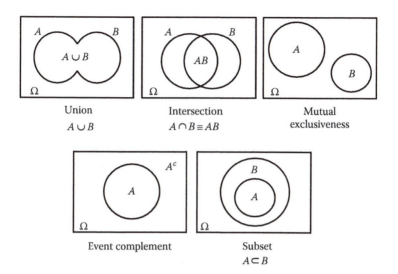

FIGURE 2.1
Venn diagrams for various event relationships.

Figure 2.1 illustrates these concepts in the form of Venn diagrams.

Operations involving the union and intersection of events follow the rules of set algebra. These rules are summarized below.

Identity laws	$A \cup \emptyset = A$	$A \cap \emptyset = \emptyset$
	$A \cup \Omega = \Omega$	$A \cap \Omega = A$
De Morgan's laws	$(A \cup B)^c = A^c \cap B^c$	$(A \cap B)^c = A^c \cup B^c$
Associative laws	$A \cup B \cup C = (A \cup B) \cup C = A \cup (B \cup C)$	
	$A \cap B \cap C = (A \cap B) \cap C = A \cap (B \cap C)$	
Distributive laws	$A \cup (B \cap C) = (A \cup B) \cap (A \cup C)$	
	$A \cap (B \cup C) = (A \cap B) \cup (A \cap C)$	
Commutative laws	$A \cup B = B \cup A$	$A \cap B = B \cap A$
Idempotency laws	$A \cup A = A$	$A \cap A = A$
Complementary laws	$A \cup A^c = \Omega$	$A \cap A^c = \emptyset$

2.3 Interpretations and Axioms of Probability

In the preceding discussion, the sample space for the toss of a die was given by $\Omega = \{1, 2, 3, 4, 5, 6\}$. If we *assume* the die is fair (which, unless otherwise noted, is assumed throughout this book) then any outcome in the sample space is as likely to appear as any other. Given this assumption, it is reasonable to conclude the proportion of time each outcome is expected to occur is $\frac{1}{6}$. Thus, the probability of each simple event in the sample space is

$$P(\{1\}) = P(\{2\}) = P(\{3\}) = P(\{4\}) = P(\{5\}) = P(\{6\}) = \frac{1}{6}$$

Similarly, suppose B is the event an odd number appears in a single toss of the die. This compound event is given by the set $B = \{1, 3, 5\}$. Since there are three ways event B can occur out of the six possible, we conclude the probability of event B is

$$P(B) = \frac{3}{6} = \frac{1}{2}$$

The following presents a view of probability known as the equally likely interpretation.

2.3.1 Equally Likely Interpretation

In this view, if a sample space Ω consists of a finite number of outcomes n, which are all equally likely to occur, then the probability of each simple event is $\frac{1}{n}$. If an event A consists of m of these n outcomes, then the probability of event A is

$$P(A) = \frac{m}{n} \tag{2.1}$$

In this equation, it is assumed the sample space consists of a *finite* number of outcomes and all outcomes are equally likely to occur. What if the sample space is uncountable? What if the sample space is finite but the outcomes are *not* equally likely?* In these situations, probability might be measured in terms of how frequently a particular outcome occurs when the experiment is repeatedly performed under identical conditions. This leads to a view of probability known as the frequency interpretation.

2.3.2 Frequency Interpretation

In this view, the probability of an event is the limiting proportion of time the event occurs in a set of n repetitions of the experiment. In particular, we write this as

$$P(A) = \lim_{n \to \infty} \frac{n(A)}{n}$$

---

* If a die is weighted in a particular way, then the outcomes of the toss are no longer considered fair, or equally likely.

where $n(A)$ is the number of times in n repetitions of the experiment the event A occurs. In this sense $P(A)$ is the limiting frequency of event A. Probabilities measured by the frequency interpretation are referred to as *objective probabilities*. There are many circumstances where it is appropriate to work with objective probabilities. However, there are limitations with this interpretation of probability. It restricts events to those that can be subjected to repeated trials conducted under *identical conditions*. Furthermore, it is not clear how many trials of an experiment are needed to obtain an event's limiting frequency.

2.3.3 Axiomatic Definition

In 1933, Russian mathematician A. N. Kolmogorov* presented a definition of probability in terms of three axioms. These axioms define probability in a way that encompasses the *equally likely and frequency interpretations* of probability. It is known as the axiomatic definition of probability. It is the view of probability adopted in this book. Under this definition it is assumed for each event A, in the sample space Ω, a real number $P(A)$ exists that denotes the probability of A. In accordance with Kolmogorov's axioms, a probability is simply a numerical value (measure) that satisfies the following:

Axiom 2.1 $0 \leq P(A) \leq 1$ for any event A in Ω

Axiom 2.2 $P(\Omega) = 1$

Axiom 2.3   For any sequence of mutually exclusive events[†] $A_1, A_2, \ldots$ defined on Ω

$$P\left(\bigcup_{i=1}^{\infty} A_i\right) = \sum_{i=1}^{\infty} P(A_i)$$

For any *finite sequence* of mutually exclusive events $A_1, A_2, \ldots,$ A_n defined on Ω

$$P\left(\bigcup_{i=1}^{n} A_i\right) = \sum_{i=1}^{n} P(A_i)$$

Axiom 2.1 states the probability of any event is a nonnegative number in the interval zero to unity. In Axiom 2.2, the sample space Ω is sometimes referred to as the *sure* or *certain event*; therefore, we have $P(\Omega)$ equal to unity. Axiom 2.3 states for any sequence of mutually exclusive events, the probability of at least one of these events occurring is the sum of the probabilities associated

---

* A. N. Kolmogorov, Grundbegriffe der Wahrscheinlichkeitsrechnung, *Ergeb. Mat. und ihrer Grenzg.*, vol. 2, no. 3, 1933. Translated into English by N. Morrison, *Foundations of the Theory of Probability*, New York (Chelsea), 1956 (Feller 1968).

† That is, $A_i \cap A_j = \emptyset$ for $i \neq j$.

with each event A_i. In Axiom 2.3, this sequence may also be finite. From these axioms come basic theorems of probability.

Theorem 2.1   *The probability event A occurs is one minus the probability it will not occur; that is, $P(A) = 1 - P(A^c)$.*

Proof.   From the complementary law $\Omega = A \cup A^c$. From Axiom 2.3 it follows that $P(\Omega) = P(A \cup A^c) = P(A) + P(A^c)$ since A and A^c are mutually exclusive events. From Axiom 2.2, we know that $P(\Omega) = 1$; therefore, $1 = P(A) + P(A^c)$ and the result $P(A) = 1 - P(A^c)$ follows.

Theorem 2.2   *The probability associated with the null event $\emptyset$ is zero*

$$P(\emptyset) = 0$$

Proof.   From Theorem 2.1 and Axiom 2.2

$$P(\emptyset) = 1 - P(\emptyset^c) = 1 - P(\Omega) = 1 - 1 = 0$$

Theorem 2.3   *If events A_1 and A_2 are mutually exclusive, then*

$$P(A_1 \cap A_2) \equiv P(A_1 A_2) = 0$$

Proof.   Since A_1 and A_2 are mutually exclusive, $A_1 \cap A_2 = \emptyset$. Thus, $P(A_1 \cap A_2) = P(\emptyset)$. From Theorem 2.2, $P(\emptyset) = 0$; therefore, $P(A_1 \cap A_2) = 0$.

Theorem 2.4   *For any two events A_1 and A_2*

$$P(A_1 \cup A_2) = P(A_1) + P(A_2) - P(A_1 \cap A_2)$$

Proof.   The event $A_1 \cup A_2$, shown in Figure 2.2, is written in terms of three mutually exclusive events, that is, $A_1 \cup A_2 = (A_1 A_2^c) \cup (A_1 A_2) \cup (A_1^c A_2)$. From Axiom 2.3, $P(A_1 \cup A_2) = P(A_1 A_2^c) + P(A_1 A_2) + P(A_1^c A_2)$.
From Figure 2.2, A_1 can be written in terms of mutually exclusive events; that is, $A_1 = (A_1 A_2^c) \cup (A_1 A_2)$; similarly $A_2 = (A_1^c A_2) \cup (A_1 A_2)$. From Axiom 2.3, it follows that $P(A_1) = P(A_1 A_2^c) + P(A_1 A_2)$ and $P(A_2) = P(A_1^c A_2) + P(A_1 A_2)$. Therefore, $P(A_1 \cup A_2)$ can be written as

$$P(A_1 \cup A_2) = P(A_1) - P(A_1 A_2) + P(A_1 A_2) + P(A_2) - P(A_1 A_2)$$

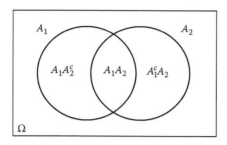

FIGURE 2.2
The partition of $A_1 \cup A_2$.

*It follows that**

$$P(A_1 \cup A_2) = P(A_1) + P(A_2) - P(A_1A_2) \blacklozenge$$

If A_1 and A_2 were mutually exclusive events, Theorem 2.4 simplifies to Axiom 2.3, that is, $P(A_1 \cup A_2) = P(A_1) + P(A_2)$ since $P(A_1A_2) \equiv P(A_1 \cap A_2) = P(\emptyset) = 0$.

Theorem 2.5   *If event A_1 is a subset of event A_2 then*

$$P(A_1) \leq P(A_2)$$

Proof.   Since A_1 is a subset of A_2, the event A_2 may be expressed as the union of two mutually exclusive events A_1 and $A_1^c A_2$. Refer to Figure 2.3. Since,

$$A_2 = A_1 \cup A_1^c A_2$$

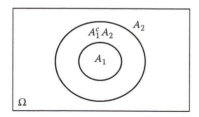

FIGURE 2.3
Event A_1 as a subset of event A_2.

---

* The symbol $\blacklozenge$ is reserved in this book to signal the completion of a proof, an example, or a case discussion.

from Axiom 2.3

$$P(A_2) = P(A_1) + P(A_1^c A_2)$$

Because $P(A_1^c A_2) \geq 0$, it follows that

$$P(A_1) \leq P(A_2) \blacklozenge$$

Example 2.1 *The sample space Ω for an experiment that consists of tossing two dice is given by the 36 possible outcomes listed in Table 2.1. The outcomes in Table 2.1 are given by the pairs (d_1, d_2),* which we assume are equally likely. Let A, B, C, and D represent the following events:*

  A: *The sum of the toss is odd*
  B: *The sum of the toss is even*
  C: *The sum of the toss is a number less than ten*
  D: *The toss yielded the same number on each die's upturned face*

Find $P(A)$, $P(B)$, $P(C)$, $P(A \cap B)$, $P(A \cup B)$, $P(B \cap C)$, and $P(B \cap C \cap D)$.

Solution The outcomes from the sample space in Table 2.1 that make up event A are

$$\{(1,2), (1,4), (1,6), (2,1), (2,3), (2,5), (3,2), (3,4), (3,6),$$
$$(4,1), (4,3), (4,5), (5,2), (5,4), (5,6), (6,1), (6,3), (6,5)\}$$

The outcomes from the sample space in Table 2.1 that make up event B are

$$\{(1,1), (1,3), (1,5), (2,2), (2,4), (2,6), (3,1), (3,3), (3,5),$$
$$(4,2), (4,4), (4,6), (5,1), (5,3), (5,5), (6,2), (6,4), (6,6)\}$$

TABLE 2.1

Sample Space for the Tossing of Two Dice

(1, 1)	(1, 2)	(1, 3)	(1, 4)	(1, 5)	(1, 6)
(2, 1)	**(2, 2)**	(2, 3)	(2, 4)	(2, 5)	(2, 6)
(3, 1)	(3, 2)	**(3, 3)**	(3, 4)	(3, 5)	(3, 6)
(4, 1)	(4, 2)	(4, 3)	**(4, 4)**	(4, 5)	(4, 6)
(5, 1)	(5, 2)	(5, 3)	(5, 4)	**(5, 5)**	(5, 6)
(6, 1)	(6, 2)	(6, 3)	(6, 4)	(6, 5)	**(6, 6)**

---

* The outcomes from tossing two dice are recorded as (d_1, d_2), where d_1 and d_2 are the numbers appearing on the upturned faces of the first and second die, respectively.

The outcomes from the sample space in Table 2.1 that make up event C are

$$\{(1,1), (1,2), (1,3), (1,4), (1,5), (1,6), (2,1), (2,2), (2,3),$$
$$(2,4), (2,5), (2,6), (3,1), (3,2), (3,3), (3,4), (3,5), (3,6),$$
$$(4,1), (4,2), (4,3), (4,4), (4,5), (5,1), (5,2), (5,3), (5,4),$$
$$(6,1), (6,2), (6,3)\}$$

The outcomes from the sample space in Table 2.1 that make up event D are

$$\{(1,1), (2,2), (3,3), (4,4), (5,5), (6,6)\}$$

Determination of $P(A)$, $P(B)$, and $P(C)$: From Equation 2.1, we can compute

$$P(A) = \frac{18}{36} = \frac{1}{2} \qquad P(B) = \frac{18}{36} = \frac{1}{2} \qquad P(C) = \frac{30}{36} = \frac{15}{18}$$

Determination of $P(A \cap B)$: Observe event A and event B are mutually exclusive, that is, they share no elements in common. Therefore, from Theorem 2.3

$$P(A \cap B) \equiv P(AB) = 0$$

Determination of $P(A \cup B)$: From Theorem 2.4.

$$P(A \cup B) = P(A) + P(B) - P(A \cap B)$$

Since $P(A \cap B) = 0$ and $P(A) = P(B) = 1/2$, it follows that $P(A \cup B) = 1$. Notice the event $(A \cup B)$ yields the sample space Ω for this experiment; by Axiom 2.2 we know $P(\Omega) = 1$.

Determination of $P(B \cap C)$: The event the sum of the toss is even and it is a number less than 10 is given by $B \cap C$. This event contains the outcomes

$$\{(1,1), (1,3), (1,5), (2,2), (2,4), (2,6), (3,1), (3,3), (3,5),$$
$$(4,2), (4,4), (5,1), (5,3), (6,2)\}$$

from which $P(B \cap C) = 14/36 = 7/18$.

Determination of $P(B \cap C \cap D)$: The event the sum of the toss is even and it is a number less than ten and the toss yielded the same number on each die's upturned face is given by $B \cap C \cap D$. This event contains the outcomes

$$\{(1,1), (2,2), (3,3), (4,4)\}$$

from which $P(B \cap C \cap D) = 4/36 = 1/9$. Notice event $B \cap C \cap D$ is a subset of event $B \cap C$. From Theorem 2.5 we expect $P(B \cap C \cap D) \leq P(B \cap C)$.

2.3.4 Measure of Belief Interpretation

From the axiomatic view, probability need only be a number satisfying the three axioms stated by Kolmogorov. Given this, it is possible for probability to reflect a "measure of belief" in an event's occurrence. For instance, a software engineer might assign a probability of 0.70 to the event "*the radar software for the Advanced Air Traffic Control System (AATCS) will not exceed 100K delivered source instructions.*" We consider this event to be nonrepeatable. It is not practical, or possible, to build the AATCS *n*-times (and under identical conditions) to determine whether this probability is indeed 0.70. When an event such as this arises, its probability may be assigned. Probabilities assigned on the basis of personal judgment, or measure of belief, are known as *subjective probabilities*.

Subjective probabilities are the most common in systems engineering projects and cost analysis problems. Such probabilities are typically assigned by expert technical opinion. The software engineer's probability assessment of 0.70 is a subjective probability. Ideally, subjective probabilities should be based on available evidence and previous experience with similar events. Subjective probabilities risk becoming suspect if they are premised on limited insights or without prior experiences. Care is also needed in soliciting subjective probabilities. They must certainly be plausible; but even more, they must be *consistent* with Kolmogorov's axioms and the theorems of probability which stem from these axioms. Consider the following:

> The XYZ Corporation has offers on two contracts A and B. Suppose the proposal team made the following subjective probability assignments...the chance of winning contract A is 40%, the chance of winning contract B is 20%, the chance of winning contract A or contract B is 60%, and the chance of winning both contract A and contract B is 10%. It turns out this set of probability assignments is *not* consistent with the axioms and theorems of probability! Why is this?* If the chance of winning contract B was changed to 30%, then this *particular set of probability assignments* would be consistent.

Kolmogorov's axioms, and the resulting theorems of probability, *do not suggest* how to assign probabilities to events; rather, they provide a way to verify the probability assignments (be they objective or subjective) are consistent.

2.3.5 Risk versus Uncertainty

There is an important distinction between the terms *risk* and *uncertainty*. Risk is the chance of loss or injury. In a situation that includes favorable and unfavorable events, risk is the *probability an unfavorable event occurs*. Uncertainty is the *indefiniteness about the outcome of a situation*. We analyze uncertainty *for the*

---

* The answer can be seen from Theorem 2.4; this is also Exercise 2.6.

purpose of measuring risk. In systems engineering, the analysis might focus on measuring the risk of failing to achieve performance objectives, overrunning the budgeted cost, or delivering the system too late to meet user needs. Conducting the analysis involves varying degrees of subjectivity. This includes defining the events of concern, as well as specifying their subjective probabilities. Given this, it is fair to ask whether it is meaningful to apply rigorous mathematical procedures to such analyses. In a speech before the 1955 Operations Research Society of America meeting, Charles Hitch addressed this question. He stated:

> "Systems analyses provide a framework which permits the judgment of experts in many fields to be combined to yield results that transcend any individual judgment. The systems analyst [cost analyst] may have to be content with better rather than optimal solutions; or with devising and costing sensible methods of hedging; or merely with discovering critical sensitivities. We tend to be worse, in an absolute sense, in applying analysis or scientific method to broad context problems; but unaided intuition in such problems is also much worse in the absolute sense. Let's not deprive ourselves of any useful tools, however short of perfection they may fail" (Hitch 1955).

2.4 Conditional Probability

In many circumstances, the probability of an event must be conditioned on knowing another event has taken place. Such a probability is known as a conditional probability. *Conditional probabilities* incorporate information about the occurrence of another event. The conditional probability of event A given an event B has occurred is denoted by $P(A \mid B)$. In Example 2.1, it was shown if a pair of dice is tossed the probability the sum of the toss is even is $1/2$; this probability is known as a *marginal* or *unconditional probability*. How would this unconditional probability change (i.e., be conditioned) if it was *known* the sum of the toss was a number less than ten? This is discussed in the following example.

> **Example 2.2** *If a pair of dice is tossed and the sum of the toss is a number less than 10, compute the probability this sum is an even number.*
>
> *Solution* Returning to Example 2.1, recall events B and C were given by
>
> > B : *The sum of the toss is even*
> >
> > C : *The sum of the toss is a number less than ten*
>
> The sample space Ω is given by the 36 outcomes in Table 2.1. In this case, we want the subset of Ω containing *only* those outcomes whose toss yielded a sum less than 10. This subset is shown in Table 2.2.

TABLE 2.2

Outcomes Associated with Event C

(1, 1)	(1, 2)	(1, 3)	(1, 4)	(1, 5)	(1, 6)
(2, 1)	**(2, 2)**	(2, 3)	(2, 4)	(2, 5)	(2, 6)
(3, 1)	(3, 2)	**(3, 3)**	(3, 4)	(3, 5)	(3, 6)
(4, 1)	(4, 2)	(4, 3)	**(4, 4)**	(4, 5)	
(5, 1)	(5, 2)	(5, 3)	(5, 4)		
(6, 1)	(6, 2)	(6, 3)			

Within Table 2.2, 14 possible outcomes are associated with the event *the sum of the toss is even, given the sum of the toss is a number less than ten.*

$$\begin{Bmatrix} (1,1), (1,3), (1,5), (2,2), (2,4), (2,6), (3,1), (3,3), (3,5) \\ (4,2), (4,4), (5,1), (5,3), (6,2) \end{Bmatrix}$$

Therefore, the probability of this event is $P(B \mid C) = \frac{14}{30}$ ◆

In Example 2.2, observe $P(B \mid C)$ was obtained directly from a subset of the sample space Ω; furthermore, $P(B \mid C) = 14/30 < P(B) = 1/2$ in Example 2.2. If A and B are events in the same sample space Ω, then $P(A \mid B)$ is the probability of event A within the subset of the sample space defined by event B. Formally, the *conditional probability of event A given event B has occurred* is defined as

$$P(A|B) = \frac{P(A \cap B)}{P(B)} \equiv \frac{P(AB)}{P(B)} \tag{2.2}$$

where $P(B) > 0$. Likewise, the *conditional probability of event B given event A has occurred* is defined as

$$P(B|A) = \frac{P(B \cap A)}{P(A)} \equiv \frac{P(BA)}{P(A)} \tag{2.3}$$

where $P(A) > 0$. In particular, relating Equation 2.3 to Example 2.2 (and referring to the computations in Example 2.1) we have

$$P(B|C) = \frac{P(B \cap C)}{P(C)} = \frac{\frac{14}{36}}{\frac{30}{36}} = \frac{14}{30}$$

Example 2.3 *A proposal team from XYZ Corporation has offers on two contracts A and B. The team made subjective probability assignments on the chances of winning these contracts. They assessed a 40% chance on the event winning contract A, a 50% chance on the event winning contract B, and a 30% chance on the event winning both contracts. Given this, what is the probability of*

a. *Winning at least one of these contracts?*

b. *Winning contract A and not winning contract B?*

c. *Winning contract A if the proposal team has won at least one contract?*

Solution

a. Winning at least one contract means winning either contract A *or* contract B or both contracts. This event is represented by the set $A \cup B$. From Theorem 2.4

$$P(A \cup B) = P(A) + P(B) - P(A \cap B)$$

therefore,

$$P(A \cup B) = 0.40 + 0.50 - 0.30 = 0.60$$

b. The event winning contract A and not winning contract B is represented by the set $A \cap B^c$. From the following Venn diagram, observe that

$$P(A) = P\left((A \cap B^c) \cup (A \cap B)\right)$$

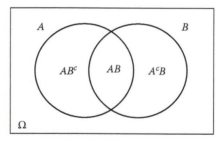

Since the events $A \cap B^c$ and $A \cap B$ are disjoint, from Theorem 2.4 we have

$$P(A) = P\left(A \cap B^c\right) + P(A \cap B)$$

This is equivalent to $P\left(A \cap B^c\right) = P(A) - P(A \cap B)$; therefore,

$$P\left(A \cap B^c\right) = P(A) - P(A \cap B) = 0.40 - 0.30 = 0.10$$

c. If the proposal team has won one of the contracts, the probability of winning contract A must be revised (or conditioned) on this information. This means we must compute $P(A \,|\, A \cup B)$. From Equation 2.2

$$P(A \,|\, A \cup B) = \frac{P(A \cap (A \cup B))}{P(A \cup B)}$$

Since $P(A) = P(A \cap (A \cup B))$ we have

$$P(A \,|\, A \cup B) = \frac{P(A \cap (A \cup B))}{P(A \cup B)} = \frac{P(A)}{P(A \cup B)} = \frac{0.40}{0.60} = \frac{2}{3} \approx 0.67 \blacklozenge$$

A consequence of conditional probability is obtained if we multiply Equations 2.2 and 2.3 by $P(B)$ and $P(A)$, respectively. This multiplication yields*

$$P(A \cap B) = P(B)P(A|B) = P(A)P(B|A) \qquad (2.4)$$

Equation 2.4 is known as the *multiplication rule*. The multiplication rule provides a way to express the probability of the intersection of two events in terms of their conditional probabilities. An illustration of this rule is presented in Example 2.4.

Example 2.4 *A box contains memory chips of which 3 are defective and 97 are nondefective. Two chips are drawn at random, one after the other, without replacement. Determine the probability*

 a. *Both chips drawn are defective.*

 b. *The first chip is defective and the second chip is nondefective.*

Solution

 a. Let A and B denote the event the first and second chips drawn from the box are *defective*, respectively. From the multiplication rule, we have

$$P(A \cap B) = P(AB) = P(A)P(B|A)$$

$$= P(\text{first chip defective}) \, P \, (\text{second chip defective}|$$

$$\text{first chip defective})$$

$$= \frac{3}{100} \left(\frac{2}{99} \right) = \frac{6}{9900}$$

 b. To determine the probability the first chip drawn is defective and the second chip is *nondefective*, let C denote the event the second chip drawn is nondefective. Thus,

$$P(A \cap C) = P(AC) = P(A)P(C|A)$$

$$= P(\text{first chip defective}) \, P \, (\text{second chip nondefective}|$$

$$\text{first chip defective})$$

$$= \frac{3}{100} \left(\frac{97}{99} \right) = \frac{291}{9900} \blacklozenge$$

In this example, sampling was performed *without replacement*. Suppose the chips sampled were *replaced*; that is, the first chip selected was replaced before the second chip was selected. In that case, the probability of a defective chip being selected on the second drawing is independent of the outcome of the first chip drawn. Specifically,

$$P \, (\text{2nd chip defective}) = P(\text{1st chip defective}) = \frac{3}{100}$$

---

* From the commutative law $P(A \cap B) = P(B \cap A)$, which is equivalent to $P(AB) = P(BA)$.

So

$$P(A \cap B) = \frac{3}{100} \left(\frac{3}{100} \right) = \frac{9}{10,000}$$

and

$$P(A \cap C) = \frac{3}{100} \left(\frac{97}{100} \right) = \frac{291}{10,000}$$

Independent Events: Two events A and B are said to be *independent* if and only if

$$P(A \cap B) = P(A)P(B) \qquad (2.5)$$

and *dependent* otherwise. The events $A_1, A_2, \ldots, A_n$ are (mutually) *independent* if and only if for every set of indices $i_1, i_2, \ldots, i_k$ between 1 and n, inclusive,

$$P\left(A_{i_1} \cap A_{i_2} \cap \ldots \cap A_{i_k}\right) = P\left(A_{i_1}\right) P\left(A_{i_2}\right) \ldots P\left(A_{i_k}\right), \quad (k = 2, \ldots, n)$$

For instance, events A_1, A_2, and A_3, are independent (or mutually independent) if the following equations are satisfied:

$$P(A_1 \cap A_2 \cap A_3) = P(A_1) P(A_2) P(A_3) \qquad (2.6)$$
$$P(A_1 \cap A_2) = P(A_1) P(A_2) \qquad (2.7)$$
$$P(A_1 \cap A_3) = P(A_1) P(A_3) \qquad (2.8)$$
$$P(A_2 \cap A_3) = P(A_2) P(A_3) \qquad (2.9)$$

It is possible to have three events A_1, A_2, and A_3 for which Equations 2.7 through 2.9 hold but Equation 2.6 does not hold. Mutual independence implies pairwise independence, in the sense that Equations 2.7 through 2.9 hold, but the converse is not true.

There is a close relationship between independent events and conditional probability. To see this, suppose events A and B are independent. This implies $P(AB) = P(A)P(B)$. From this, Equations 2.2 and 2.3 become, respectively, $P(A|B) = P(A)$ and $P(B|A) = P(B)$. Thus, when two events are independent, the occurrence of one event has no impact on the probability the other event occurs.

To illustrate the concept of independence, suppose a fair die is tossed. Let A be the event an odd number appears. Let B be the event one of these numbers $\{2, 3, 5, 6\}$ appears, then $P(A) = 1/2$ and $P(B) = 2/3$. Since $A \cap B$ is the event represented by the set $\{3, 5\}$, we can readily state $P(A \cap B) = 1/3$. Therefore, $P(A \cap B) = P(A)P(B)$ and we conclude events A and B are independent. Dependence can be illustrated by tossing two fair dice, as described in Example 2.1. In that example, A was the event the sum of the toss is odd and B was the event the sum of the toss is even. In the solution to Example 2.1, it was shown $P(A \cap B) = 0$ and $P(A)$ and $P(B)$ were each $1/2$. Since $P(A \cap B) \neq P(A)P(B)$ we would conclude events A and B are dependent, in this case.

It is important not to confuse the meaning of independent events with mutually exclusive events. If events A and B are mutually exclusive, the event A *and* B is empty; that is, $A \cap B = \emptyset$. This implies $P(A \cap B) = P(\emptyset) = 0$. If events A and B are *independent* with $P(A) \neq 0$ and $P(B) \neq 0$, then A and B cannot be mutually exclusive since $P(A \cap B) = P(A)P(B) \neq 0$.

Theorem 2.6   *For any two independent events A_1 and A_2*

$$P(A_1 \cup A_2) = 1 - P\left(A_1^c\right) P\left(A_2^c\right)$$

Proof.   From Theorem 2.1 we can write

$$P(A_1 \cup A_2) = 1 - P\left((A_1 \cup A_2)^c\right)$$

From De Morgan's law (Section 2.2) $(A_1 \cup A_2)^c = A_1^c \cap A_2^c$; therefore,

$$P(A_1 \cup A_2) = 1 - P\left(A_1^c \cap A_2^c\right) \equiv 1 - P\left(A_1^c A_2^c\right)$$

Since events A_1 and A_2 are independent,

$$P(A_1 \cup A_2) = 1 - P\left(A_1^c\right) P\left(A_2^c\right) \blacklozenge \qquad (2.10)$$

To prove this theorem, we used a result that if A_1 and A_2 are independent then A_1^c and A_2^c are also independent. Showing this is left as an exercise for the reader. Extending Theorem 2.6, it can be shown that if $A_1, A_2, \ldots, A_n$ are independent, then

$$P(A_1 \cup A_2 \cup A_3 \cup \ldots \cup A_n) = 1 - P\left(A_1^c A_2^c A_3^c \ldots A_n^c\right)$$
$$= 1 - P\left(A_1^c\right) P\left(A_2^c\right) P\left(A_3^c\right) \ldots P\left(A_n^c\right) \qquad (2.11)$$

2.5 Bayes' Rule

Suppose we have a collection of events A_i representing possible conjectures about a topic. Furthermore, suppose we have some initial probabilities associated with the "truth" of these conjectures. Bayes' rule* provides a way

---

* Named in honor of Thomas Bayes (1702–1761), an English minister and mathematician.

to update (revise) initial probabilities when new information about these conjectures is evidenced.

Bayes' rule is a consequence of conditional probability. Suppose we partition a sample space Ω into a finite collection of three mutually exclusive events (see Figure 2.4). Define these events as A_1, A_2, and A_3, where $A_1 \cup A_2 \cup A_3 = \Omega$. Let B denote an arbitrary event contained in Ω. From Figure 2.4 we can write the event B as

$$B = (A_1 \cap B) \cup (A_2 \cap B) \cup (A_3 \cap B) \tag{2.12}$$

Since the events $(A_1 \cap B)$, $(A_2 \cap B)$, and $(A_3 \cap B)$ are mutually exclusive, we can apply Axiom 2.3 and write

$$P(B) = P(A_1 \cap B) + P(A_2 \cap B) + P(A_3 \cap B)$$

From the multiplication rule given in Equation 2.4, $P(B)$ can be expressed in terms of conditional probability as

$$P(B) = P(A_1) P(B|A_1) + P(A_2) P(B|A_2) + P(A_3) P(B|A_3) \tag{2.13}$$

Equation 2.13 is known as the total probability law. Its generalization is

$$P(B) = \sum_{i=1}^{n} P(A_i) P(B|A_i) \tag{2.14}$$

where $\Omega = \bigcup_{i=1}^{n} A_i$ and $A_i \cap A_j = \emptyset$ and $i \neq j$. The conditional probability for each event A_i given event B has occurred is

$$P(A_i|B) = \frac{P(A_i \cap B)}{P(B)} = \frac{P(A_i) P(B|A_i)}{P(B)} \tag{2.15}$$

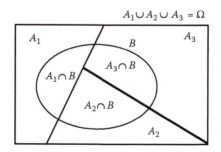

FIGURE 2.4
Partitioning Ω into three mutually exclusive sets.

When the total probability law is applied to Equation 2.15, we have

$$P(A_i \mid B) = \frac{P(A_i)\,P(B \mid A_i)}{\sum_{i=1}^{n} P(A_i)\,P(B \mid A_i)} \tag{2.16}$$

Equation 2.16 is known as *Bayes' rule.*

Example 2.5 *ChipyTech Corporation has three divisions D_1, D_2, and D_3. Each manufactures a specific type of microprocessor chip. From the total annual output of chips produced by the corporation, D_1 manufactures 35%, D_2 manufactures 20%, and D_3 manufactures 45%. Data collected from the quality control group indicate 1% of the chips from D_1 are defective, 2% of the chips from D_2 are defective, and 3% of the chips from D_3 are defective. Suppose a chip was randomly selected from the total annual output produced and it was found to be defective. What is the probability it was manufactured by D_1? By D_2? By D_3?*

Solution Let A_i denote the *event* the selected chip was produced by division D_i for $i = 1, 2, 3$. Let B denote the event the selected chip is defective. To determine the probability the defective chip was manufactured by D_i we must compute the conditional probability $P(A_i|B)$. From the information provided, we have

$$P(A_1) = 0.35, P(A_2) = 0.20, \text{ and } P(A_3) = 0.45$$
$$P(B|A_1) = 0.01, P(B|A_2) = 0.02, \text{ and } P(B|A_3) = 0.03$$

The total probability law and Bayes' rule will be used to determine $P(A_i|B)$ for each $i = 1, 2, 3$. Recall from Equation 2.13 $P(B)$ can be written as

$$P(B) = P(A_1)\,P(B|A_1) + P(A_2)\,P(B|A_2) + P(A_3)\,P(B|A_3)$$
$$P(B) = 0.35(0.01) + 0.20(0.02) + 0.45(0.03) = 0.021$$

and from Bayes' rule we can write

$$P(A_i|B) = \frac{P(A_i)\,P(B|A_i)}{\sum_{i=1}^{n} P(A_i)\,P(B|A_i)} = \frac{P(A_i)\,P(B|A_i)}{P(B)}$$

from which

$$P(A_1|B) = \frac{P(A_1)\,P(B|A_1)}{P(B)} = \frac{0.35(0.01)}{0.021} = 0.167$$

$$P(A_2|B) = \frac{P(A_2)\,P(B|A_2)}{P(B)} = \frac{0.20(0.02)}{0.021} = 0.190$$

$$P(A_3|B) = \frac{P(A_3)\,P(B|A_3)}{P(B)} = \frac{0.45(0.03)}{0.021} = 0.643$$

Table 2.3 provides a comparison of $P(A_i)$ with $P(A_i|B)$ for each $i = 1, 2, 3$. The probabilities given by $P(A_i)$ are the probabilities the selected chip will have been produced by division D_i before it is randomly selected and before

TABLE 2.3

Comparison of $P(A_i)$ and $P(A_i|B)$

i	$P(A_i)$	$P(A_i|B)$
1	0.35	0.167
2	0.20	0.190
3	0.45	0.643

it is known whether or not the chip is defective. Therefore, $P(A_i)$ are the *prior*, or *a priori* (before the fact), probabilities. The probabilities given by $P(A_i|B)$ are the probabilities the selected chip was produced by division D_i after it is known the selected chip is defective. Therefore, $P(A_i|B)$ are the *posterior*, or *a posteriori* (after the fact), probabilities. Bayes' rule provides a means for the computation of posterior probabilities from the known prior probabilities $P(A_i)$ and the conditional probabilities $P(B|A_i)$ for a particular situation or experiment.

Bayes' rule established a philosophy in probability theory that became known as *Bayesian inference* and *Bayesian decision theory*. These areas play an important role in the application of probability theory to cost and systems engineering problems. In Equation 2.16, we may think of A_i as representing possible states of nature to which an analyst or systems engineer assigns subjective probabilities. These subjective probabilities are the prior probabilities, which are often premised on personal judgments based on past experience. In general, Bayesian methods offer a powerful way to revise, or update, probability assessments as new (or refined) information becomes available.

Exercises

2.1 State the interpretation of probability implied by the following:

a. The probability tails appears on the toss of a fair coin is $1/2$.

b. After recording the outcomes of 50 tosses of a fair coin, the probability tails appears is 0.54.

c. It is with certainty the coin is fair!

d. The probability is 60% that the stock market will close 500 points above yesterday's closing count.

e. The design team believes there is less than a 5% chance the new microchip will require more than 12,000 gates.

2.2 A sack contains 20 marbles exactly alike in size but different in color. Suppose the sack contains 5 blue marbles, 3 green marbles, 7 red marbles, 2 yellow marbles, and 3 black marbles. Picking a single marble from the sack and then replacing it, what is the probability of choosing the following:

    a. Blue marble?

    b. Green marble?

    c. Red marble?

    d. Yellow marble?

    e. Black marble?

    f. Non-blue marble

    g. Red or non-red marble?

2.3 If a fair coin is tossed, what is the probability of not obtaining heads? What is the probability of the event: (heads or not heads)?

2.4 Show the probability of the event: (A or A complement) is *always* unity.

2.5 Suppose two tetrahedrons (4-sided polygons) are randomly tossed. Assuming the tetrahedrons are weighted fair, determine the set of all possible outcomes Ω. Assume each face is numbered 1, 2, 3, and 4.

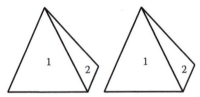

Two tetrahedrons for Exercise 2.5

Let the sets A, B, C, and D represent the following events:

A: *The sum of the toss is even*

B: *The sum of the toss is odd*

C: *The sum of the toss is a number less than 6*

D: *The toss yielded the same number on each upturned face*

    a. Find $P(A)$, $P(B)$, $P(C)$, $P(A \cap B)$, $P(A \cup B)$, $P(B \cup C)$, and $P(B \cap C \cap D)$.

    b. Verify $P(A \cup B)^c = P(A^c \cap B^c)$ (De Morgan's Law).

2.6 The XYZ Corporation has offers on two contracts A and B. Suppose the proposal team made the following subjective probability assessments: the chance of winning contract A is 40%, the chance of winning contract B is 20%, the chance of winning contract A or contract B is 60%, the chance of winning both contracts is 10%.

    a. Explain why the given set of probability assignments is *inconsistent* with the axioms of probability.

    b. What must $P(B)$ equal such that it and the set of other assigned probabilities specified are consistent with the axioms of probability?

2.7 Suppose a coin is balanced such that tails appears three times more frequently than heads. Show the probability of obtaining tails with such a coin is 3/4. What would you expect this probability to be if the coin was fair; that is, equally balanced?

2.8 Suppose the sample space of an experiment is given by $\Omega = A \cup B$. Compute $P(A \cap B)$ if $P(A) = 0.25$ and $P(B) = 0.80$.

2.9 If A and B are disjoint subsets of Ω show that

   a. $P(A^c \cup B^c) = 1$

   b. $P(A^c \cap B^c) = 1 - [P(A) + P(B)]$

2.10 Two missiles are launched. Suppose there is a 75% chance missile A hits the target and a 90% chance missile B hits the target. If the probability missile A hits the target is *independent* of the probability missile B hits the target, determine the probability missile A or missile B hits the target. Find the probability needed for missile A such that if the probability of missile B hitting the target remains at 90%, the probability missile A or missile B hits the target is 0.99.

2.11 Suppose A and B are independent events. Show that

   a. The events A^c and B^c are independent.

   b. The events A and B^c are independent.

   c. The events A^c and B are independent.

2.12 Suppose A and B are independent events with $P(A) = 0.25$ and $P(B) = 0.55$. Determine the probability

   a. At least one event occurs.

   b. Event B occurs but event A does not occur.

2.13 Suppose A and B are independent events with $P(A) = r$ and the probability that "at least A or B occurs" is s. Show the only value for $P(B)$ is $(s - r)(1 - r)^{-1}$.

2.14 In Exercise 2.5, suppose event C has occurred. Enumerate the set of remaining possible outcomes. From this set compute $P(B)$. Compare this with $P(B \mid C)$ where $P(B \mid C)$ is determined from the definition of conditional probability.

2.15 At a local sweet shop, 10% of all customers buy ice cream, 2% buy fudge, and 1% buy both ice cream and fudge. If a customer selected at random bought fudge, what is the probability the customer bought an ice cream? If a customer selected at random bought ice cream, what is the probability the customer bought fudge?

2.16 For any two events A and B, show that $P(A \mid A \cap (A \cap B)) = 1$.

2.17 A production lot contains 1000 microchips of which 10% are defective. Two chips are successively drawn at random without replacement. Determine the probability

   a. Both chips selected are nondefective.

   b. Both chips are defective.

    c. The first chip is defective and the second chip is nondefective.

    d. The first chip is nondefective and the second chip is defective.

2.18 Suppose the sampling scheme in Exercise 2.17 was with replacement; that is, the first chip is returned to the lot before the second chip is drawn. Show how the probabilities computed in Exercise 2.17 are changed.

2.19 Spare power supply units for a communications terminal are provided to the government from three different suppliers A_1, A_2, and A_3. Thirty percent come from A_1, 20% come from A_2, and 50% come from A_3. Suppose these units occasionally fail to perform according to their specifications and the following has been observed: 2% of those supplied by A_1 fail, 5% of those supplied by A_2 fail, and 3% of those supplied by A_3 fail. What is the probability any one of these units provided to the government will perform *without* failure?

2.20 In a single day, ChipyTech Corporation's manufacturing facility produces 10,000 microchips. Suppose machines A, B, and C individually produce 3000, 2500, and 4500 chips daily. The quality control group has determined the output from machine A has yielded 35 defective chips, the output from machine B has yielded 26 defective chips, and the output from machine C has yielded 47 defective chips.

    a. If a chip was selected at random from the daily output, what is the probability it is defective?

    b. What is the probability a randomly selected chip was produced by machine A? By machine B? By machine C?

    c. Suppose a chip *was* randomly selected from the day's production of 10,000 microchips and it was found to be defective. What is the probability it was produced by machine A? By machine B? By machine C?

References

Feller, W. 1968. *An Introduction to Probability Theory and Its Applications*, vol. 1, 3rd rev. edn. New York: John Wiley & Sons, Inc.

Hitch, C. J. 1955. *An Appreciation of Systems Analysis*, p.699. Santa Monica, CA: The RAND Corporation.

3

Distributions and the Theory of Expectation

3.1 Random Variables and Probability Distributions

Consider the experiment of tossing two fair dice described in Example 2.1. Suppose x represents the sum of the toss. Define X as a variable that takes on only values given by x. If the sum of the toss is 2 then $X = 2$; if the sum of the toss is 3 then $X = 3$; if the sum of the toss is 7 then $X = 7$. Numerical values of X are associated with *events* defined from the sample space Ω for this experiment, which was given in Table 2.1. In particular,

$X = 2$ is associated with only this simple event $\{(1, 1)\}$*
$X = 3$ is associated with only these two simple events $\{(1, 2)\}, \{(2, 1)\}$
$X = 7$ is associated with only these six simple events $\{(1, 6)\}, \{(2, 5)\},$
$\qquad \{(3, 4)\}, \{(4, 3)\}, \{(5, 2)\}, \{(6, 1)\}$

Here, we say X is a random variable. This is illustrated in Figure 3.1. Formally, a *random variable* is a real-valued function defined over a sample space. The sample space is the *domain* of a random variable. Traditionally, random variables are denoted by capital letters such as X, W, and Z.

The event $X = x$ is equivalent to

$$\{X = x\} \equiv \{\omega \in \Omega \mid X(\omega) = x\}$$

This represents a subset of Ω consisting of all sample points ω such that $X(\omega) = x$. In Figure 3.1, the event $\{X = 3\}$ is equivalent to

$$\{X = 3\} \equiv \{(1, 2), (2, 1)\}$$

The probability of the event $\{X = x\}$ is equivalent to

$$P(\{X = x\}) \equiv P(\{\omega \in \Omega \mid X(\omega) = x\})$$

---

* The outcomes from tossing two dice are recorded as (d_1, d_2), where d_1 and d_2 are the numbers appearing on the upturned faces of the first and second die, respectively. Therefore, in this discussion, $x = d_1 + d_2$.

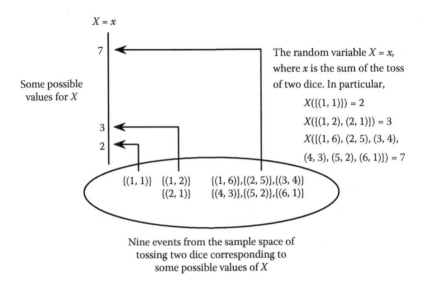

Nine events from the sample space of
tossing two dice corresponding to
some possible values of X

FIGURE 3.1
Some possible values of a random variable.

In Figure 3.1, the probability of the event $\{X = 3\}$ is equivalent to

$$P(\{X = 3\}) \equiv P(\{(1,2), (2,1)\}) = \frac{2}{36}$$

For convenience, the notation $P(\{X = x\}) \equiv P(X = x)$ is adopted in this book.

Random variables can be characterized as discrete or continuous. A random variable is *discrete* if its set of possible values is finite or countably infinite. A random variable is *continuous* if its set of possible values is uncountable.

3.1.1 Discrete Random Variables

Consider again the simple experiment of tossing a pair of fair dice. Let the random variable X represent the sum of the toss. The sample space Ω for this experiment consists of the 36 outcomes given in Table 2.1. The random variable X is discrete since the *only* possible values are $x = 2, 3, 4, 5, 6, \ldots, 12$. The function that describes the probabilities associated with the event $\{X = x\}$, for all *feasible values* of x, is shown in Figure 3.2. This function is known as the probability function of X.

The *probability function* of a *discrete random variable* X is defined as

$$p_X(x) = P(X = x) \tag{3.1}$$

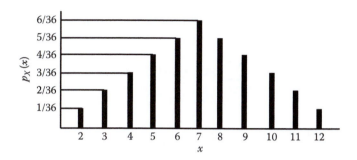

FIGURE 3.2
Probability function for the sum of two dice tossed.

The probability function is also referred to as the *probability mass function* or the *frequency function* of X. The probability function associates probabilities to events described by distinct (single) points of interest. Over all *feasible* (possible) values of x, probability functions satisfy, by the axioms of probability, the following conditions:

$$\text{a) } p_X(x) \geq 0 \quad \text{b) } \sum_x p_X(x) = 1$$

If x is *not a feasible* value of X then

$$p_X(x) = P(X = x) = P(\emptyset) = 0$$

It is often of interest to determine probabilities associated with events of the form $\{X \leq x\}$. For instance, suppose we wanted the probability that the sum of the numbers resulting from the toss of two fair dice will not exceed seven. This is equivalent to computing $P(X \leq 7)$; in this instance, we have $P(X \leq 7) = P(\{X = 2\} \cup \{X = 3\} \cup \ldots \cup \{X = 7\})$. Thus, X can take a value not exceeding seven if and only if X takes on one of the values $2, 3, \ldots, 7$. Since the events $\{X = 2\}, \{X = 3\}, \ldots, \{X = 7\}$ are mutually exclusive, from Axiom 2.3 and Figure 3.2 we have

$$P(X \leq 7) = P(X = 2) + P(X = 3) + \cdots + P(X = 7) = \frac{21}{36}$$

The function that produces probabilities for events of the form $\{X \leq x\}$ is known as the cumulative distribution function (CDF). Formally, if X is a discrete random variable then its CDF is defined by

$$F_X(x) = P(X \leq x) = \sum_{t \leq x} p_X(t) \quad (-\infty < x < \infty) \tag{3.2}$$

In terms of this definition, we would write $P(X \leq 7)$ as

$$F_X(7) = P(X \leq 7) = \sum_{t \leq 7} p_X(t) = p_X(2) + p_X(3) + \cdots + p_X(7) = \frac{21}{36}$$

where, from Equation 3.1, $p_X(x) = P(X = x)$ for $x = 2, 3, \ldots, 7$.

The CDF for the random variable with probability function in Figure 3.2 is pictured in Figure 3.3. Notice the CDF is a "staircase" or "step" function. This is characteristic of CDFs for *discrete random variables*. The height of the "step" along the CDF is the probability the value associated with that step occurs. For instance, in Figure 3.3, the probability that $X = 3$ is the height of the step (jump) between $X = 2$ and $X = 3$; that is, $P(X = 3) = \frac{3}{36} - \frac{1}{36} = \frac{2}{36}$.

If X is a discrete random variable and a is any real number that is a feasible (or possible) value of X, then $P(X = a) = p_X(a)$ is equal to the height of the step (jump) of $F_X(x)$ at $x = a$.

The following presents theorems for determining probabilities from the CDF of a discrete random variable X. In the theorems below, a and b are real numbers with $a < b$.

Theorem 3.1   *The probability of $\{X > a\}$ is $1 - F_X(a)$.*

Proof.   Let A denote the event $\{X > a\}$; then $A^c = \{X \leq a\}$; from Theorem 2.1 and the definition given by Equation 3.2, it immediately follows that

$$P(X > a) = 1 - P(X \leq a) = 1 - F_X(a)$$

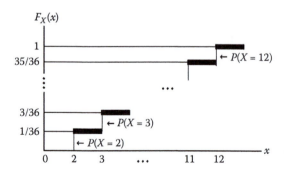

FIGURE 3.3
Cumulative distribution function for the sum of two dice tossed.

Theorem 3.2   *The probability of $\{X \geq a\}$ is $1 - F_X(a) + P(X = a)$.*

Proof. *We can write the event $\{X \geq a\}$ as the union of two mutually exclusive events $\{X = a\}$ and $\{X > a\}$; that is,*

$$\{X \geq a\} = \{X = a\} \cup \{X > a\}$$

$$\{X = a\} \qquad \{X > a\}$$

From Theorems 2.4 and 3.1 we have

$$P(X \geq a) = P(\{X = a\} \cup \{X > a\}) = P(X = a) + P(X > a)$$
$$= P(X = a) + 1 - F_X(a) \equiv 1 - F_X(a) + P(X = a)$$

Theorem 3.3   *The probability of $\{X < a\}$ is $F_X(a) - P(X = a)$.*

Proof. *This is a direct consequence of Theorems 3.1 and 3.2. The proof is left as an exercise for the reader.*

Theorem 3.4   *The probability of $\{a < X \leq b\}$ is $F_X(b) - F_X(a)$.*

Proof. *We can write the event $\{X \leq b\}$ as the union of two mutually exclusive events $\{X \leq a\}$ and $\{a < X \leq b\}$; that is,*

$$\{X \leq b\} = \{X \leq a\} \cup \{a < X \leq b\}$$

$$a \qquad b$$
$$\{X \leq a\} \qquad \{a < X \leq b\}$$

From Theorem 2.4

$$P(X \leq b) = P(\{X \leq a\} \cup \{a < X \leq b\}) = P(X \leq a) + P(a < X \leq b)$$

Thus,

$$F_X(b) = F_X(a) + P(a < X \leq b)$$

Therefore

$$P(a < X \leq b) = F_X(b) - F_X(a)$$

Theorem 3.5   *The probability of $\{a < X < b\}$ is $F_X(b) - F_X(a) - P(X = b)$.*

Proof. We can write the event $\{X < b\}$ as the union of two mutually exclusive events $\{X \le a\}$ and $\{a < X < b\}$; that is,

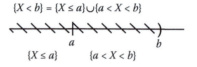

$$\{X < b\} = \{X \le a\} \cup \{a < X < b\}$$

$$\underset{\{X \le a\} \qquad\qquad \{a < X < b\}}{\underset{a \qquad\qquad\qquad b}{}}$$

From Theorem 2.4

$$P(X < b) = P(\{X \le a\} \cup \{a < X < b\}) = P(X \le a) + P(a < X < b)$$

It follows that

$$P(X < b) - P(X \le a) = P(a < X < b)$$

From Theorem 3.3, $P(X < b) = F_X(b) - P(X = b)$; since $P(X \le a) = F_X(a)$ we have $P(a < X < b) = F_X(b) - F_X(a) - P(X = b)$, which was to be shown.

Theorem 3.6   *The probability of $\{a \le X < b\}$ is*

$$F_X(b) - F_X(a) + P(X = a) - P(X = b).$$

Proof. We can write the event $\{a \le X < b\}$ as the union of two mutually exclusive events $\{X = a\}$ and $\{a < X < b\}$; that is,

$$\{a \le X < b\} = \{X = a\} \cup \{a < X < b\}$$

From Theorem 2.4

$$P(a \le X < b) = P(\{X = a\} \cup \{a < X < b\}) = P(X = a) + P(a < X < b)$$

From Theorem 3.5, $P(a < X < b) = F_X(b) - F_X(a) - P(X = b)$; therefore,

$$P(a \le X < b) = F_X(b) - F_X(a) + P(X = a) - P(X = b)$$

Theorem 3.7   *The probability of $\{a \le X \le b\}$ is $F_X(b) - F_X(a) + P(X = a)$.*

Proof. We can write the event $\{a \le X \le b\}$ as the union of three mutually exclusive events $\{X = a\}$, $\{a < X < b\}$, and $\{X = b\}$; that is,

$$\{a \le X \le b\} = \{X = a\} \cup \{a < X < b\} \cup \{X = b\}$$

From Axiom 2.3 and Theorem 3.5

$$
\begin{aligned}
P(a \le X \le b) &= P(\{X = a\} \cup \{a < X < b\} \cup \{X = b\}) \\
&= P(X = a) + P(a < X < b) + P(X = b) \\
&= P(X = a) + [F_X(b) - F_X(a) - P(X = b)] + P(X = b) \\
&= F_X(b) - F_X(a) + P(X = a) \; \blacklozenge
\end{aligned}
$$

The following presents the first of many case discussions in this book. The discussion addresses how a corporation might assess the chance of making a profit on a new electronics product.

Case Discussion 3.1:[*] ChipyTech Corporation is a major producer and supplier of electronics products to industry worldwide. They are planning to bring a new product to the market. Management needs to know the product's potential for profit and loss during its first year on the market. In addition, they want to know the chance of *not making* a profit the first year. Suppose profit (Park and Jackson 1984) is given by Equation 3.3

$$Profit = (U_{Price} - U_{Cost})V \tag{3.3}$$

where U_{Price} is a discrete random variable that represents the product's unit price, U_{Cost} is a discrete random variable that represents the unit cost to manufacture the product, and V is a discrete random variable that represents the product's *sales* volume for year 1, which is assumed to be nonzero.

A profit exists when $U_{Price} > U_{Cost}$, a loss exists when $U_{Price} < U_{Cost}$, and *no profit* exists when $U_{Price} \le U_{Cost}$. For purposes of this case discussion, we will assume U_{Price}, U_{Cost}, and V are *independent* random variables.

Suppose the corporation's sales, price, and cost histories for similar products have been analyzed. Further, suppose interviews were carefully conducted with subject matter experts from the engineering and marketing departments of ChipyTech. From the interviews and the historical data, possible values for the product's unit price, unit cost, and sales volume were established along with their respective probabilities of occurrence. Figure 3.4 presents these values for U_{Price}, U_{Cost}, and V.

---

[*] Adapted and expanded from an example in Park, W. R. and D. E. Jackson. 1984. *Cost Engineering Analysis—A Guide to Economic Evaluation of Engineering Projects*, 2nd edn. New York: John Wiley & Sons, Inc.

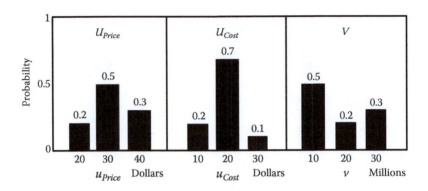

FIGURE 3.4
Possible values for U_{Price}, U_{Cost}, and V.

To find the dollar range on the product's profit or loss potential, we first list all possible combinations of U_{Price}, U_{Cost}, and V. This list is shown in Table 3.1. Since U_{Price}, U_{Cost}, and V are given to be independent random variables,* the probability that any combination of them will occur is

$$P\left(\{U_{Price} = u_{Price}\} \cap \{U_{Cost} = u_{Cost}\} \cap \{V = v\}\right)$$
$$= P\left(U_{Price} = u_{Price}\right) P\left(U_{Cost} = u_{Cost}\right) P(V = v) \qquad (3.4)$$

where the values for $P\left(U_{Price} = u_{Price}\right)$, $P\left(U_{Cost} = u_{Cost}\right)$, and $P(V=v)$ are given in Figure 3.4. For example, the probability the new product will have a unit price of 20 dollars and a unit cost of 10 dollars and a sales volume of 10 million (the first year) is

$$P\left(\{U_{Price} = 20\} \cap \{U_{Cost} = 10\} \cap \{V = 10\}\right)$$
$$= P\left(U_{Price} = 20\right) P\left(U_{Cost} = 10\right) P(V = 10) = 0.020 \qquad (3.5)$$

Table 3.1 summarizes the possible values for *Profit*. Table 3.1 also shows the probability *Profit* takes a value according to a specific combination of U_{Price}, U_{Cost}, and V. From Table 3.1, observe there is a potential loss of as much as 300 ($M) and a potential gain of as much as 900 ($M). How probable are these extremes? What is the chance the corporation will *not make* a profit the first year? The following discussion addresses these questions.

From Table 3.1 it can be seen there is less than a 1% chance (i.e., 0.6%) the new product will realize a loss of 300 ($M) during its first year on the

---

* When random variables are independent, their associated events are independent. This is discussed further in Chapter 5.

TABLE 3.1

Possible Profits and Their Probabilities

U_{Price} (\$)	U_{Cost} (\$)	V (Millions)	Profit (\$M)	Probability
20	10	10	100	0.020
20	10	20	200	0.008
20	10	30	300	0.012
20	**20**	**10**	**0**	**0.070**
20	**20**	**20**	**0**	**0.028**
20	**20**	**30**	**0**	**0.042**
20	**30**	**10**	**−100**	**0.010**
20	**30**	**20**	**−200**	**0.004**
20	**30**	**30**	**−300**	**0.006**
30	10	10	200	0.050
30	10	20	400	0.020
30	10	30	600	0.030
30	20	10	100	0.175
30	20	20	200	0.070
30	20	30	300	0.105
30	**30**	**10**	**0**	**0.025**
30	**30**	**20**	**0**	**0.010**
30	**30**	**30**	**0**	**0.015**
40	10	10	300	0.030
40	10	20	600	0.012
40	10	30	900	0.018
40	20	10	200	0.105
40	20	20	400	0.042
40	20	30	600	0.063
40	30	10	100	0.015
40	30	20	200	0.006
40	30	30	300	0.009
Total Probability			1	

market. Similarly, the maximum profit of 900 (\$M) has just under a 2% chance (i.e., 1.8%) of occurring.

The corporation will *not make* a profit (i.e., *Profit* ≤ 0) when $U_{Price} \leq U_{Cost}$. There are nine events in Table 3.1 (shown by the bold-faced figures) that produce *Profit* ≤ 0. Let these events be represented by $A_1, A_2, \ldots, A_9$, where

$$A_1 = \{\{U_{Price} = 20\} \cap \{U_{Cost} = 20\} \cap \{V = 10\}\}$$
$$A_2 = \{\{U_{Price} = 20\} \cap \{U_{Cost} = 20\} \cap \{V = 20\}\}$$
$$\vdots$$
$$A_9 = \{\{U_{Price} = 30\} \cap \{U_{Cost} = 30\} \cap \{V = 30\}\}$$

These events are mutually exclusive. Therefore, from Axiom 2.3 the probability that *Profit* ≤ 0 is

$$P(Profit \leq 0) = P\left(\overset{9}{\underset{i=1}{\cup}} A_i \right) = \sum_{i=1}^{9} P(A_i) = 0.210 \qquad (3.6)$$

where each $P(A_i)$ is given in Table 3.1.

Table 3.1 can also be used to develop the *probability function* for the random variable *Profit*. Since *Profit* is given (in this discussion) to be a discrete random variable, its probability function is

$$p_{Profit}(x) = P(Profit = x) \qquad (3.7)$$

where feasible values of x are given in Table 3.1. Figure 3.5 is the graph of $p_{Profit}(x)$. Among the many useful aspects of the probability function is identifying the value of x associated with the highest probability of occurrence. In Figure 3.5, a profit of 200 ($M) has the highest probability of occurrence. A number of other computations can be determined from $p_{Profit}(x)$. For example, from Figure 3.5 we have

$$P(Profit \leq 0) = p_{Profit}(-300) + p_{Profit}(-200)$$
$$+ p_{Profit}(-100) + p_{Profit}(0) = 0.210$$

Notice that $P(Profit \leq 0)$ is really the value of the *cumulative distribution function* for *Profit* at $x = 0$. From Equation 3.2, the CDF of *Profit* is

$$F_{Profit}(x) = P(Profit \leq x) = \sum_{t \leq x} p_{Profit}(t)$$

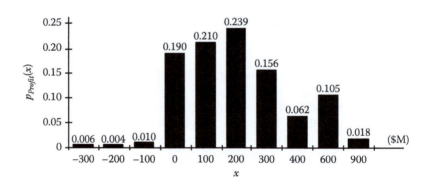

FIGURE 3.5
Probability function for profit—Case Discussion 3.1.

Equation 3.8 presents $F_{Profit}(x)$ for this case discussion.

$$F_{Profit}(x) = \begin{cases} 0 & \text{if } -300 < x \\ 0.006 & \text{if } -300 \leq x < -200 \\ 0.010 & \text{if } -200 \leq x < -100 \\ 0.020 & \text{if } -100 \leq x < 0 \\ 0.210 & \text{if } 0 \leq x < 100 \\ 0.420 & \text{if } 100 \leq x < 200 \\ 0.659 & \text{if } 200 \leq x < 300 \\ 0.815 & \text{if } 300 \leq x < 400 \\ 0.877 & \text{if } 400 \leq x < 600 \\ 0.982 & \text{if } 600 \leq x < 900 \\ 1 & \text{if } 900 \geq x \end{cases} \tag{3.8}$$

The probability that ChipyTech will *not make* a profit can now be read directly from the CDF (Equation 3.8), specifically

$$F_{Profit}(0) = P(Profit \leq 0) = 0.210$$

A graph of $F_{Profit}(x)$ is presented with Example 3.3 (Figure 3.12). From Equation 3.8, the probability *Profit* will fall within other intervals of interest can be determined. From Theorems 3.2 through 3.7, with reference to Figure 3.5 and Equation 3.8, we have the following:

$$P(Profit \geq 200) = 1 - F_{Profit}(200) + P(Profit = 200)$$

$$= 1 - 0.659 + 0.239 = 0.580$$

$$P(Profit < 200) = F_{Profit}(200) - P(Profit = 200)$$

$$= 0.659 - 0.239 = 0.420$$

$$P(200 < Profit \leq 600) = F_{Profit}(600) - F_{Profit}(200)$$

$$= 0.982 - 0.659 = 0.323$$

$$P(200 < Profit < 600) = F_{Profit}(600) - F_{Profit}(200) - P(Profit = 600)$$

$$= 0.982 - 0.659 - 0.105 = 0.218$$

$$P(200 \leq Profit < 600) = F_{Profit}(600) - F_{Profit}(200)$$

$$+ P(Profit = 200) - P(Profit = 600) = 0.457$$

$$P(200 \leq Profit \leq 600) = F_{Profit}(600) - F_{Profit}(200)$$

$$+ P(Profit = 200) = 0.562$$

In summary, Case Discussion 3.1 illustrates how fundamental probability concepts such as the axioms, independence, the probability function, and the CDF can provide decision-makers insights on profits and their associated probabilities.

3.1.2 Continuous Random Variables

Mentioned in the beginning of this chapter, a random variable is continuous if its set of possible values is uncountable. For instance, suppose T is a random variable representing the duration (in hours) of an electronic device. If the possible values of T are given by $\{t : 0 \leq t \leq 2500\}$, then T is a *continuous random variable*.

In general, we say X is a *continuous random variable* if there exists a *nonnegative function* $f_X(x)$, defined on the real line, such that for any interval A

$$P(X \in A) = \int_A f_X(x) \, dx$$

The function $f_X(x)$ is called the probability density function (PDF) of X. Unlike the probability function for a discrete random variable, the PDF *does not* directly produce a probability—$f_X(a)$ does not produce $p_X(a)$, defined by Equation 3.1. Here, the probability that X is contained in any subset of the real line is determined by integrating $f_X(x)$ over that subset. Since X *must* assume some value on the real line, it will always be true that

$$\int_{-\infty}^{\infty} f_X(x) \, dx \equiv P(X \in (-\infty, \infty)) = 1$$

In this case, the CDF of the random variable X is defined as

$$F_X(x) = P(X \leq x) = P(X \in (-\infty, x]) = \int_{-\infty}^{x} f_X(t) \, dt \qquad (3.9)$$

A useful way to view Equation 3.9 is shown by Figure 3.6; if we assume $f_X(x)$ is a PDF, then from calculus we can interpret the probabilities of the events $\{X \leq a\}$ and $\{a \leq X \leq b\}$ as the areas of the indicated regions in Figure 3.6.

When X is a *continuous random variable*, the probability $X = a$ is zero; this is because

$$P(X = a) = P(a \leq X \leq a) = \int_a^a f_X(x) \, dx = 0 \qquad (3.10)$$

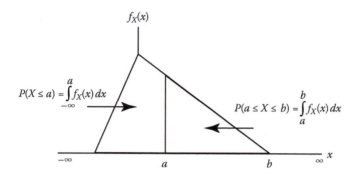

FIGURE 3.6
A probability density function.

From this it is seen the inclusion or exclusion of an interval's endpoints does not affect the probability X falls in the interval; thus, if a and b are any two real numbers

$$P(a < X \leq b) = P(a < X < b)$$
$$= P(a \leq X < b) = P(a \leq X \leq b) = F_X(b) - F_X(a) \qquad (3.11)$$

when X is a *continuous random variable*. Referring back to Equation 3.9, note that $F_X(x)$ is determined from $f_X(x)$ by integration. From calculus, it follows that $f_X(x)$ is determined from $F_X(x)$ by differentiation; that is,

$$f_X(x) = \frac{d(F_X(x))}{dx}$$

provided the derivative exists at all but a finite number of points.

3.1.3 Properties of $F_X(x)$

For *any* discrete or continuous random variable, the value of $F_X(x)$ at any x must be a number in the interval $0 \leq F_X(x) \leq 1$. The function $F_X(x)$ is always continuous from the right. It is nondecreasing as x increases; that is, if $x_1 \leq x_2$ then $F_X(x_1) \leq F_X(x_2)$. Last,

$$\lim_{x \to -\infty} F_X(x) = 0 \text{ and } \lim_{x \to \infty} F_X(x) = 1$$

Example 3.1 *Let I be a continuous random variable* that represents the size of a software application being developed for a data reduction task. Let I be expressed*

---

* In this example, and in many that follow, software size I is *treated* as a continuous random variable. In reality, the number of delivered source instructions for a software application is

as the number of delivered source instructions (DSI). Suppose the state of the technical information about the application's functional and performance require- ments is very sparse. Given this, suppose subject matter experts have assessed the "true" size of the application will fall somewhere in the interval [1000, 5000]. Furthermore, because of the sparseness of available information, suppose their size assessment is such that I could take any value in [1000, 5000] with constant (uniform) probability density.

    a. Compute the PDF and the CDF of I.

    b. Determine a value x such that $P(I \leq x) = 0.80$.

Solution

    a. Figure 3.7 presents a function with the property that its value is c (a constant) at any point in the interval [1000, 5000]. For this function to be a probability density, it is necessary to find c such that its area is one. It will then be true that all subintervals of [1000, 5000] that are the same in length will occur with equal, or constant, probability (an exercise for the reader).

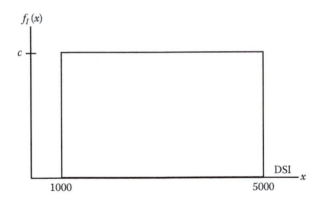

FIGURE 3.7
Probability density function for Example 3.1.

---

a positive integer—for example, *it takes 4553 source instructions to pre-process the data stream passing into the radar's primary processor.* If, for example, we treat software size as a discrete random variable, then each distinct value (assessed by subject matter experts as "possible") also requires an assessment of its probability of occurrence. Although this is a valid way to describe such a random variable, it is not clear how many distinct values (and their associated probabilities) are needed to adequately capture the overall distribution of possible values. In practice, a continuous distribution is often used to describe the range of possible values for a random variable such as software size. This enables subject matter experts to focus on the "shape" that best describes the distribution of probability, rather than assessing individual probabilities associated to each distinct possible value. If needed, the resulting continuous distribution could later be translated into a discrete form.

From Figure 3.7, $f_I(x)$ can be written as

$$f_I(x) = \begin{cases} c, & \text{if } 1000 \le x \le 5000 \\ 0, & \text{otherwise} \end{cases} \tag{3.12}$$

For $f_I(x)$ to be a PDF, we need to find c such that

$$\int_{-\infty}^{\infty} f_I(x)\, dx = \int_{1000}^{5000} c\, dx = 4000c = 1$$

Therefore $c = \frac{1}{4000}$. Thus, the PDF of the random variable I is

$$f_I(x) = \begin{cases} \dfrac{1}{4000}, & \text{if } 1000 \le x \le 5000 \\ 0, & \text{otherwise} \end{cases} \tag{3.13}$$

To determine the CDF, we must evaluate the integral

$$F_I(x) = P(I \le x) = \int_{-\infty}^{x} f_I(t)\, dt \quad \text{for} \quad -\infty < x < \infty$$

as x moves across the interval $-\infty < x < \infty$. From Equation 3.9, and the PDF in Equation 3.13, we can write the CDF as

$$F_I(x) = \begin{cases} 0, & \text{if } x < 1000 \\ \displaystyle\int_{1000}^{x} \dfrac{1}{4000}\, dt = (x - 1000)/4000, & \text{if } 1000 \le x < 5000 \\ 1, & \text{if } x \ge 5000 \end{cases}$$

$$\tag{3.14}$$

Note that $F_I(x)$ is a straight line, as illustrated in Figure 3.8.

b. The value of x such that $P(I \le x) = 0.80$ is obtained from Equation 3.14 by solving

$$\frac{x - 1000}{4000} = 0.80$$

for x. The solution is $x = 4200$. Therefore, there is an 80% chance the "true" software size will be less than or equal to 4200 DSI.

Example 3.2 *Suppose the PDF for I in Example 3.1 is now defined by the two regions shown in Figure 3.9.*

a. *Find c such that $f_I(x)$ in Figure 3.9 is a PDF.*
b. *Determine $F_I(x)$.*
c. *Compute $P(I \le 2000)$, $P(2000 < I < 4000)$, $P(2000 < I \le 5000)$.*

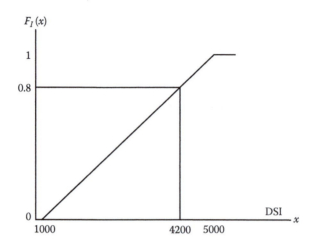

FIGURE 3.8
The cumulative distribution function for Example 3.1.

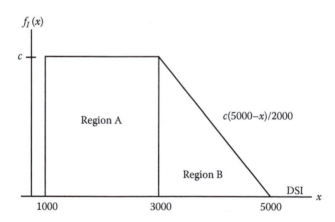

FIGURE 3.9
Probability density function for Example 3.2.

Solution

a. From Figure 3.9 it can be determined that

$$f_I(x) = \begin{cases} c, & \text{if } 1000 \leq x < 3000 \\ c(5000 - x)/2000, & \text{if } 3000 \leq x \leq 5000 \end{cases} \tag{3.15}$$

For $f_I(x)$ to be a PDF there must exist a constant c such that

$$\int_{-\infty}^{\infty} f_I(x)\, dx = 1$$

This implies c is the solution to

$$\int_{1000}^{3000} c\, dx + \int_{3000}^{5000} c((5000 - x)/2000)\, dx = 1$$

from which $c = 1/3000$. Thus, the PDF is

$$f_I(x) = \begin{cases} 1/3000, & \text{if } 1000 \leq x < 3000 \\ (5000 - x)/6(10^6), & \text{if } 3000 \leq x \leq 5000 \end{cases} \quad (3.16)$$

b. To determine the CDF $F_I(x)$, we must evaluate

$$F_I(x) = P(I \leq x) = \int_{-\infty}^{x} f_I(t)\, dt \quad \text{for} \quad -\infty < x < \infty \quad (3.17)$$

as x moves across the interval $-\infty < x < \infty$. From the PDF given in Equation 3.16, $F_I(x)$ is

$$F_I(x) = \begin{cases} 0, & \text{if } x < 1000 \\ \int_{1000}^{x} \dfrac{1}{3000}\, dt, & \text{if } 1000 \leq x < 3000 \\ \int_{1000}^{3000} \dfrac{1}{3000}\, dt + \int_{3000}^{x} (5000 - t)/6(10^6)\, dt, & \text{if } 3000 \leq x < 5000 \\ 1, & \text{if } x \geq 5000 \end{cases}$$

which is equal to

$$F_I(x) = \begin{cases} 0, & \text{if } x < 1000 \\ (x - 1000)/3000, & \text{if } 1000 \leq x < 3000 \\ 2/3 - (x - 7000)(x - 3000)/12(10^6), & \text{if } 3000 \leq x < 5000 \\ 1, & \text{if } x \geq 5000 \end{cases} \quad (3.18)$$

c. Probabilities can be determined from Equation 3.18. The probability I is less than or equal to 2000 DSI is

$$P(I \leq 2000) = F_I(2000) = \frac{1}{3} = 0.333$$

The probability I will fall between 2000 and 4000 DSI is

$$P(2000 < I < 4000) = F_I(4000) - F_I(2000) = \frac{7}{12} = 0.583$$

The probability I will fall between 2000 and 5000 DSI is

$$P(2000 < I \leq 5000) = F_I(5000) - F_I(2000) = \frac{2}{3} = 0.667\blacklozenge$$

A graph of the CDF for this example is given in Figure 3.10. When examining such a CDF, it is often useful to determine the value of x associated with $F_X(x) = 0.50$. In Figure 3.10, this value is 2500 (an exercise for the reader). A value of 2500 DSI for I has an equal probability of being larger or smaller.

This leads to the definition of an important measure about a distribution function known as the median. If X is a random variable with distribution function $F_X(x)$, a number x satisfying *both*

$$P(X \le x) \ge \frac{1}{2} \quad \text{and} \quad P(X \ge x) \ge \frac{1}{2}$$

is called the *median of* X. This will be denoted by $Med(X)$. Using Theorem 3.2, the above inequalities combine to yield the expression (Rohatgi 1976)

$$\frac{1}{2} \le F_X(x) \le \frac{1}{2} + P(X = x) \tag{3.19}$$

If X is a continuous random variable, we know $P(X = x) = 0$ for all x; therefore, from Expression 3.19, the median of X is the number x satisfying

$$F_X(x) = \frac{1}{2} \tag{3.20}$$

When X is a continuous random variable its distribution function $F_X(x)$ is monotonically increasing, as seen in Figure 3.10; therefore, a unique value of x exists such that Equation 3.20 is satisfied. When X is a *discrete* random

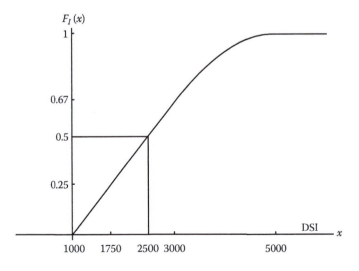

FIGURE 3.10
Cumulative distribution function for Example 3.2.

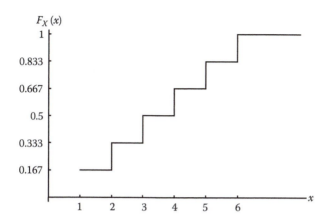

FIGURE 3.11
CDF of a random variable with uncountably many medians.

variable, $Med(X)$ may not be unique. For instance, in Figure 3.11, every point in the interval $3 \le x < 4$ is a median of X. From Figures 3.10 and 3.11 we see that *every distribution function has at least one median*.

The median is one measure among a class of measures about a distribution known as *fractiles*. In general, the value x_α is called the α-fractile of X if $P(X \le x_\alpha) = \alpha$. The median is the 0.50-fractile of X; its value is given by $x_{0.50}$. In Figure 3.10, we have $x_{0.50} = 2500$. Other α-fractiles of common interest are $x_{0.25}$ and $x_{0.75}$. Fractiles are one way to express *percentiles* of a distribution. In general, the α-fractile of X is the $\alpha(100)$th percentile of X. For instance, the median is the 50th percentile of X.

3.2 Expectation of a Random Variable

When looking at the possible values of a random variable, a useful value to determine is its expectation. The expectation of a random variable is also known as its mean. The *expectation* (or *mean*) of a *discrete* random variable X is defined as

$$E(X) \equiv \mu_X = \sum_x x p_X(x) \tag{3.21}$$

The expectation* of a random variable is the summation of all its possible values weighted by the probabilities associated with these values. The

---

* The expectation $E(X)$ for a discrete random variable X exists if and only if the summation in Equation 3.21 is absolutely convergent; that is, if and only if $\sum_x |x| \, p_X(x) < \infty$.

terms expectation and mean (usually denoted by the Greek symbol μ) are synonymous.

Example 3.3 *Return to Case Discussion 3.1 and determine the following:*

    a. $P(Profit \geq E(Profit))$

    b. $P(Profit = Med(Profit))$

Solution

  a. First determine $E(Profit)$. From Case Discussion 3.1, the probability function for *Profit* is given in Figure 3.5. Since *Profit* was defined by a discrete random variable, from Equation 3.21 we have

$$E(Profit) = \sum_{i=1}^{10} x_i p_{Profit}(x_i)$$

$$= -300(0.006) + (-200)(0.004) + (-100)(0.010)$$

$$+ 0(0.190) + 100(0.210) + 200(0.239) + 300(0.156)$$

$$+ 400(0.062) + 600(0.105) + 900(0.018)$$

$$= 216$$

Therefore, the expected profit is 216 ($M). From Theorem 3.2, the probability *Profit* will be greater than or equal to its expected value is

$$P(Profit \geq E(Profit)) = 1 - F_{Profit}(E(Profit)) + P(Profit = E(Profit))$$

or

$$P(Profit \geq 216) = 1 - F_{Profit}(216) + P(Profit = 216)$$

From Equation 3.8, $F_{Profit}(216) = 0.659$; however, $P(Profit = 216) = 0$ since the point $x = 216$ is not a feasible (possible) value for *Profit*; so

$$P(Profit \geq 216) = 1 - 0.659 + 0 = 0.341$$

  b. First determine $Med(Profit)$. The median of *Profit* can be found by expression (3.19). Referring to Equation 3.8 and Figure 3.5, it can be seen that $x = 200$ satisfies both

$$P(Profit \leq x) \geq \frac{1}{2} \quad and \quad P(Profit \geq x) \geq \frac{1}{2}$$

From Equation 3.8

$$P(Profit \leq 200) = F_{Profit}(200) = 0.659 \geq 1/2$$

Therefore, the first inequality $P(Profit \leq x) \geq 1/2$ is true when $x = 200$. It now remains to verify that $P(Profit \geq x) \geq 1/2$ when $x = 200$. From Theorem 3.2

$$P(Profit \geq 200) = 1 - F_{Profit}(200) + P(Profit = 200)$$

$$= 1 - 0.659 + 0.239 = 0.580 \geq 1/2$$

Therefore, the second inequality is also true. It is left as an exercise for the reader to show that $x = 200$ is the *only* median of *Profit*, in this case. To complete part (b) we need to determine $P(Profit = Med(Profit))$. Since it was established that $Med(Profit) = 200$, it can be readily seen from Figure 3.5

$$P(Profit = Med(Profit)) = P(Profit = 200) = p_{Profit}(200) = 0.239$$

This result could also be obtained from the CDF of *Profit*. Recall $P(X = a) = p_X(a)$ is the height of the jump of $F_X(x)$ at $x = a$, where a is a feasible value of X. From Equation 3.8, the height of the jump of $F_{Profit}(x)$ at $x = 200$ is $0.659 - 0.420 = 0.239$. Figure 3.12 illustrates this probability and presents the CDF for *Profit*, as described in Case Discussion 3.1.◆

Example 3.4 *Suppose the probability function of the cost to develop an inspection system for radomes is given below.*

a. *What is the expected cost?*
b. *What is the 0.95-fractile of Cost?*

Cost ($M)	40	65	80	95	105
Probability Function for *Cost*	0.30	0.20	0.25	0.20	0.05

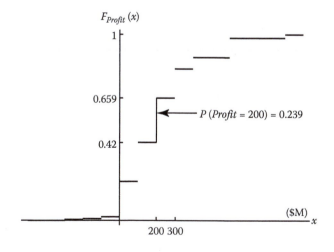

FIGURE 3.12
Cumulative distribution function for profit—defined in Case Discussion 3.1 and Example 3.3.

Solution

a. From the information in this table and Equation 3.21

$$E(Cost) = 40(0.30) + 65(0.20) + 80(0.25) + 95(0.20) + 105(0.05)$$
$$= 69.25$$

Therefore, the expected cost of the inspection system is 69.25 ($M).

b. We will use the CDF to determine the 0.95-fractile of *Cost*. The following table expresses the probability function and the distribution function of *Cost*. From this table, 95 ($M) is the 0.95-fractile of *Cost*; that is, $P(Cost \leq 95) = 0.95$.

Cost ($M)	Probability Function	Cumulative Probability
40	0.30	0.30
65	0.20	0.50
80	0.25	0.75
95	0.20	0.95
105	0.05	1.00

This discussion focused on determining the expected value of a random variable for the discrete case. If X is a *continuous* random variable, the expectation* (or the mean) of X is defined as

$$E(X) \equiv \mu_X = \int_{-\infty}^{\infty} x f_X(x)\, dx \qquad (3.22)$$

Example 3.5 *Using Equation 3.22, compute $E(I)$ in Example 3.1.*

Solution In Example 3.1, the PDF of I was

$$f_I(x) = \begin{cases} \dfrac{1}{4000}, & \text{if } 1000 \leq x \leq 5000 \\ 0, & \text{otherwise} \end{cases}$$

from Equation 3.22

$$E(I) = \int_{-\infty}^{\infty} x f_I(x)\, dx = \int_{1000}^{5000} x \frac{1}{4000}\, dx = 3000 \text{ DSI}$$

---

* The expectation $E(X)$ for a continuous random variable X exists if and only if the integral in Equation 3.22 is absolutely convergent; that is, if and only if $\int_{-\infty}^{\infty} |x| f_X(x)\, dx < \infty$.

Therefore, the expected (mean) size $E(I)$ of the software application described in Example 3.1 is 3000 DSI. In Figure 3.13, notice $E(I)$ falls exactly between the interval $[1000, 5000]$. In Chapter 4, we will see when $f_X(x)$ is described by a *rectangular region*, within an interval $[a, b]$, then $E(X) = (a + b)/2$.

Example 3.6 *Compute $E(I)$ for the PDF in Example 3.2.*

Solution In Example 3.2, the PDF of I was

$$f_I(x) = \begin{cases} 1/3000, & \text{if } 1000 \le x < 3000 \\ (5000 - x)/6(10^6), & \text{if } 3000 \le x \le 5000 \end{cases}$$

Using Equation 3.22

$$E(I) = \int_{-\infty}^{\infty} x f_I(x)\, dx = \int_{1000}^{3000} x \frac{1}{3000}\, dx + \int_{3000}^{5000} x((5000 - x)/6(10^6))\, dx$$

$$= 2555.56 \approx 2556 \text{ DSI}$$

Therefore, the expected (or mean) size $E(I)$ of the software application described in Example 3.2 is approximately 2556 DSI. A graph illustrating the location of $E(I)$, in this example, is shown in Figure 3.14.

Example 3.7 *Let Cost denote the unit production cost of a transmitter synthesizer unit (TSU) for a communications terminal. Suppose there is uncertainty in the fabrication, assembly, inspection, and test hours per TSU. Because of this, suppose production engineering assessed that Cost is best described by the PDF in Figure 3.15. Determine*

    a. *$E(Cost)$*

    b. *$P(Cost > E(Cost))$*

    c. *$Med(Cost)$*

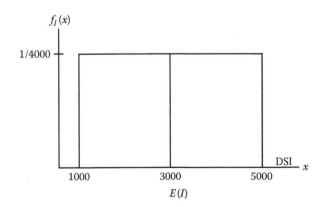

FIGURE 3.13
The expectation of I for Example 3.5.

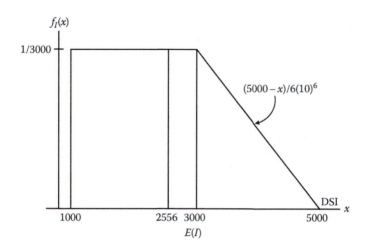

FIGURE 3.14
The expectation of *I* for Example 3.6.

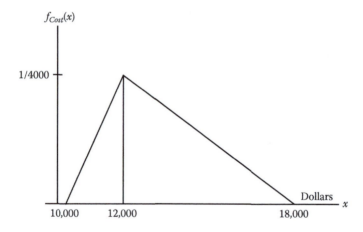

FIGURE 3.15
PDF for *Cost* in Example 3.7.

Solution

a. To compute $E(Cost)$, it is necessary to determine the mathematical form of the PDF in Figure 3.15. It is left to the reader to verify Equation 3.23 is indeed the PDF.

$$f_{Cost}(x) = \begin{cases} (x - 10{,}000)/8(10^6), & \text{if } 10{,}000 \leq x < 12{,}000 \\ (18{,}000 - x)/24(10^6), & \text{if } 12{,}000 \leq x \leq 18{,}000 \end{cases}$$

$$(3.23)$$

From Equation 3.22

$$E(Cost) \equiv \mu_{Cost} = \int_{-\infty}^{\infty} x f_{Cost}(x)\, dx$$

For this example

$$E(Cost) = \int_{10,000}^{12,000} x((x - 10,000)/8(10^6))\, dx$$

$$+ \int_{12,000}^{18,000} x((18,000 - x)/24(10^6))\, dx = 13,333.3$$

Thus, the expected (mean) cost of the TSU is approximately 13,333 dollars.

b. To compute $P(Cost > E(Cost))$, recall from Theorem 2.1

$$P(Cost > E(Cost)) = 1 - P(Cost \le E(Cost))$$

$$= 1 - F_{Cost}(E(Cost)) = 1 - F_{Cost}(13,333.3)$$

From Equation 3.9

$$F_{Cost}(13,333.3) = \int_{-\infty}^{13,333.3} f_{Cost}(t)\, dt$$

$$= \int_{10,000}^{12,000} ((t - 10,000)/8(10^6))\, dt$$

$$+ \int_{12,000}^{13,333.3} ((18,000 - t)/24(10^6))\, dt = 0.54629$$

Therefore

$$P(Cost > E(Cost)) = 1 - 0.54629 = 0.45371$$

c. From Equation 3.20, the median of *Cost* is

$$Med(Cost) = P(Cost \le x) = 0.50$$

We need to find x such that

$$F_{Cost}(x) = P(Cost \le x) = \int_{-\infty}^{x} f_{Cost}(t)\, dt = 0.50$$

In Figure 3.15, the area under the curve between $10,000 \le x < 12,000$ accounts for only 25% of the total area (which must equal unity) between $10,000 \le x \le 18,000$; that is,

$$P(Cost \le 12,000) = \int_{10,000}^{12,000} ((t - 10,000)/8(10^6))\, dt = 0.25$$

Therefore, the value of x that satisfies $P(Cost \leq x) = 0.50$ must be to the right of $x = 12,000$. To find this value we need to solve the equation below; specifically, we must find x such that

$$\int_{10,000}^{12,000} ((t - 10,000)/8(10^6))\, dt + \int_{12,000}^{x} ((18,000 - t)/24(10^6))\, dt = 0.50$$

This expression simplifies to solving

$$\int_{12,000}^{x} ((18,000 - t)/24(10^6))\, dt = 0.25$$

for x. It turns out the only feasible value for x is 13,101; showing this is left for the reader. Therefore, we say the median cost of the transmitter synthesizer unit is 13,101; that is, $Med(Cost) = 13,101$ dollars.* ◆

Thus far, we have discussed the expectation (or mean) and the median of a random variable. Another value of interest is the mode. The mode of a random variable X, denoted by $Mode(X)$, is the value of X that occurs most frequently. It is often referred to as the most likely or most probable value of X. Formally, we say that a is the *mode of X* if

$$p_X(a) = \max_t p_X(t) \text{ when } X \text{ is a } discrete \text{ random variable}$$

$$f_X(a) = \max_t f_X(t) \text{ when } X \text{ is a } continuous \text{ random variable}$$

The mode of a random variable is not necessarily unique. The random variable described by the rectangular PDF in Figure 3.7 does not have a *unique* mode. However, in Example 3.7, $x = 12,000$ is the *unique* mode of the random variable *Cost*. The mean, median, and mode of a random variable are collectively known as *measures of central tendency*. Figure 3.16 illustrates these measures for the PDF in Example 3.7.

The term average is often used in the same context as the expected value (or mean) of a random variable. The following theorem explains this context.

Theorem 3.8   *Let X be a random variable with mean E(X). If an experiment is repeated n-times under identical conditions and X_i is the random variable X associated with the ith round of the experiment, then*

$$P\left(\lim_{n \to \infty} \frac{1}{n} \sum_{i=1}^{n} X_i = E(X) \right) = 1$$

---

* Mentioned in the preface, the numerical precision shown in this example, and elsewhere in this book, is strictly for teaching purposes. Rounding results to a sensible level of precision is always applied in practice, particularly in the practice of cost analysis.

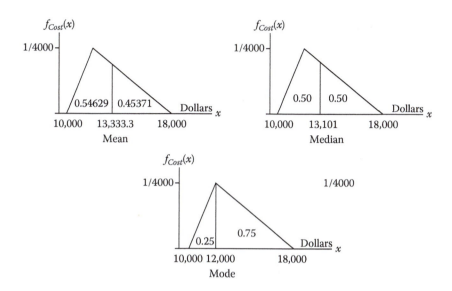

FIGURE 3.16
Central tendency measures for the PDF in Example 3.7.

Theorem 3.8 is known as the *strong law of large numbers*. It states that for sufficiently large n, it is virtually certain the average of the observed values of $X_1, X_2, \ldots, X_n$ will be approximately the same as the expected value of X. For example, it can be shown the expected value associated with tossing a fair six-sided die is 3.5. This does not mean we expect to obtain 3.5 on a toss; rather, the average value of many repeated tosses is expected to be approximately 3.5.

3.2.1 Expected Value of a Function

The need to determine the expected value of a function arises frequently in practice. For instance, in cost analysis the effort Eff_{SW} (staff months) to develop software of size I might be given by*

$$\mathit{Eff}_{SW} = 2.8I^{1.2} \tag{3.24}$$

We might ask "*What is the expected software development effort?*" Assuming I is a continuous random variable, from Equation 3.22 we could write the expected software development effort as

$$E(\mathit{Eff}_{SW}) = \int_{-\infty}^{\infty} u f_{\mathit{Eff}_{SW}}(u)\, du \tag{3.25}$$

---

* Boehm, B. W. 1981. *Software Engineering Economics*. Englewood Cliffs, NJ: Prentice-Hall, Inc. In Equation 3.24, I is in thousands of DSI.

To use Equation 3.25 we need the PDF of Eff_{SW}. As we shall see in Chapter 5, this can be difficult for certain kinds of functions. Is there another approach to computing $E(Eff_{SW})$? Note that Eff_{SW} is a function of I.

$$Eff_{SW} = 2.8I^{1.2} = g(I) \tag{3.26}$$

It follows that

$$E(Eff_{SW}) = E(g(I)) \tag{3.27}$$

The following proposition presents a general way to determine $E(Eff_{SW})$ from $E(g(I))$, where $E(g(I))$ is determined from the PDF of I.

Proposition 3.1   *If X is a random variable and $g(x)$ is a real-valued function defined for all x that are feasible (possible) values of X, then*

$$E(g(X)) = \begin{cases} \sum_{x} g(x) p_X(x), & \text{if } X \text{ is discrete} \\ \int_{-\infty}^{\infty} g(x) f_X(x) \, dx, & \text{if } X \text{ is continuous} \end{cases} \tag{3.28}$$

In this equation, the summation and integral must be absolutely convergent. Applying Proposition 3.1 to the discussion on Eff_{SW}, we have

$$E(Eff_{SW}) = E(g(I)) = \int_{-\infty}^{\infty} g(x) f_I(x) \, dx \tag{3.29}$$

Thus, the only information needed to determine $E(Eff_{SW})$ is the function $g(I)$ and $f_I(x)$, the PDF of I. For now, further discussion of this problem is deferred to Chapter 5. In particular, Case Discussion 5.2 presents the determination of $E(Eff_{SW})$ in detail.

Theorem 3.9   *If a and b are real numbers, then $E(aX + b) = aE(X) + b$*

Proof. Let $g(X) = aX + b$; if X is a discrete random variable, then from Equation 3.28

$$E(aX + b) = \sum_{x} (ax + b) p_X(x) = \sum_{x} ax p_X(x) + \sum_{x} b p_X(x)$$

$$= a \sum_{x} x p_X(x) + b \sum_{x} p_X(x) = aE(X) + b \cdot 1 = aE(X) + b$$

If X is a continuous random variable, then from Equation 3.28

$$E(aX + b) = \int_{-\infty}^{\infty} (ax + b)f_X(x)\,dx = a \int_{-\infty}^{\infty} xf_X(x)\,dx + b \int_{-\infty}^{\infty} f_X(x)\,dx$$

$$= aE(X) + b \cdot 1 = aE(X) + b \blacklozenge$$

Directly from this proof it can be shown the expected value of a constant is the constant itself; that is, $E(b) = b$. From Theorem 3.9, it can also be seen that $E(aX) = aE(X)$, where a is a real number. Showing these two results is an exercise for the reader.

Thus far, we have addressed the expectation (or mean) of a random variable. A quantity known as the variance measures its spread or dispersion (deviation) around the mean. The *variance* of a random variable X is

$$Var(X) \equiv \sigma_X^2 = E\left[(X - E(X))^2\right] \equiv E\left[(X - \mu_X)^2\right] \qquad (3.30)$$

The positive square root of $Var(X)$ is known as the *standard deviation* of X, which is denoted by σ_X.

$$\sigma_X = \sqrt{Var(X)} \qquad (3.31)$$

Example 3.8 *Let X represent the sum of the toss of a pair of fair dice.*

    a. *Determine the expected sum.*

    b. *Determine the variance of the sum.*

Solution In this example, X is a discrete random variable.

    a. From Equation 3.21 and Figure 3.2, the expected sum is

$$E(X) = \frac{1}{36}(2) + \frac{2}{36}(3) + \frac{3}{36}(4) + \frac{4}{36}(5) + \frac{5}{36}(6) + \frac{6}{36}(7)$$

$$+ \frac{5}{36}(8) + \frac{4}{36}(9) + \frac{3}{36}(10) + \frac{2}{36}(11) + \frac{1}{36}(12) = \frac{252}{36} = 7$$

    b. From part (a) we can write $Var(X) = E[(X - 7)^2]$. If we let $g(X) = (X - 7)^2$ then from Equation 3.28

$$E[g(X)] = E[(X - 7)^2] = \sum_x (x - 7)^2 p_X(x) \quad x = 2, \ldots, 12 \qquad (3.32)$$

From Figure 3.2, $p_X(2) = \frac{1}{36}, p_X(3) = \frac{2}{36}, \ldots, p_X(12) = \frac{1}{36}$. Working through the computation, Equation 3.32 is equal to 5.833; therefore,

$$Var(X) = E[g(X)] = E[(X - 7)^2] = \sum_x (x - 7)^2 p_X(x) = 5.833$$

The variance computed in Example 3.8 could be interpreted as follows: The average value of the square of the deviations from the expected sum ($E(X) = 7$) of many repeated tosses of two dice is 5.833. In this case, what is the standard deviation of X? From the definition of $Var(X)$ in Equation 3.30, we can deduce the following theorems.

Theorem 3.10 $Var(X) = E(X^2) - (\mu_X)^2$

Proof.   The proof follows from the definition of $Var(X)$ and the properties of expectation, as presented in Theorem 3.9.

$$Var(X) = E[(X - E(X))^2]$$
$$= E(X^2 - 2XE(X) + (E(X))^2) = E(X^2 - 2X\mu_X + (\mu_X)^2)$$
$$= E(X^2) - E(2X\mu_X) + E(\mu_X)^2$$
$$= E(X^2) - 2\mu_X E(X) + (\mu_X)^2$$
$$= E(X^2) - 2(\mu_X)^2 + (\mu_X)^2$$
$$= E(X^2) - (\mu_X)^2$$

Theorem 3.10 is a convenient alternative for computing the variance of a random variable. It is left as an exercise for the reader to use this theorem to verify $Var(X) = 5.833$, where X is the random variable in Example 3.8.

Theorem 3.11   *If a and b are real numbers, then*

$$Var(aX + b) = a^2 Var(X)$$

Proof.   The proof follows directly from the definition of $Var(X)$ and Theorem 3.9; that is,

$$Var(X) = E[(aX + b - E(aX + b))^2]$$
$$= E[(aX + b - aE(X) - b)^2]$$
$$= E[(aX - aE(X))^2]$$
$$= E[(a(X - E(X)))^2]$$
$$= E[a^2(X - E(X))^2]$$
$$= a^2 E[(X - E(X))^2]$$
$$= a^2 Var(X)$$

This theorem demonstrates the variance of a random variable described by the *linear* function $aX + b$ is unaffected by the constant term b.

Example 3.9 *For the communication terminal's transmitter synthesizer unit (TSU) described in Example 3.7, compute*

a. *Var(Cost) and σ_{Cost} using Theorem 3.10*
b. *Determine $P(|Cost - \mu_{Cost}| \leq \sigma_{Cost})$*

Solution

a. From Example 3.7, the PDF for *Cost* is

$$f_{Cost}(x) = \begin{cases} (x - 10{,}000)/8(10^6) & 10{,}000 \leq x < 12{,}000 \\ (18{,}000 - x)/24(10^6) & 12{,}000 \leq x \leq 18{,}000 \end{cases}$$

From Theorem 3.10 we have

$$Var(Cost) = E(Cost^2) - (\mu_{Cost})^2$$

From part (a) in Example 3.7, $\mu_{Cost} = E(Cost) = 13{,}333.3$; therefore,

$$Var(Cost) = E(Cost^2) - (13{,}333.3)^2$$

From Equation 3.28 we can write

$$E(Cost^2) = \int_{10{,}000}^{12{,}000} x^2((x - 10{,}000)/8(10^6))\, dx$$
$$+ \int_{12{,}000}^{18{,}000} x^2((18{,}000 - x)/24(10^6))\, dx$$
$$= 1.80667(10^8) \quad (\$)^2$$

Therefore

$$Var(Cost) = \sigma_{Cost}^2 = 1.80667(10^8) - (13{,}333.3)^2$$
$$= 2.88889(10^6) \quad (\$)^2$$

from which

$$\sigma_{Cost} = \sqrt{Var(Cost)} = 1699.67 \approx 1700 \ (\$)$$

The variance squares the units that define the random variable. Since $\$^2$ is not a useful way to look at *Cost*, the standard deviation σ_{Cost}, which is in dollar units, is usually a better way to interpret this deviation.

b. Probabilities associated with intervals* expressed in terms of the mean and standard deviation can be computed. For some positive real number k

$$P(|Cost - \mu_{Cost}| \leq k\sigma_{Cost})$$
$$= P(\mu_{Cost} - k\sigma_{Cost} \leq Cost \leq \mu_{Cost} + k\sigma_{Cost})$$

From Equation 3.11, we can express this probability in terms of F_{Cost} as

$$P(|Cost - \mu_{Cost}| \leq k\sigma_{Cost})$$
$$= F_{Cost}(\mu_{Cost} + k\sigma_{Cost}) - F_{Cost}(\mu_{Cost} - k\sigma_{Cost})$$

For part (b) we need $k = 1$; from part (a) $\mu_{Cost} = 13{,}333.3$ and $\sigma_{Cost} = 1700$; thus,

$$P(|Cost - \mu_{Cost}| \leq \sigma_{Cost}) = P(11{,}633.3 \leq Cost \leq 15{,}033.3)$$
$$= F_{Cost}(15{,}033.3) - F_{Cost}(11{,}633.3)$$

where

$$F_{Cost}(15{,}033.3) = \int_{-\infty}^{15{,}033.3} f_{Cost}(t)\, dt$$

$$= \int_{10{,}000}^{12{,}000} ((t - 10{,}000)/8(10^6))\, dt$$

$$+ \int_{12{,}000}^{15{,}033.3} ((18{,}000 - t)/24(10^6))\, dt = 0.817$$

and

$$F_{Cost}(11{,}633.3) = \int_{-\infty}^{11{,}633.3} f_{Cost}(t)\, dt$$

$$= \int_{10{,}000}^{11{,}633.3} ((t - 10{,}000)/8(10^6))\, dt = 0.167$$

So

$$P(|Cost - \mu_{Cost}| \leq \sigma_{Cost}) = 0.817 - 0.167 = 0.65 \blacklozenge$$

---

* Probability intervals are often given in the form $P(|X - a| \leq b)$ or $P(|X - a| > b)$, where a and b are any two real numbers. In general, $P(|X - a| \leq b) = P(-b \leq X - a \leq b) = P(a - b \leq X \leq a + b)$; furthermore, $P(|X - a| > b) = 1 - P(|X - a| \leq b) = 1 - P(a - b \leq X \leq a + b)$.

The TSU cost falls within ± 1 ($k = 1$) standard deviation (σ) around its expected (or mean) cost with probability 65%. The range of values for x associated with this probability is shown in Figure 3.17. This range is sometimes referred to as the 1-sigma interval.

A random variable can be standardized when its mean and variance are known. A *standardized* random variable has zero mean and unit variance. To see this, suppose X is a random variable with mean μ_X and variance σ_X^2. The *standard form* of X is the random variable $Y = (X - \mu_X)/\sigma_X$. From Theorems 3.9 and 3.11, it can be shown Y has zero mean and unit variance; that is,

$$E(Y) = E\left(\frac{X - \mu_X}{\sigma_X}\right) = \frac{1}{\sigma_X}E(X - \mu_X) = \frac{1}{\sigma_X}[E(X) - \mu_X] = 0$$

$$Var(Y) = Var\left(\frac{X - \mu_X}{\sigma_X}\right) = \frac{1}{\sigma_X^2}Var(X - \mu_X) = \frac{1}{\sigma_X^2}Var(X) = \frac{\sigma_X^2}{\sigma_X^2} = 1$$

Referring to Example 3.9, we have

$$E(Y) = E\left(\frac{(X - 13{,}333.3)}{1700}\right) = 0$$

$$Var(Y) = Var\left(\frac{(X - 13{,}333.3)}{1700}\right) = 1$$

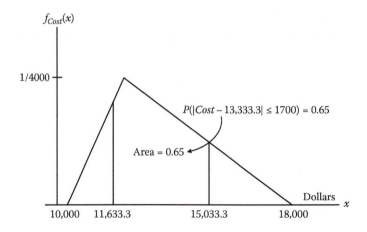

FIGURE 3.17
1-Sigma interval for the TSU cost.

3.3 Moments of Random Variables

Moments provide important information about the distribution function of a random variable. Such information includes the random variable's mean and variance, as well as the shape of its distribution function. Suppose X is a random variable and k is any positive integer. The expectation $E(X^k)$ is called the *kth moment of X*, which is given by Equation 3.33. In general, we say the kth moment of X is

$$E(X^k) = \begin{cases} \sum\limits_{x} x^k p_X(x), & \text{if } X \text{ is discrete} \\ \int\limits_{-\infty}^{\infty} x^k f_X(x)\, dx, & \text{if } X \text{ is continuous} \end{cases} \tag{3.33}$$

In Equation 3.33, the summation and integral must be absolutely convergent. The mean is the first moment of X. It is the "balance point" or the "center of gravity" of the probability mass (or density) function. This is in contrast to the median. If the random variable is discrete, the median divides the entire mass of the distribution function into two equal parts; each part contains the mass $1/2$. If the random variable is continuous, the median divides the entire area under the density function into equal parts. Each part contains an area equal to $1/2$ (refer to Figure 3.16).

The second moment of the random variable $(X - \mu_X)$ is $E[(X - \mu_X)^2]$. From Equation 3.30 this is the variance of X, which provides a measure of the dispersion of X about its mean. What do higher moments of a random variable reveal about the shape of its distribution function?

Let Y be the standardized random variable of X; that is, $Y = (X - \mu_X)/\sigma_X$. The third and fourth moments of Y are known as the coefficients of *skewness* and *kurtosis*. These coefficients are given by Equations 3.34 and 3.35, respectively.

$$\gamma_1 = E(Y^3) = E\left[\left(\frac{X - \mu_X}{\sigma_X}\right)^3\right] \tag{3.34}$$

$$\gamma_2 = E(Y^4) = E\left[\left(\frac{X - \mu_X}{\sigma_X}\right)^4\right] \tag{3.35}$$

Skewness, given by γ_1, is a measure of the symmetry of the distribution function of X about the mean of X. If this function has a long tail to the left, then γ_1 is usually negative and we say the distribution function is negatively skewed. If this function has a long tail to the right, then γ_1 is usually positive and we say the distribution function is positively skewed.

In cost analysis, it is common to see distributions with $\gamma_1 > 0$. In such distributions, the probability of exceeding the mode (often associated with the point estimate) is greater than the probability of falling below the mode. Experience suggests this is due to a variety of reasons. These include changing requirements, understating a project's true technical complexity, or planning the project against unrealistic cost or schedule objectives. Positively skewed distributions are often used to represent uncertainty in system definition variables, such as weight or software size. Point estimates for these variables, particularly in the early phases of a system's design, typically have a high probability of being exceeded.

If the distribution function of X is symmetric about the mean of X, then $\gamma_1 = 0$. The distribution function of X is symmetric about $x = a$ if

$$P(X \geq a + x) = P(X \leq a - x) \text{ for all } x \tag{3.36}$$

From Theorem 3.2, Equation 3.36 can be written as

$$F_X(a - x) = 1 - F_X(a + x) + P(X = a + x) \tag{3.37}$$

If Equation 3.37 is true for all x, we say the distribution function $F_X(x)$ is symmetric with a as the *center of symmetry*. If the center of symmetry is the origin, then $a = 0$ and

$$F_X(-x) = 1 - F_X(x) + P(X = x) \tag{3.38}$$

If X is a continuous random variable, Equation 3.38 simplifies to

$$F_X(-x) = 1 - F_X(x) \tag{3.39}$$

The distribution function of a continuous random variable X is symmetric with center a, if and only if

$$f_X(a - x) = f_X(a + x) \text{ for all } x \tag{3.40}$$

If $F_X(x)$ is a symmetric distribution, the center of symmetry is *always the median*. In certain symmetric distributions the mean or the mode may also equal the median. If the distribution function of a *continuous* random variable X is symmetric *and the mean of X exists*, then the median and mean of X are equal and they both locate the center of symmetry. The Cauchy distribution* is a symmetric distribution whose mean does not exist (i.e., it is not well

---

* The Cauchy distribution is given by $f_X(x) = \{\pi b [1 + ((x - a)/b)^2]\}^{-1}$. The moments of X do not exist; however, X has a unique median and a unique mode, which both fall at $x = a$. In the Cauchy distribution, the median and the mode are equal; they also locate the center of symmetry.

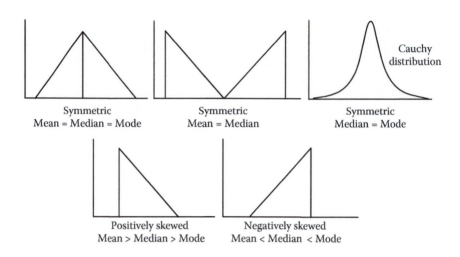

FIGURE 3.18
Illustrative symmetric and skewed distributions.

defined). It has a unique median and a unique mode that equal each other. In the Cauchy distribution, both the median and the mode locate the center of symmetry. Figure 3.18 illustrates these and other cases of symmetric and skewed distributions.

Kurtosis, given by γ_2 (Equation 3.35), measures the peakedness of a random variable's distribution function around its mean. The kurtosis of a distribution function is usually compared with the value $\gamma_2 = 3$, which is the kurtosis of a standardized normal probability distribution (discussed in Chapter 4). If $\gamma_2 > 3$, the distribution function of X has greater kurtosis (less peaked) than the normal probability distribution. If $\gamma_2 < 3$, the distribution function of X has less kurtosis (more peaked) than the normal probability distribution.

If we don't know exactly how a random variable is distributed, but we have knowledge about its mean, variance, skewness, and kurtosis, we can often guess its overall shape. In some instances, only the mean and variance of a random variable are needed to uniquely specify the form of its distribution.

3.4 Probability Inequalities Useful in Cost Analysis

Thus far, we have shown how probabilities can be computed from the distribution function of a random variable. However, circumstances frequently exist when the underlying distribution is unknown. This section presents inequalities that provide bounds on the probability of an event independent of the form of the underlying distribution function.

The *Markov inequality*, due to A. A. Markov (1856–1922), can be used to compute an upper bound on the probability of an event when X is nonnegative and only its mean is known. The *Chebyshev inequality*, derived by P. L. Chebyshev (1821–1894), bounds the probability that a random variable takes a value within k standard deviations around its mean. Chebyshev's inequality will be shown to be a consequence of Markov's inequality. Before discussing the details of these inequalities, we will first discuss the expected value of an *indicator function*.

The Indicator Function: For a random variable X, the indicator function of the event $A = \{X \geq a\}$ is

$$I_A(X) = \begin{cases} 1, & \text{if event } \{X \geq a\} \text{ occurs} \\ 0, & \text{if event } \{X \geq a\} \text{ does not occur} \end{cases}$$

The expected value of $I_A(X)$ is the probability the event A occurs. This can be seen from the following argument. From Equation 3.21, we can write

$$E(I_A(X)) = 1 \cdot P(X \geq a) + 0 \cdot [1 - P(X \geq a)] = P(A)$$

Markov's Inequality: If X is a nonnegative random variable whose mean μ is positive, then $P(X \geq c\mu) \leq c^{-1}$ for any constant $c > 0$.

Proof. The random variable X is given to be nonnegative with positive mean μ. Since $c > 0$ it follows that $c\mu > 0$. Let

$$I_A(X) = \begin{cases} 1, & \text{if event } \{X \geq c\mu\} \text{ occurs} \\ 0, & \text{if event } \{X \geq c\mu\} \text{ does not occur} \end{cases}$$

where A is the event $\{X \geq c\mu\}$. From this it follows that

$$I_A(X) \leq \frac{X}{c\mu}$$

The expected value of $I_A(X)$ is

$$E(I_A(X)) \leq \frac{1}{c\mu}E(X)$$

Since $E(X) = \mu$ and $E(I_A(X)) = P(A)$ it follows immediately that

$$P(A) = P(X \geq c\mu) \leq 1/c \blacklozenge$$

Markov's inequality states the probability X takes a value greater than or equal to c times its mean cannot exceed $1/c$. For instance, if $c = 2$ then $P(X \geq 2\mu)$ can never exceed $1/2$. If $c = 1$ then $P(X \geq \mu)$ is bounded by unity, which is consistent with the first axiom of probability (Chapter 2). Markov's inequality is meaningless if c is less than one. Markov's inequality may also be written as

$$P(X \geq a) \leq \frac{1}{a}E(X)$$

where X is nonnegative and $a > 0$; this result follows immediately from the Markov inequality proof (showing this is left as an exercise for the reader).

From a cost analysis perspective, Markov's inequality provides decision-makers an upper bound on the probability that *Cost* is greater than c times its mean. For instance, suppose the mean cost of a system is determined to be 100 million dollars ($M). Regardless of the underlying distribution function for *Cost*, Markov's inequality guarantees the probability that *Cost* takes a value greater than 200 ($M) can never exceed $1/2$.

The probability bound yielded by Markov's inequality is quite conservative. To illustrate this, suppose the random variable *Cost* is described by the PDF in Figure 3.19. This is a lognormal probability distribution* with mean 100 ($M) and standard deviation 25 ($M); it is slightly skewed to the right.

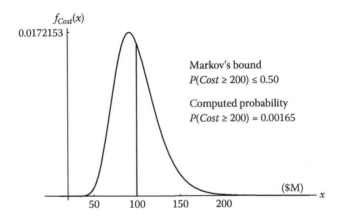

FIGURE 3.19
A lognormal PDF for *Cost* with mean 100 ($M).

---

* The lognormal distribution is often used in cost and economic analyses. It will be fully discussed in Chapter 4, with additional applications provided in the subsequent chapters.

As seen in Figure 3.19, the Markov bound is substantially larger than the computed probability of 0.00165 (shown in Example 4.8). Such a wide disparity is not surprising since Markov's inequality relies only on the mean of a random variable. In systems engineering, decision-makers typically need more insight into the probability that *Cost* is likely to be exceeded than that provided by Markov's inequality. If values for the mean and variance of *Cost* are available, then Chebyshev's inequality provides probability bounds that improve on those obtained from Markov's inequality.

Chebyshev's Inequality: If X is a random variable with finite mean μ and variance σ^2, then for $k \geq 1$

$$P(\mu - k\sigma < X < \mu + k\sigma) \geq 1 - \frac{1}{k^2} \tag{3.41}$$

Proof. Recall that

$$P(|X - a| \geq b) = 1 - P(|X - a| < b) = 1 - P(a - b < X < a + b) \tag{3.42}$$

where a and b are real numbers. Suppose we let $a = \mu$ and $b = k\sigma$. Then

$$P(|X - a| \geq b) = P(|X - \mu| \geq k\sigma)$$

Now $(X - \mu)^2 \geq k^2\sigma^2$ if and only if $|X - \mu| \geq k\sigma$; from Markov's inequality

$$P((X - \mu)^2 \geq k^2\sigma^2) \leq \frac{1}{k^2\sigma^2} E((X - \mu)^2) \tag{3.43}$$

Since $E((X - \mu)^2) = \sigma^2$, inequality (3.43) reduces to $P((X - \mu)^2 \geq k^2\sigma^2) \leq \frac{1}{k^2}$, which is equivalent to

$$P(|X - \mu| \geq k\sigma) \leq \frac{1}{k^2} \tag{3.44}$$

or

$$\frac{1}{k^2} \geq P(|X - \mu| \geq k\sigma)$$

From Equation 3.42

$$\frac{1}{k^2} \geq P(|X - \mu| \geq k\sigma) = 1 - P(\mu - k\sigma < X < \mu + k\sigma)$$

therefore

$$P(\mu - k\sigma < X < \mu + k\sigma) \geq 1 - \frac{1}{k^2} \blacklozenge$$

Chebyshev's inequality states that for any random variable X, the probability that X will assume a value within k standard deviations of its mean is at least $1 - 1/k^2$. From Equation 3.41, the probability a random variable takes a value within 2 standard deviations of its mean will always be at least 0.75. If X is a continuous random variable, at least 95% of the *area under any probability density function* will always fall within 4.5 standard deviations of the mean.

Like Markov's inequality, probabilities produced by Chebyshev's inequality are also conservative, but to a lesser extent. To illustrate this, consider once again the random variable *Cost* with mean 100 ($M), standard deviation 25 ($M), and the PDF given in Figure 3.19. It can be *computed* that the interval

$$[\mu - 2\sigma, \mu + 2\sigma] = [50, 150] \ (\$M)$$

accounts for nearly 96% (refer to Example 4.8) of the total probability (area) under $f_{Cost}(x)$. This computed probability is in contrast to Chebyshev's inequality (Equation 3.41), which indicates the interval [50, 150] ($M) accounts for at least 75% of the total probability.

Various forms of Chebyshev's inequality are given below; in each form $a > 0$.

A. $P(|X - \mu| \geq k\sigma) \leq \dfrac{1}{k^2}$

B. $P(|X - \mu| < a) \geq 1 - \dfrac{\sigma^2}{a^2}$

C. $P(|X - \mu| \geq a) \leq \dfrac{\sigma^2}{a^2}$

D. $P(X - \mu \geq a) \leq P(|X - \mu| \geq a) \leq \dfrac{\sigma^2}{a^2}$

Suppose $\mu = 100$, $\sigma = 25$, and $a = 100$. From Form D of Chebyshev's inequality we have

$$P(Cost - 100 > 100) \leq \frac{(25)^2}{(100)^2}$$

$$\Rightarrow P(Cost > 200) \leq \frac{1}{16} = 0.0625$$

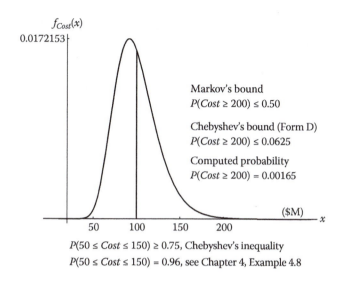

$f_{Cost}(x)$

0.0172153

Markov's bound
$P(Cost \geq 200) \leq 0.50$

Chebyshev's bound (Form D)
$P(Cost \geq 200) \leq 0.0625$

Computed probability
$P(Cost \geq 200) = 0.00165$

($M)

x

50   100   150   200

$P(50 \leq Cost \leq 150) \geq 0.75$, Chebyshev's inequality
$P(50 \leq Cost \leq 150) = 0.96$, see Chapter 4, Example 4.8

FIGURE 3.20
Some probability bounds on *Cost*.

Thus, the probability *Cost* will exceed twice its mean will not be more than $\frac{1}{16}$ (or 0.0625). From the previous discussion, Markov's inequality revealed this bound could not be more than $\frac{1}{2}$. Although these results are consistent, Form D of Chebyshev's inequality provides a significant refinement on the probability bound for this event. This is not surprising since additional information about the random variable *Cost*, specifically its variance, is taken into account. Because of this, Chebyshev's inequality will always provide a tighter probability bound than that produced by Markov's inequality. Figure 3.20 summarizes this discussion and contrasts these probability bounds for the PDF given in Figure 3.19.

The probability inequalities presented here share the common characteristic that their bounds are valid for *any type* of distribution function. Although these bounds are conservative, they do offer decision-makers probabilities that are *independent of the underlying distribution*. When inequalities such as Chebyshev's are used in conjunction with an assumed or approximated distribution, decision-makers are provided alternative ways to view the probability associated with the same event.

3.5 Cost Analysis Perspective

In cost uncertainty analysis, two important statistical measures to determine are the expected (mean) cost and the standard deviation of cost. A classical way to view the relationship between a mean and a standard deviation is presented in Figure 3.21. Shown is a special distribution known as the *normal probability distribution* (Chapter 4).

The normal distribution is symmetric about its mean. It has the property that its mode and median equal its mean. In particular, the 1-sigma interval

$$[\mu - \sigma, \mu + \sigma]$$

will always account for slightly more than 68% of the total area under a *normal* PDF. Similarly, the 2-sigma interval

$$[\mu - 2\sigma, \mu + 2\sigma]$$

will always account for slightly more than 95% of the total area under a *normal* PDF.

Although the mean is an important statistical measure that contributes many useful insights about the underlying distribution, it is just a single value among infinitely many that define the curve. Alone, the mean provides no direct view into the variability implicit to the distribution. For this reason, analysts and decision-makers must consider the mean and the standard deviation jointly. Figure 3.22 illustrates this point. Comparing just the difference in the mean costs between system design alternatives A and B, it may appear to a decision-maker alternative B is the better choice.

However, when the dispersion σ in cost is considered and the 1-sigma interval is determined for each alternative, the decision-maker may very well select alternative A instead. Specifically, the 1-sigma interval for alternative A (from Figure 3.22) is

$$[\mu - \sigma, \mu + \sigma] = [81, 99] \ (\$M)$$

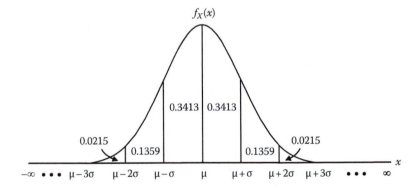

FIGURE 3.21
Areas under the normal probability distribution.

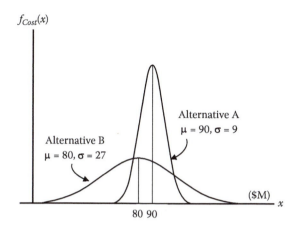

FIGURE 3.22
Comparing the mean costs of alternatives.

The 1-sigma interval for alternative B is

$$[\mu - \sigma, \mu + \sigma] = [53, 107] \ (\$M)$$

Thus, for the same level of confidence implied by the 1-sigma interval (68%) choosing alternative B implies accepting three times the variability in cost (54 ($M)) than that associated with alternative A (18 ($M)). Clearly, this result would not have been seen if comparing the mean costs was the sole criterion for selecting an alternative.

This discussion illustrates the usefulness of another statistic known as the *coefficient of dispersion*. Defined by Equation 3.45, the coefficient of dispersion D is the ratio of the standard deviation to the mean.

$$D = \frac{\sigma}{\mu} \qquad (3.45)$$

Consider again Figure 3.22. The coefficient of dispersion for alternative A is 0.10. This implies the value of Cost at one standard deviation above the mean will be 10% higher than the mean of Cost, which is 90 ($M) for alternative A. Similarly, the coefficient of dispersion for alternative B is 0.3375. This implies the value of Cost at one standard deviation above its mean will be nearly 34% higher than the mean of Cost, which is 80 ($M) for alternative B. Clearly, a significantly higher cost penalty exists at 1-sigma above the mean under alternative B than for alternative A. A decision-maker might consider this cost risk to be unacceptable. Although the cost mean for alternative A is 10 ($M) higher than the cost mean for alternative B, its significantly lower cost variance (i.e., less cost risk) may be the acceptable trade-off.

Exercises

3.1 Let X denote the sum of the toss of two fair dice. Determine the following using the probability function in Figure 3.2 and the appropriate theorems in Section 3.1.

a. $P(X < 7)$

b. $P(X > 7)$

c. $P(X \geq 7)$

d. $P(10 \leq X \leq 12)$

e. $P(10 \leq X < 12)$

f. $P(10 < X < 12)$

3.2 Suppose the probability function for the development and production cost of a microchip is given below. Determine the following:

a. The CDF of *Cost*

b. $P(Cost \leq 35)$

c. $P(Cost > 25)$

d. $P(Cost \geq 25)$

e. $P(20 \leq Cost < 35)$

f. $P(20 < Cost < 35)$

g. $P(Cost < 35)$

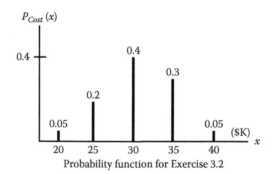

Probability function for Exercise 3.2

3.3 For any random variable X, show that $P(X < a) = F_X(a) - P(X = a)$.

3.4 Refer to Case Discussion 3.1 and answer the following:

a. Find $p_{Profit}(x)$ and $F_{Profit}(x)$ if $P(V = 5) = 0.1$, $P(V = 15) = 0.8$, and $P(V = 20) = 0.1$, where V is the sales volume (in millions).

b. With what probability does *Profit* $= 0$?

3.5 Suppose the profit function to sell 10,000 electronic widgets, with a unit price of $10 per widget, is given by *Profit* $= (10)^4(10 - U_{Cost})$, where U_{Cost} is a discrete random variable that represents the unit cost (in dollars) of each widget. If U_{Cost} can take one of the values in the set $\{4, 7, 10\}$, where u_{Cost} represents one of these values, find the constant c such that $p_{Profit}(u_{Cost}) = c\,Profit$ is a *probability function*.

3.6 Suppose *Cost* is a continuous random variable whose possible values are given by the interval $20 \le x \le 70$, where x is in dollars million (\$M).

a. Find c such that the function below is a PDF.

b. Compute $P(Cost \le 30)$, $P(30 < Cost < 70)$, $P(Cost = 30)$.

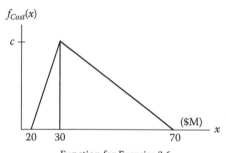

Function for Exercise 3.6

3.7 Show that $f_{Cost}(x)$ in Exercise 3.6 is the derivative of $F_{Cost}(x)$, where

$$F_{Cost}(x) = \begin{cases} 0, & \text{if } x < 20 \\[2mm] \dfrac{1}{500}(x-20)^2, & \text{if } 20 \le x < 30 \\[2mm] \dfrac{1}{5} + \dfrac{1}{50}\left[40 - \dfrac{(x-70)^2}{40}\right], & \text{if } 30 \le x < 70 \\[2mm] 1, & \text{if } x \ge 70 \end{cases}$$

3.8 For the PDF in Example 3.1 (Figure 3.7), show that all subintervals of $[1000, 5000]$ that are the same in length will occur with equal probability.

3.9 a. Given the probability function in Exercise 3.2, determine $Med(Cost)$.

b. From the *Profit* probability function in Case Discussion 3.1, show that $x = 200$ is the *only* value of x that satisfies the relationship

$$(1/2) \le F_{Profit}(x) \le (1/2) + P(Profit = x)$$

c. In Example 3.2, show that $Med(I) = 2500$ DSI.

3.10 Suppose the uncertainty in the size I of a software application is expressed by the PDF in the following figure.

a. Determine $F_I(x)$.

b. Compute $P(I \le 50{,}000)$, $P(40{,}000 \le I \le 60{,}000)$, $P(50{,}000 \le I \le 65{,}000)$.

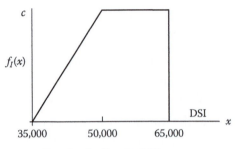

Function for Exercise 3.10

3.11 In Exercise 3.10, show that $Med(I) = 53,750$ DSI.

3.12 Find the expected number of workstations purchased per month and the standard deviation if the probability function for the monthly demand is given in the following table.

Workstations Purchased per Month	14	9	36	6	4
Probability	0.23	0.15	0.42	0.10	0.10

Probability Function for Exercise 3.12

3.13 From Case Discussion 3.1, the profit on a new electronics product manufactured and sold by ChipyTech Corporation was given by

$$Profit = (U_{Price} - U_{Cost})V$$

Suppose the product's sales volume V for its first year on the market is set at 30 million. Suppose the probability functions of U_{Price} and U_{Cost} are given in the figure below. Assume U_{Price} and U_{Cost} are independent.

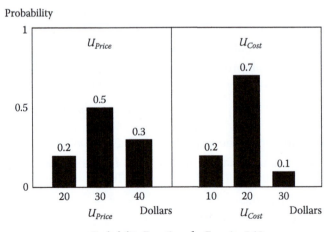

Probability Functions for Exercise 3.13

Compute

a. $p_{Profit}(x)$ and $F_{Profit}(x)$

b. $E(Profit)$

c. $Var(Profit)$

d. $P(Profit = E(Profit))$

e. $P(Profit < E(Profit))$

f. The probability of making *no profit*

3.14 A random variable X takes the value 1 with probability p and the value 0 with probability $1 - p$. Show that $E(X) = p$ and $Var(X) = p(1 - p)$.

3.15 From Exercise 3.10, compute the following:

a. $E(I)$

b. σ_I

c. $P(|I - E(I)| > \sigma_I)$

3.16 Let Y be a random variable with the probability function given in the following table. Compute

a. $E(3Y + 1)$

b. $Var(3Y + 1)$

y	1	2	3	4	5
$P(Y = y)$	1/4	1/8	1/4	1/4	1/8

Probability Function for Exercise 3.16

3.17 Suppose $E(X) = 4$ and $Var(X) = E(X)/2$. Find the expectation and variance of the random variable $(1 - 2X)/2$.

3.18 a. If X has mean μ_X show that $E(X - \mu_X)$ is always zero.

b. If a and b are constants, show that $E(b) = b$ and $E(aX) = aE(X)$.

3.19 a. Let X represent the value of the toss of a fair six-sided die. Show that $E(X) = 3.5$. Determine $Var(X)$.

b. If X is a random variable representing the sum of the toss of a pair of fair six-sided dice, use Theorem 3.10 to verify that $Var(X) = 5.833$.

3.20 Find a general formula for the kth moment of a continuous random variable X with density function $f_X(x) = (b - a)^{-1}$, where $a \leq x \leq b$.

3.21 Suppose X is a continuous random variable with $f_X(x) = 1$ in the interval $0 \leq x \leq 1$. Show that the coefficient of skewness for $f_X(x)$ is zero.

3.22 If the PDF of X is given by

$$f_X(x) = \frac{1}{\sqrt{2\pi}\sigma} e^{-\frac{1}{2}[(x-\mu)^2/\sigma^2]}$$

show that X is symmetric with center equal to μ.

3.23 If a is a constant, show that Markov's inequality can also be written in the form $P(X \geq a) \leq a^{-1}E(X)$.

3.24 Let N_W be a random variable representing the number of widgets produced in a month. Suppose the expected number of widgets produced by a manufacturer during a month is 2000.

   a. Find an upper bound on the probability this month's production will exceed 3200 widgets.

   b. Suppose the standard deviation of a month's production is known to be 35 widgets. Find a and b such that the number of widgets produced this month falls in the interval $a < N_W < b$ with probability at least 0.75.

3.25 Suppose *Cost* is a random variable with $E(Cost) = 3$ and $Var(Cost) = 1$. Use Chebyshev's inequality to compute a lower bound on

   a. $P(2 < Cost < 4)$

   b. $P(|Cost - 3| < 5)$

References

Park, W. R. and D. E. Jackson. 1984. *Cost Engineering Analysis—A Guide to Economic Evaluation of Engineering Projects*, 2nd edn. New York: John Wiley & Sons, Inc.

Rohatgi, V. K. 1976. *An Introduction to Probability Theory and Mathematical Statistics*. New York: John Wiley & Sons, Inc.

4

Special Distributions for Cost Uncertainty Analysis

In probability theory there is a class of distribution functions known as special distributions. Special distributions are those that occur frequently in the theory and application of probability. A well-known special distribution is the Bernoulli distribution, a discrete distribution whose probability function is given by Equation 4.1. The Bernoulli distribution can be used to study a random variable X representing the outcome of an experiment that succeeds, $\{X = 1\}$, with probability p or fails, $\{X = 0\}$, with probability $(1 - p)$.

$$p_X(x) = P(X = x) = \begin{cases} p, & \text{if } x = 1 \\ 1 - p, & \text{if } x = 0 \end{cases} \qquad (4.1)$$

Another well-known special distribution is the normal distribution, a continuous distribution discussed later in this chapter. Special distributions have been well-studied over the years and are fully described in a two-volume text by Johnson and Kotz (1969). To avoid an extended exposition on the entire class of special distributions, this chapter focuses on a subset of them that frequently arises in cost uncertainty analysis.

4.1 Trapezoidal Distribution

The trapezoidal distribution is illustrated in Figure 4.1. It is rarely presented in traditional, or classical, texts on probability theory. Despite this, the trapezoidal distribution is highly useful and flexible for many situations in cost uncertainty analysis. Seen in Figure 4.1, it can model a random variable whose probability density function (PDF) increases in the interval $a \leq x < m_1$, remains constant across the interval $m_1 \leq x < m_2$, then decreases to zero in the interval $m_2 \leq x \leq b$.

Mathematically, a trapezoidal distribution can arise from the sum of two independent continuous random variables whose PDFs are constants over different closed intervals of the real line.* In cost uncertainty analysis, the

---

* Independent random variables are discussed in Chapter 5.

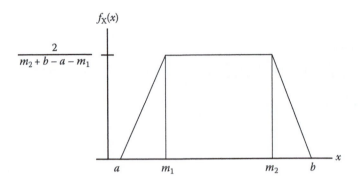

FIGURE 4.1
Trapezoidal probability density function.

trapezoidal distribution is primarily used to *directly* specify a range of possible values for a random variable. For instance, suppose an experienced software engineer was asked to assess the number of DSI needed to build a particular software application. The engineer may have solid technical reasons why this number would be less than $x = b$ DSI or be greater than $x = a$ DSI. However, the engineer may believe it is more likely the number of DSI will fall in an interval of constant density between m_1 and m_2. This can be represented by a trapezoidal distribution, as shown in Figure 4.1.

A random variable X is said to have a *trapezoidal distribution* if its PDF is given by Equation 4.2 (Young and Young 1995)

$$f_X(x) = \begin{cases} \dfrac{2}{(m_2 + b - a - m_1)} \cdot \dfrac{1}{m_1 - a}(x - a), & \text{if } a \leq x < m_1 \\[3mm] \dfrac{2}{(m_2 + b - a - m_1)}, & \text{if } m_1 \leq x < m_2 \\[3mm] \dfrac{2}{(m_2 + b - a - m_1)} \cdot \dfrac{1}{b - m_2}(b - x), & \text{if } m_2 \leq x \leq b \end{cases} \quad (4.2)$$

where $-\infty < a < m_1 < m_2 < b < \infty$. A trapezoidal PDF is illustrated in Figure 4.1. The numbers a and b represent the minimum and maximum possible values of X, respectively. Note that $f_X(x) = 0$ if $x < a$ or $x > b$. The mode of X is not unique. It is any value of x in the interval $m_1 \leq x \leq m_2$. For the remainder of this book, a random variable X with PDF given by Equation 4.2 will be implied by the following expression:

$$X \sim Trap(a, m_1, m_2, b)^*$$

---

* The symbol $\sim$ means "is distributed as." In this case, we say X is distributed as a trapezoidal random variable with parameters a, m_1, m_2, and b. We might also say X is a trapezoidal random variable with PDF given by Equation 4.2.

The cumulative distribution function (CDF) of X is given by Equation 4.3 (Young and Young 1995). A graph of $F_X(x)$ is shown in Figure 4.2.

$$F_X(x) = \begin{cases} 0, & \text{if } x < a \\[2mm] \dfrac{1}{(m_2 + b - a - m_1)} \cdot \dfrac{1}{m_1 - a}(x - a)^2, & \text{if } a \leq x < m_1 \\[2mm] \dfrac{1}{(m_2 + b - a - m_1)}(2x - a - m_1), & \text{if } m_1 \leq x < m_2 \\[2mm] 1 - \dfrac{1}{(m_2 + b - a - m_1)} \cdot \dfrac{1}{b - m_2}(b - x)^2, & \text{if } m_2 \leq x < b \\[2mm] 1, & \text{if } x \geq b \end{cases} \quad (4.3)$$

The CDF is linear in the interval $m_1 \leq x < m_2$, where the density function is constant, and quadratic in the intervals $a \leq x < m_1$ and $m_2 \leq x < b$.

Theorem 4.1 *If X is a trapezoidal random variable, then*

$$E(X) = \frac{((m_2 + b)^2 - m_2 b) - ((a + m_1)^2 - a m_1)}{3(m_2 + b - a - m_1)}$$

$$Var(X) = \frac{(m_2^2 + b^2)(m_2 + b) - (a^2 + m_1^2)(a + m_1)}{6(m_2 + b - a - m_1)} - [E(X)]^2$$

Example 4.1 *Let X represent the uncertainty in the number of delivered source instructions (DSI) of a new software application. Suppose this uncertainty*

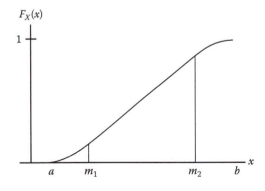

FIGURE 4.2
The trapezoidal cumulative distribution function.

is expressed as the trapezoidal density function in Figure 4.3. Determine the following:

 a. $E(X)$

 b. $Med(X)$

 c. $P(X \leq E(X) + \sigma_X)$

Solution

 a. It is given that $X \sim Trap(25{,}000, 28{,}000, 35{,}000, 37{,}500)$; therefore, we have $a = 25{,}000$, $m_1 = 28{,}000$, $m_2 = 35{,}000$, $b = 37{,}500$. Substituting these values into the expectation formula in Theorem 4.1 yields

$$E(X) = \frac{((m_2 + b)^2 - m_2 b) - ((a + m_1)^2 - am_1)}{3(m_2 + b - a - m_1)}$$

$$= 31{,}363.24786 \approx 31{,}363 \text{ DSI}$$

Since we need σ_X in part (c) of this example, we will compute $Var(X)$ at this point; from Theorem 4.1 we have

$$\sigma_X = \sqrt{Var(X)}$$

$$= \sqrt{\frac{(m_2^2 + b^2)(m_2 + b) - (a^2 + m_1^2)(a + m_1)}{6(m_2 + b - a - m_1)} - [31{,}363.24786]^2}$$

$$= 2925.26 \approx 2925 \text{ DSI}$$

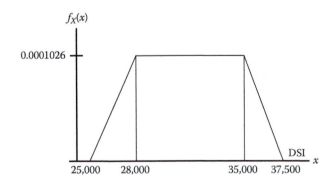

FIGURE 4.3

Trapezoidal probability density function for Example 4.1.

b. To compute $Med(X)$, the median size of the software application, we need to find x such that $F_X(x) = 1/2$. It can be shown (left for the reader) that

$$P(25,000 \leq X \leq 28,000) = \frac{2}{13} < \frac{1}{2}$$

$$P(25,000 \leq X \leq 35,000) = \frac{2}{13} + \frac{28}{39} = \frac{34}{39} > \frac{1}{2}$$

Thus, the median of X will fall in the region of constant probability density; this is equivalent to finding x along the CDF of X such that

$$\frac{1}{(35,000 + 37,500 - 25,000 - 28,000)}(2x - 25,000 - 28,000) = \frac{1}{2}$$

Solving this yields $x = 31,375$; thus $Med(X) = 31,375$ DSI.

c. To determine $P(X \leq E(X) + \sigma_X)$ we have from part (a) the result

$$E(X) + \sigma_X = 31,363 + 2,925 = 34,288 \text{ DSI}$$

The value $x = 34,288$ falls in the linear region of $F_X(x)$; from Equation 4.3

$$P(X \leq E(X) + \sigma_X) = P(X \leq 34,288) = F_X(34,288) = 0.798$$

Thus, there is nearly an 80% probability the amount of code to build the new software application will not exceed 34,288 DSI.

4.1.1 Uniform Distribution

The uniform distribution can be considered a special case of the trapezoidal distribution.* In Figure 4.1, as $(m_1 - a)$ and $(b - m_2)$ approach zero (in the limit), the trapezoidal distribution approaches a distribution with uniform (or constant) probability density, shown in Figure 4.4.

A random variable X is said to have a *uniform distribution* (or *rectangular distribution*) if its PDF is constant and is given by

$$f_X(x) = \frac{1}{b - a}, \text{ if } a \leq x \leq b \tag{4.4}$$

where $-\infty < a < b < \infty$. The numbers a and b are the minimum and maximum possible values of X, respectively. Note that $f_X(x) = 0$ if $x < a$ or $x > b$. A random variable described by a uniform PDF has the following interesting property. If the unit interval $0 \leq x \leq 1$ is the range of values for X, then

---

* It is also a special case of the beta distribution, which is discussed later in this chapter.

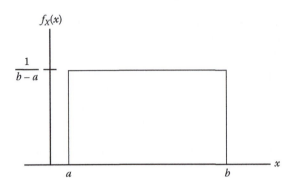

FIGURE 4.4
The uniform probability density function.

$f_X(x) = 1$ and the probability X falls in any subinterval $a' \leq x \leq b'$ of the unit interval is simply the length of that subinterval; specifically,

$$P(a' \leq X \leq b') = \int_{a'}^{b'} 1 \, dx = b' - a'$$

For the remainder of this book, a random variable X with PDF given by Equation 4.4 will be implied by the expression

$$X \sim Unif(a, b)$$

The CDF of X is given by Equation 4.5.

$$F_X(x) = \begin{cases} 0, & \text{if } x < a \\ \dfrac{1}{b-a}(x-a), & \text{if } a \leq x < b \\ 1, & \text{if } x \geq b \end{cases} \qquad (4.5)$$

A graph of $F_X(x)$ is shown in Figure 4.5. Since the density function of X is constant in the interval $a \leq x \leq b$, the cumulative distribution is strictly a *linear function* of x in the interval $a \leq x \leq b$.

The uniform distribution has no skew and no unique mode. From a cost analysis perspective, such random variables might be the number of DSI required for a new software application (refer to Example 3.1), the weight of a new electronic device, or an unknown contractor's software productivity rate. In practice, the uniform distribution is used when a random variable is best described *only* by its extreme possible values. In cost analysis, this occurs most often in the very early stages of a system's design.

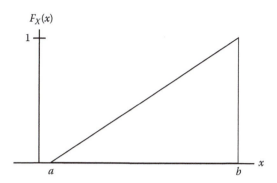

FIGURE 4.5
The uniform cumulative distribution function.

Theorem 4.2 *If X is a uniform random variable then*

$$E(X) = \frac{1}{2}(a + b)$$

$$Var(X) = \frac{1}{12}(b - a)^2$$

Example 4.2 *If X has a uniform distribution, show that Med(X) = E(X).*

Solution Since $X \sim Unif(a, b)$ we know that

$$F_X(x) = \frac{1}{b-a}(x - a) \text{ if } a \leq x < b$$

Since X is a continuous random variable, we know X has a unique median. The median of X will be the value x such that

$$F_X(x) = \frac{1}{b-a}(x - a) = \frac{1}{2}$$

Solving the expression for x yields $x = (a + b)/2$, which is $Med(X)$. From Theorem 4.2 we see that $Med(X) = (a+b)/2 = E(X)$, when $X \sim Unif(a, b)$.

4.1.2 Triangular Distribution

The Triangular distribution can also be considered a special case of the trapezoidal distribution. In a trapezoidal distribution, if $m_1 = m_2 = m$, then it becomes a triangular distribution, such as the one shown in Figure 4.6.

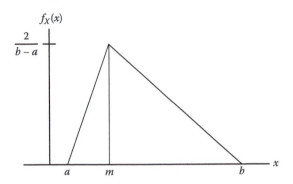

FIGURE 4.6
Triangular probability density function.

A random variable X is said to have a *triangular distribution* if its PDF is given by

$$f_X(x) = \begin{cases} \dfrac{2(x-a)}{(b-a)(m-a)}, & \text{if } a \le x < m \\[3mm] \dfrac{2(b-x)}{(b-a)(b-m)}, & \text{if } m \le x \le b \end{cases} \quad (4.6)$$

where $-\infty < a < m < b < \infty$. The numbers a, m, and b represent the minimum, the mode (most likely), and the maximum possible values of X, respectively. Note that $f_X(x) = 0$ if $x < a$ or $x > b$. In cost analysis, the mode m is often regarded as the point estimate.*

For the remainder of this book, a random variable X with PDF given by Equation 4.6 will be implied by the expression

$$X \sim Trng(a, m, b)$$

The CDF of X is given by Equation 4.7.

$$F_X(x) = \begin{cases} 0, & \text{if } x < a \\[3mm] \dfrac{(x-a)^2}{(b-a)(m-a)}, & \text{if } a \le x < m \\[3mm] 1 - \dfrac{(b-x)^2}{(b-a)(b-m)}, & \text{if } m \le x < b \\[3mm] 1, & \text{if } x \ge b \end{cases} \quad (4.7)$$

A graph is shown in Figure 4.7.

---

* Associating the point estimate (defined in Chapter 1) to the mode of a distribution is traditional in cost analysis; however, there are no strict reasons for doing so. An analyst might judge the point estimate is best represented by the median, or by the mean, of a distribution.

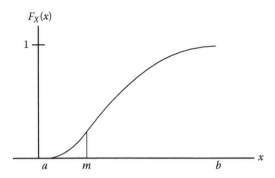

FIGURE 4.7
The triangular cumulative distribution function.

The CDF is a quadratic function of x in the intervals $a \leq x < m$ and $m \leq x < b$. The location of m relative to a and b determines how much probability there is on either side of m. This is illustrated by the three triangular distributions in Figure 4.8.

As seen in Figure 4.8, the closer the mode is to the variable's maximum possible value b, the less likely the variable will exceed its mode. The closer the mode is to the variable's minimum possible value a, the more likely the variable will exceed its mode. For this reason the triangular distribution is often favored as a subjective probability distribution. Only three values a, m, and b are needed to specify the distribution. From these values, subject matter experts focus the distribution in a way that appropriately reflects the overall subjective distribution of probability for the variable under consideration.

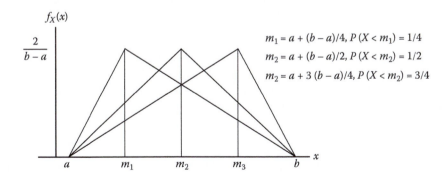

$$m_1 = a + (b - a)/4, P\,(X < m_1) = 1/4$$
$$m_2 = a + (b - a)/2, P\,(X < m_2) = 1/2$$
$$m_2 = a + 3\,(b - a)/4, P\,(X < m_2) = 3/4$$

FIGURE 4.8
A family of triangular probability density functions. (From Evans, M. et al., *Statistical Distributions*, 2nd edn., John Wiley & Sons, Inc., New York, 1993.)

Theorem 4.3 *If X is a triangular random variable then*

$$E(X) = (a + m + b)/3$$

$$Var(X) = \frac{1}{18} \left\{ (m - a)(m - b) + (b - a)^2 \right\}$$

Example 4.3 *In Example 3.7, the uncertainty in the unit production cost of a transmitter synthesizer unit (TSU) for a communications terminal was given by the PDF in Figure 3.15. Use Theorem 4.3 to show that $E(Cost) = 13{,}333.3$ $\$$ and $Var(Cost) = 2.89(10^6)$ $\$^2$.*

Solution Referring to Example 3.7, we see the PDF for *Cost* can be written in the form given by Equation 4.6 with $a = 10{,}000$, $m = 12{,}000$, and $b = 18{,}000$. Substituting these values into the expected value and variance formulas given in Theorem 4.3 yields

$$E(Cost) = (a + m + b)/3 = (10 + 12 + 18)10^3/3 = 13{,}333.3\$$$

$$Var(Cost) = \frac{1}{18} \left\{ (12 - 10)(12 - 18) + (18 - 10)^2 \right\} (10^6) = 2.89(10^6)\$^2$$

4.2 Beta Distribution

The beta distribution, like the distributions discussed in Section 4.1, can be used to describe a random variable whose range of possible values is bounded by an interval of the real line. A random variable X is said to have a *beta distribution* if its PDF is given by

$$f_X(x) = \begin{cases} \dfrac{1}{b-a} \cdot \dfrac{\Gamma(\alpha+\beta)}{\Gamma(\alpha)\Gamma(\beta)} \left(\dfrac{x-a}{b-a}\right)^{\alpha-1} \left(\dfrac{b-x}{b-a}\right)^{\beta-1} & a < x < b \\ 0 & \text{otherwise} \end{cases} \tag{4.8}$$

where α and β ($\alpha > 0$ and $\beta > 0$) determine the shape of the density function and $\Gamma(\alpha)$ is the gamma function of the argument α.*

Beta distributions are in *standard form* when they are defined over the unit interval. A random variable Y is said to have a *standard beta distribution* if its PDF is given by

$$f_Y(y) = \begin{cases} \dfrac{\Gamma(\alpha+\beta)}{\Gamma(\alpha)\Gamma(\beta)} (y)^{\alpha-1}(1-y)^{\beta-1} & 0 < y < 1 \\ 0 & \text{otherwise} \end{cases} \tag{4.9}$$

---

* In general, $\Gamma(\alpha) = \int_0^\infty t^{\alpha-1}e^{-t}dt$. If α is a positive integer, then $\Gamma(\alpha) = (\alpha - 1)!$.

For the remainder of this book, the random variables X and Y with density functions given by Equations 4.8 and 4.9 will be implied by the expressions $X \sim Beta(\alpha, \beta, a, b)$ and $Y \sim Beta(\alpha, \beta)$, respectively. The transformation* of $X \sim Beta(\alpha, \beta, a, b)$ to its standard form $Y \sim Beta(\alpha, \beta)$ is done by letting $y = (x - a)/(b - a)$.

Graphs of the standard beta PDF for various α and β are illustrated in Figures 4.9 and 4.10. Figure 4.9 illustrates several possible shapes associated with the standard beta density function. When $\alpha = \beta$, it is symmetric about $y = 0.5$, which is the median of Y. When $\alpha = \beta$, the median, mean, and

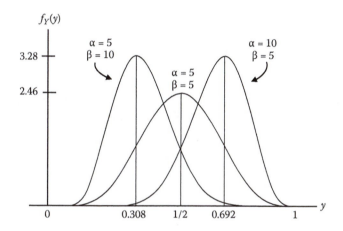

FIGURE 4.9
A family of standard beta probability density functions.

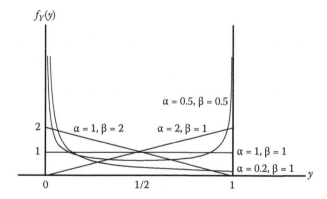

FIGURE 4.10
More standard beta probability density functions.

* Transformations of random variables are formally discussed in Chapter 5.

mode of Y are equal. If $\alpha > 1$ and $\beta > 1$, then the mode of Y is unique and occurs at

$$y = \frac{1 - \alpha}{2 - \alpha - \beta} \qquad (4.10)$$

Figure 4.10 illustrates some other shapes associated with the standard beta density. For instance, the beta density is **U** shaped if $\alpha < 1$ and $\beta < 1$. If $\alpha = 1$ and $\beta = 1$ the beta density becomes the $Unif(0, 1)$ (uniform) density function. A $Beta(1, 2)$ density is a right-skewed triangular PDF, while a $Beta(2, 1)$ is a left-skewed triangular PDF.

As seen in Figure 4.9 and Figure 4.10, the beta density can take a wide variety of shapes. This characteristic makes the beta density among the most diverse of the special distributions for describing (or modeling) a random variable whose range of possible values is bounded by an interval of the real line.

In general, from the transformation $y = (x - a)/(b - a)$ it can be shown the CDF of X can be found from the CDF of Y according to

$$F_X(x) = F_Y\left(\frac{x - a}{b - a}\right) = F_Y(y)$$

However, a closed form expression for the CDF of Y (given by Equation 4.11) does not exist.

$$F_Y(y) = \int_0^y f_Y(t)\,dt \ \text{ if } 0 < y < 1 \qquad (4.11)$$

Values for $F_Y(y)$ are determined through a numerical integration procedure. A number of software applications, such as *Mathematica*® (Wolfram 1991), are available for numerically computing the integral given by Equation 4.11. A family of graphs for $F_Y(y)$ is presented in Figure 4.11. These CDFs are the integrals of the three beta densities given in Figure 4.9.

Theorem 4.4   *If $Y \sim Beta(\alpha, \beta)$ and $X \sim Beta(\alpha, \beta, a, b)$, then*

$$E(Y) = \frac{\alpha}{\alpha + \beta} \qquad (4.12)$$

$$E(X) = a + (b - a)E(Y) \qquad (4.13)$$

$$Var(Y) = \frac{\alpha\beta}{(\alpha + \beta + 1)(\alpha + \beta)^2} \qquad (4.14)$$

$$Var(X) = (b - a)^2 Var(Y) \qquad (4.15)$$

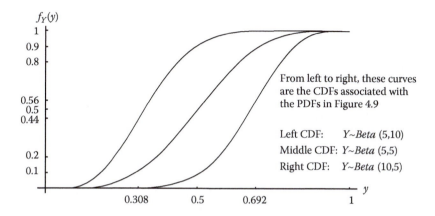

FIGURE 4.11
A family of standard beta cumulative distribution functions.

If the mean and variance of Y are known, it can be shown from Theorem 4.4, the shape parameters of the beta distribution are uniquely determined by

$$\alpha = E(Y)\left[\frac{E(Y)(1 - E(Y))}{Var(Y)} - 1\right] \qquad (4.16)$$

$$\beta = \alpha\left(\frac{1 - E(Y)}{E(Y)}\right) \qquad (4.17)$$

Last, if $Y \sim Beta(\alpha, \beta)$, then $1 - Y \sim Beta(\beta, \alpha)$. Discuss how this property is seen in Figures 4.9 and 4.11.

> **Example 4.4** *Suppose the activity time X (in minutes) to complete the assembly of a microcircuit is beta distributed in the interval $4 < x < 9$, with shape parameters $\alpha = 5$ and $\beta = 10$. Determine $P(X \leq Mode(X))$.*
>
> *Solution* From Equation 4.10
>
> $$Mode(Y) = \frac{1 - \alpha}{2 - \alpha - \beta} = \frac{1 - 5}{2 - 15} = \frac{4}{13} \approx 0.308$$
>
> where Y is the standard beta density of X. This is in terms of the unit interval, that is, if $Y \sim Beta(5, 10)$ then $Mode(Y) = 0.308$. From the transformation $y = (x - a)/(b - a)$, the value $y = 0.308$ in the unit interval is equivalent to the value $x = 5.54$ in the interval $4 < x < 9$ (where $a = 4$ and $b = 9$); therefore, $Mode(X) = 5.54$. To determine $P(X \leq Mode(X))$ we have
>
> $$P(X \leq Mode(X)) = P\left(\frac{X - a}{b - a} \leq \frac{Mode(X) - a}{b - a}\right) = P(Y \leq Mode(Y))$$

Since $Y \sim Beta(5, 10)$ we have

$$f_Y(y) = \frac{\Gamma(15)}{\Gamma(5)\Gamma(10)} (y)^4 (1 - y)^9 \quad 0 < y < 1$$

From numerical integration it can be shown that

$$P(Y \leq Mode(Y)) = F_Y(0.308) = \int_0^{0.308} \frac{\Gamma(15)}{\Gamma(5)\Gamma(10)} (y)^4 (1 - y)^9 \, dy \approx 0.44$$

Since $P(X \leq Mode(X)) = P(Y \leq Mode(Y))$, we conclude that

$$P(X \leq Mode(X)) = 0.44$$

Therefore, with a probability of 0.44 the assembly time of the microcircuit will be less than or equal to 5.54 min. Discuss why this probability is also seen in Figure 4.11.

4.3 Normal Distribution

The distributions presented in Sections 4.1 and 4.2 can be thought of as finite distributions. Random variables described by *finite distributions* have values that are restricted to a bounded interval of the real line. The trapezoidal, uniform, triangular, and beta distributions are examples of finite distributions. In contrast to these, a random variable described by a normal distribution is unbounded. Its values fall in the open interval given by the entire real line. The normal distribution is the first of two *infinite distributions* we will discuss in this chapter.

The trapezoidal, uniform, triangular, and beta PDFs are frequently used in cost analysis to *directly specify* the uncertainty in the value of a variable. Typically, such variables are inputs for deriving cost.* These variables might include the number of new DSI for a software function, the weight of a future hardware item (e.g., a satellite), or the time required to assemble a new electronic device. The normal distribution *could* be used in the same way; however, from a cost analysis perspective, the normal distribution most often characterizes the *underlying distribution function of a derived cost*. In this sense, the normal distribution can reflect the shape of an "output" distribution—particularly one generated from a summation of "input" distributions, like those discussed in Sections 4.1 and 4.2. For instance, suppose the random variable *Cost* is derived from the sum of the cost of each work breakdown structure cost element X_i ($i = 1, \ldots, n$) in a system. Specifically, if

$$Cost = X_1 + X_2 + X_3 + \cdots + X_n \tag{4.18}$$

---

* This is illustrated in the discussion associated with Figure 1.4.

then under certain conditions (discussed in Chapters 5 and 6) the normal distribution will characterize the underlying distribution function of *Cost*.

A random variable X is said to be *normally distributed* if its PDF is given by

$$f_X(x) = \frac{1}{\sqrt{2\pi}\,\sigma} e^{-\frac{1}{2}\left[(x-\mu)^2/\sigma^2\right]} \qquad (4.19)$$

where $-\infty < x < \infty$ and $\sigma > 0$. Equation 4.19 is also known as the *Gaussian* distribution, named after the German mathematician Karl Friedrich Gauss (1777–1855). For the remainder of this book, a random variable X with PDF given by Equation 4.19 will be implied by the expression $X \sim N(\mu, \sigma^2)$. The normal PDF is *uniquely defined* by two parameters μ and σ^2. Theorem 4.6 will show these parameters are the mean and variance of X, respectively. A graph of the normal PDF is presented in Figure 4.12.

The normal distribution is symmetric about its mean μ. It has the property that its mode and median equal its mean. The numbers in Figure 4.12 are the areas under the curve within the indicated intervals. Specifically,

$$P(\mu - \sigma \leq X \leq \mu + \sigma) = \int_{\mu-\sigma}^{\mu+\sigma} f_X(x)\,dx = 0.6826 \qquad (4.20)$$

where $f_X(x)$ is given by Equation 4.19. Similarly,

$$P(\mu - 2\sigma \leq X \leq \mu + 2\sigma) = 0.9544 \qquad (4.21)$$

$$P(\mu - 3\sigma \leq X \leq \mu + 3\sigma) = 0.9973 \qquad (4.22)$$

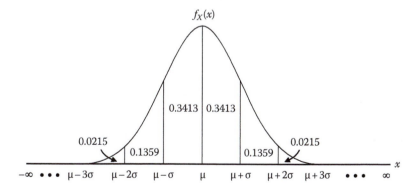

FIGURE 4.12
The normal probability density.

Thus, when X is *normally distributed*, the probability X falls within $\pm 1\sigma$ from its mean is always 0.6826; the probability X falls within $\pm 2\sigma$ from its mean is always 0.9544; the probability X falls within $\pm 3\sigma$ from its mean is always 0.9973.

The peak of the normal PDF is governed only by the variance of X. Furthermore, $Mode(X)$ occurs at $x = \mu$. The PDF evaluated at $x = \mu$ is equal to $0.399/\sigma$. Decreasing σ increases the maximum height of the normal PDF and the concentration of probability around the mean μ. This is illustrated in Figure 4.13.

If $X \sim N(\mu, \sigma^2)$ and $Z = (X - \mu)/\sigma$, the standard form of X, it can be shown (Theorem 4.5) that Z has a normal distribution with mean 0 and variance 1. The density function of Z is known as the *standard normal density*, which is given by the following equation:

$$f_Z(z) = \frac{1}{\sqrt{2\pi}}e^{-z^2/2} \quad \text{where} \ -\infty < z < \infty \tag{4.23}$$

For the remainder of this book, a random variable Z with PDF given by Equation 4.23 will be implied by the expression $Z \sim N(0,1)$. A graph of $f_Z(z)$ is shown in Figure 4.14. The peak of the *standard normal density* occurs at $z = 0$, which is $Mode(Z)$. Since $Var(Z) = 1$ the standard normal PDF evaluated at $Mode(Z)$ is equal to 0.399.

Closed form expressions for the CDFs $F_X(x)$ and $F_Z(z)$ do not exist. However, from the transformation $z = (x - \mu)/\sigma$ it can be shown that

$$F_X(x) = F_Z((x - \mu)/\sigma) = F_Z(z) \tag{4.24}$$

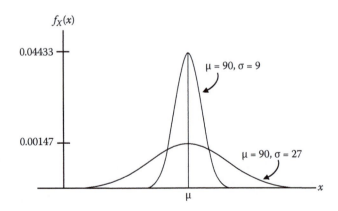

FIGURE 4.13
A comparison of the heights of two normal PDFs.

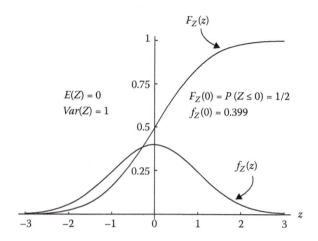

FIGURE 4.14
The standard normal PDF and CDF.

where

$$F_Z(z) = P(Z \le z) = \int_{-\infty}^{z} f_Z(y)\,dy$$

and $f_Z(y)$ is given by Equation 4.23. Thus, values for $F_X(x)$ can be obtained from values for $F_Z(z)$ by a numerical integration of $f_Z(y)$. The results of such an integration are summarized in Table A.1. A graph of $F_Z(z)$ is also shown in Figure 4.14.

Since the standard normal is symmetric about $z = 0$, $P(Z \le -k) = P(Z > k)$. In terms of the CDF of Z, this is equivalent to $F_Z(-k) = 1 - F_Z(k)$. In particular, if $X \sim N(\mu, \sigma^2)$, then the probability X is within $\pm k\sigma$ of the mean of X is

$$P(\mu - k\sigma \le X \le \mu + k\sigma) = P(-k \le Z \le k)$$
$$= F_Z(k) - F_Z(-k) = F_Z(k) - [1 - F_Z(k)] = 2F_Z(k) - 1 \qquad (4.25)$$

Example 4.5 *Using Table A.1, show that $P(\mu - \sigma \le X \le \mu + \sigma) = 0.6826$.*

Solution From Equation 4.25, we see that $k = 1$ in this case. So,

$$P(\mu - \sigma \le X \le \mu + \sigma) = P(-1 \le Z \le 1) = 2F_Z(1) - 1$$

From Table A.1, $F_Z(1) = 0.8413$; therefore,

$$P(\mu - \sigma \le X \le \mu + \sigma) = P(-1 \le Z \le 1) = 2(0.8413) - 1 = 0.6826 \blacklozenge$$

If $X \sim N(\mu, \sigma^2)$ then the probability statements about X can be written in terms of its standard form Z. From Equation 4.24, we have the general relationship

$$P(a \leq X \leq b) = F_Z\left(\frac{b - \mu}{\sigma}\right) - F_Z\left(\frac{a - \mu}{\sigma}\right) \qquad (4.26)$$

Example 4.6 *In Figure 1.6, the distribution function of a system's cost was normal with mean 110.42 ($M) and standard deviation 21.65 ($M). Given this, determine $P(100 \leq Cost \leq 140)$.*

Solution We are given $Cost \sim N(110.42, (21.65)^2)$. In terms of Equation 4.26

$$P(100 \leq X \leq 140) = F_Z\left(\frac{140 - 110.42}{21.65}\right) - F_Z\left(\frac{100 - 110.42}{21.65}\right)$$

$$= F_Z(1.37) - F_Z(-0.48)$$

Since $F_Z(-k) = 1 - F_Z(k)$, we have $F_Z(-0.48) = 1 - F_Z(0.48)$; therefore,

$$P(100 \leq X \leq 140) = F_Z(1.37) - [1 - F_Z(0.48)]$$

From Table A.1, $F_Z(1.37) = 0.91465$ and $F_Z(0.48) = 0.68439$. So,

$$P(100 \leq X \leq 140) = 0.599 \approx 0.60$$

Thus, there is nearly a 60% chance the system's cost will fall between 100 and 140 million dollars.

Example 4.7 *Suppose the uncertainty in a system's cost is described by the normal PDF shown in Figure 4.15. Suppose there is a 5% chance the system's cost will not exceed 30.34 ($M) and an 85% chance its cost will not exceed*

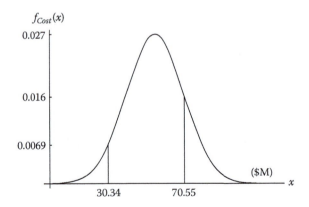

FIGURE 4.15
PDF for Example 4.7.

70.55 ($M). From this information determine the mean and standard deviation of the system's cost.

Solution We are given $P(Cost \leq 30.34) = 0.05$ and $P(Cost \leq 70.55) = 0.85$. Expressing the random variable *Cost* in standard form we have

$$P\left(Z \leq \frac{30.34 - \mu}{\sigma}\right) = 0.05 \text{ and } P\left(Z \leq \frac{70.55 - \mu}{\sigma}\right) = 0.85$$

where μ and σ are the mean and standard deviation of *Cost*, respectively. We will first work with the probability

$$P\left(Z \leq \frac{30.34 - \mu}{\sigma}\right) = 0.05$$

From Table A.1, $P(Z \leq 1.645) = 0.95$; it follows that

$$1 - P(Z \leq 1.645) = 0.05$$

This is equivalent to $P(Z > 1.645) = 0.05$. Since the standard normal distribution is symmetric about $z = 0$, $P(Z > 1.645) = P(Z \leq -1.645)$; therefore, we have

$$\frac{30.34 - \mu}{\sigma} = -1.645 \tag{4.27}$$

A similar reasoning applies to the other probability. From Table A.1

$$P\left(Z \leq \frac{70.55 - \mu}{\sigma}\right) = 0.85$$

is true when

$$\frac{70.55 - \mu}{\sigma} = 1.04 \tag{4.28}$$

Solving Equations 4.27 and 4.28 simultaneously for μ and σ yields

$$\mu \approx 55 \text{ ($M)}$$
$$\sigma \approx 15 \text{ ($M)}$$

Theorem 4.5 *If $X \sim N(\mu, \sigma^2)$, then $Z \sim N(0, 1)$ where $Z = (X - \mu)/\sigma$.*

Proof. Since $X \sim N(\mu, \sigma^2)$, we have

$$F_X(x) = P(X \leq x) = \int_{-\infty}^{x} \frac{1}{\sqrt{2\pi}\,\sigma} e^{-\frac{1}{2}\left[(t-\mu)^2/\sigma^2\right]}\, dt$$

By the definition of a CDF, we also have

$$F_Z(z) = P(Z \leq z) = P\left(\frac{X - \mu}{\sigma} \leq z\right) = P(X \leq z\sigma + \mu)$$

$$= \int_{-\infty}^{z\sigma+\mu} \frac{1}{\sqrt{2\pi}\,\sigma} e^{-\frac{1}{2}\left[(x-\mu)^2/\sigma^2\right]} dx \tag{4.29}$$

If we let $y = (x - \mu)/\sigma$, then $\sigma\,dy = dx$; substituting this change of variable into Equation 4.29 yields

$$F_Z(z) = \int_{-\infty}^{z} \frac{1}{\sqrt{2\pi}\,\sigma} e^{-\frac{1}{2}y^2} \sigma\,dy = \int_{-\infty}^{z} \frac{1}{\sqrt{2\pi}} e^{-\frac{1}{2}y^2} dy \tag{4.30}$$

Equation 4.30 is the CDF of the standard normal density; thus,

$$f_Z(z) = \frac{1}{\sqrt{2\pi}} e^{-\frac{1}{2}z^2}$$

Therefore, $Z \sim N(0, 1)$. This implies $E(Z) = 0$ and $Var(Z) = 1$.

Theorem 4.6   If $X \sim N(\mu, \sigma^2)$, then $E(X) = \mu$ and $Var(X) = \sigma^2$.

Proof.   Since $X \sim N(\mu, \sigma^2)$, we have

$$E(X) = \int_{-\infty}^{\infty} x \cdot \frac{1}{\sqrt{2\pi}\,\sigma} e^{-\frac{1}{2}\left[(x-\mu)^2/\sigma^2\right]} dx$$

By the change of variable $z = (x - \mu)/\sigma$, we have

$$E(X) = \int_{-\infty}^{\infty} (z\sigma + \mu) \cdot \frac{1}{\sqrt{2\pi}\,\sigma} e^{-\frac{1}{2}z^2} \sigma\,dz$$

which simplifies to

$$E(X) = \sigma \int_{-\infty}^{\infty} z \cdot \frac{1}{\sqrt{2\pi}} e^{-\frac{1}{2}z^2} dz + \mu \int_{-\infty}^{\infty} \frac{1}{\sqrt{2\pi}} e^{-\frac{1}{2}z^2} dz \tag{4.31}$$

The first integral in Equation 4.31 is E(Z). This integral is equal to zero since the integral exists and its integrand is an odd function; that is,

$$E(Z) = \int_{-\infty}^{\infty} z \cdot \frac{1}{\sqrt{2\pi}} e^{-\frac{1}{2}z^2} dz = 0$$

The second integral in Equation 4.31 is unity since it is the integral of the standard normal density function. Therefore, Equation 4.31 simplifies to

$$E(X) = \sigma E(Z) + \mu \cdot 1 = \sigma \cdot 0 + \mu = \mu$$

To show that Var(X) = σ^2, recall that Var(X) = $E(X^2) - (E(X))^2$. We know that

$$E(X^2) = \int_{-\infty}^{\infty} x^2 \cdot \frac{1}{\sqrt{2\pi}\,\sigma} e^{-\frac{1}{2}[(x-\mu)^2/\sigma^2]} dx$$

From the family of integrals of exponential functions, presented in Appendix A, note that

$$E(X^2) = \int_{-\infty}^{\infty} x^2 \cdot \frac{1}{\sqrt{2\pi}\,\sigma} e^{-\frac{1}{2}[(x-\mu)^2/\sigma^2]} dx = \mu^2 + \sigma^2$$

therefore, Var(X) = $\mu^2 + \sigma^2 - (\mu)^2 = \sigma^2$.

4.4 Lognormal Distribution

The lognormal probability distribution is the last of the infinite distributions we will discuss in this book. It has broad applicability in engineering, economics, and cost analysis. In engineering, the failure rates of mechanical or electrical components often follow a lognormal distribution. In economics, the random variation between the production cost of goods to capital and labor costs is frequently modeled after the lognormal distribution; the classical example is the Cobb–Douglas production function, given by Equation 4.32.

$$Q = a W_1^{a_1} W_2^{a_2} \tag{4.32}$$

In this equation, the production cost of goods Q is a function of capital cost W_1 and labor cost W_2; the terms a, a_1, and a_2 are real numbers. Under certain conditions, Q can be shown to have a lognormal probability distribution.

In cost analysis, Abramson and Young (1997) observed that the lognormal can approximate the probability distribution of a system's total cost—particularly when the cost distribution is positively skewed. Empirical studies by Garvey and Taub (Garvey 1996, Garvey and Taub 1997) identify circumstances where the lognormal can approximate the combined (joint) distribution of a program's total cost and schedule.*

The lognormal distribution has a close relationship with the normal distribution. If X is a *nonnegative random variable* where the natural logarithm of X, denoted by $\ln X$, follows the normal distribution, then X is said to have a lognormal distribution. This is illustrated in Figure 4.16. On the left-side of Figure 4.16, the random variable X has a lognormal PDF, with $E(X) = 100$ and $Var(X) = 625$. On the right-side is the representation of X in logarithmic space. In logarithmic space, X has a normal PDF, with $E(\ln X) = 4.57486$ and $Var(\ln X) = 0.0606246$. How the latter two values are determined is discussed in Theorem 4.8.

Under certain conditions (discussed in Chapter 5), normal distribution can arise from a summation of many random variables (as illustrated by Equation 4.18); the lognormal distribution can arise from a multiplicative combination of many random variables, as illustrated by Equation 4.32.

A random variable X is said to be *lognormally distributed* if its PDF is given by

$$f_X(x) = \frac{1}{\sqrt{2\pi}\,\sigma_Y}\frac{1}{x}e^{-\frac{1}{2}\left[(\ln x - \mu_Y)^2/\sigma_Y^2\right]} \tag{4.33}$$

where $0 < x < \infty$, $\sigma_Y > 0$, $\mu_Y = E(\ln X)$, and $\sigma_Y^2 = Var(\ln X)$. For the remainder of this book, a random variable X with PDF given by Equation 4.33 will be implied by the expression $X \sim LogN(\mu_Y, \sigma_Y^2)$. The parameters

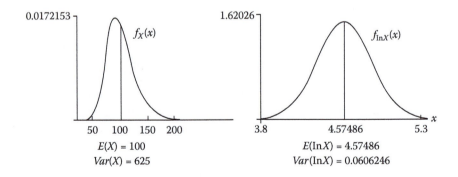

FIGURE 4.16
PDFs of X and $\ln X$, where $X \sim LogN(100,625)$ and $\ln X \sim N(4.57486, 0.0606246)$.

---

* This is fully discussed in detail Chapter 7.

μ_Y and σ_Y^2 are the mean and variance of the normally distributed random variable $Y = \ln X$, which is the logarithmic representation of X (refer to Figure 4.16). Graphs of a family of lognormal PDFs are presented in Figure 4.17. Notice the lognormal PDF is positively skewed and values for x are always nonnegative.

Theorem 4.7 *If X is a lognormal random variable, then $E(X) = \mu_X = e^{\mu_Y + \frac{1}{2}\sigma_Y^2}$ and $\text{Var}(X) = \sigma_X^2 = e^{2\mu_Y + \sigma_Y^2}(e^{\sigma_Y^2} - 1)$.*

Proof. Since X has a lognormal distribution, the PDF of X is given by Equation 4.33; therefore,

$$E(X) = \int_0^\infty x f_X(x)dx = \int_0^\infty x \cdot \frac{1}{\sqrt{2\pi}\,\sigma_Y} \frac{1}{x} e^{-\frac{1}{2}\left[(\ln x - \mu_Y)^2/\sigma_Y^2\right]} dx \qquad (4.34)$$

Equation 4.34 simplifies to

$$E(X) = \int_0^\infty \frac{1}{\sqrt{2\pi}\,\sigma_Y} e^{-\frac{1}{2}\left[(\ln x - \mu_Y)^2/\sigma_Y^2\right]} dx \qquad (4.35)$$

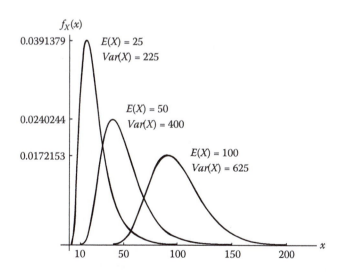

FIGURE 4.17
A family of lognormal probability density functions.

If we set $y = \ln x - \mu_Y$, *then* $-\infty < y < \infty$, $x = e^y e^{\mu_Y}$, *and* $dx = e^y e^{\mu_Y} dy$. *Substituting this into Equation 4.35, we have*

$$E(X) = \int_{-\infty}^{\infty} \frac{1}{\sqrt{2\pi}\,\sigma_Y} e^{-\frac{1}{2}[y^2/\sigma_Y^2]} e^y e^{\mu_Y} dy \tag{4.36}$$

$$E(X) = e^{\mu_Y} \int_{-\infty}^{\infty} \frac{1}{\sqrt{2\pi}\,\sigma_Y} e^{-\frac{1}{2}[(y^2 - 2\sigma_Y^2 y)/\sigma_Y^2]} dy$$

$$E(X) = e^{\mu_Y} \int_{-\infty}^{\infty} \frac{1}{\sqrt{2\pi}\,\sigma_Y} e^{-\frac{1}{2\sigma_Y^2}[(y-\sigma_Y^2)^2 - \sigma_Y^4]} dy$$

$$E(X) = e^{\mu_Y} \int_{-\infty}^{\infty} \frac{1}{\sqrt{2\pi}\,\sigma_Y} e^{-\frac{1}{2\sigma_Y^2}[(y-\sigma_Y^2)^2]} e^{\frac{1}{2}\sigma_Y^2} dy$$

$$E(X) = e^{\mu_Y + \frac{1}{2}\sigma_Y^2} \int_{-\infty}^{\infty} \frac{1}{\sqrt{2\pi}\,\sigma_Y} e^{-\frac{1}{2\sigma_Y^2}[(y-\sigma_Y^2)^2]} dy = e^{\mu_Y + \frac{1}{2}\sigma_Y^2} \tag{4.37}$$

The integral in Equation 4.37 is unity since it is the PDF of a $N(\sigma^2, \sigma^2)$ random variable. This result can be generalized to the rth moment of X. It is left to the reader to show that

$$E(X^r) = e^{r\mu_Y + \frac{1}{2}\sigma_Y^2 r^2} \tag{4.38}$$

To show that $Var(X) = e^{2\mu_Y + \sigma_Y^2}(e^{\sigma_Y^2} - 1)$, recall that

$$Var(X) = E(X^2) - (E(X))^2 \tag{4.39}$$

Substituting Equations 4.37 and 4.38 (with $r = 2$) into Equation 4.39, it is easily shown that $Var(X) = e^{2\mu_Y + \sigma_Y^2}(e^{\sigma_Y^2} - 1)$. ◆

This theorem can be illustrated by referring to Figure 4.16, where $\mu_Y = E(\ln X) = 4.57486$ and $\sigma_Y^2 = Var(\ln X) = 0.0606246$. From Theorem 4.7

$$E(X) = e^{\mu_Y + \frac{1}{2}\sigma_Y^2} = e^{4.57486 + \frac{1}{2}(0.0606246)} = 100$$

$$Var(X) = e^{2\mu_Y + \sigma_Y^2}(e^{\sigma_Y^2} - 1) = e^{2(4.57486) + 0.0606246}(e^{0.0606246} - 1) = 625$$

Thus, when X is a lognormal random variable, its mean and variance are defined in terms of the normally distributed random variable $Y = \ln X$. The same is true about the mode and median of X; in particular, if X is a lognormal random variable, then

$$Mode(X) = e^{\mu_Y - \sigma_Y^2} \tag{4.40}$$

$$Median(X) = e^{\mu_Y} \tag{4.41}$$

In Figure 4.16,

$$Mode(X) = e^{4.57486 - 0.0606246} = 91.307$$

$$Median(X) = e^{4.57486} = 97.014$$

The lognormal PDF peaks at the value

$$f_X(Mode(X)) = \frac{1}{\sqrt{2\pi}} \frac{1}{\sigma_Y} (e^{\frac{1}{2}\sigma_Y^2 - \mu_Y}) \tag{4.42}$$

Showing this is left as an exercise for the reader.

In cost analysis applications of the lognormal distribution, we typically do not have values for $E(\ln X)$ and $Var(\ln X)$ (where X might represent the cost of a system or a particular work breakdown structure cost element). How do we specify the distribution function of a lognormal random variable X when only $E(X)$ and $Var(X)$ are known? Theorem 4.8 addresses this question. Theorem 4.8 presents transformation formulas for determining $E(\ln X)$ and $Var(\ln X)$ when only $E(X)$ and $Var(X)$ are known.

Theorem 4.8 *If X is a lognormal random variable with mean* $E(X) = \mu_X$ *and* $Var(X) = \sigma_X^2$, *then*

$$\mu_Y = E(\ln X) = \frac{1}{2} \ln \left[\frac{(\mu_X)^4}{(\mu_X)^2 + \sigma_X^2} \right] \tag{4.43}$$

and

$$\sigma_Y^2 = Var(\ln X) = \ln \left[\frac{(\mu_X)^2 + \sigma_X^2}{(\mu_X)^2} \right] \tag{4.44}$$

Proof. From Theorem 4.7 we have

$$\mu_X = e^{\mu_Y + \frac{1}{2}\sigma_Y^2}$$

$$\ln \mu_X = \mu_Y + \frac{1}{2} \sigma_Y^2 \tag{4.45}$$

$$2 \ln \mu_X = 2\mu_Y + \sigma_Y^2 \tag{4.46}$$

We will first establish Equation 4.44 in Theorem 4.8 and then use that result to establish Equation 4.43. From Theorem 4.7

$$Var(X) = \sigma_X^2 = e^{2\mu_Y + \sigma_Y^2}(e^{\sigma_Y^2} - 1)$$

$$\ln(e^{\sigma_Y^2} - 1) = \ln \sigma_X^2 - (2\mu_Y + \sigma_Y^2)$$

$$\ln(e^{\sigma_Y^2} - 1) = \ln \sigma_X^2 - 2\ln \mu_X$$

$$\ln(e^{\sigma_Y^2} - 1) = \ln \left(\frac{\sigma_X^2}{\mu_X^2}\right)$$

$$e^{\sigma_Y^2} = \frac{\sigma_X^2}{\mu_X^2} + 1$$

Therefore $\sigma_Y^2 = Var(\ln X) = \ln\left[\dfrac{(\mu_X)^2 + \sigma_X^2}{(\mu_X)^2}\right]$. *To establish Equation 4.43, write*

$$\mu_Y = \ln \mu_X - \frac{1}{2}\sigma_Y^2$$

From Equation 4.44 we have

$$\mu_Y = \ln \mu_X - \frac{1}{2}\ln\left[\frac{(\mu_X)^2 + \sigma_X^2}{(\mu_X)^2}\right]$$

$$\mu_Y = \frac{1}{2}\left(2\ln \mu_X - \ln\left[\frac{(\mu_X)^2 + \sigma_X^2}{(\mu_X)^2}\right]\right)$$

$$\mu_Y = \frac{1}{2}\left(\ln(\mu_X)^2 - \ln\left[\frac{(\mu_X)^2 + \sigma_X^2}{(\mu_X)^2}\right]\right)$$

Therefore, $\mu_Y = E(\ln X) = \dfrac{1}{2}\ln\left[\dfrac{(\mu_X)^4}{(\mu_X)^2 + \sigma_X^2}\right]$ ♦

Using Theorem 4.8 the parameters μ_Y and σ_Y^2, which uniquely specify the lognormal PDF, can be determined from $E(X)$ and $Var(X)$. In Figure 4.17, the left-most PDF has $E(X) = 25$ and $Var(X) = 225$; from Theorem 4.8 this is equivalent to a lognormal PDF with parameters $\mu_Y = 3.06513$ and $\sigma_Y^2 = 0.307485$. The middle PDF (in Figure 4.17) has $E(X) = 50$ and $Var(X) = 400$; from Theorem 4.8 this is equivalent to a lognormal PDF with parameters $\mu_Y = 3.83781$ and $\sigma_Y^2 = 0.14842$. The right-most PDF (in Figure 4.17) has

$E(X) = 100$ and $Var(X) = 625$; from Theorem 4.8 this is equivalent to a lognormal PDF with parameters $\mu_Y = 4.57486$ and $\sigma_Y^2 = 0.0606246$. Thus, the equations for the three PDFs in Figure 4.17, from left to right, are as follows:

$$f_X(x) = \frac{1}{\sqrt{2\pi}\,(0.554513)}\frac{1}{x}e^{-\frac{1}{2}\left[(\ln x - 3.06513)^2/0.307485\right]}$$

$$f_X(x) = \frac{1}{\sqrt{2\pi}\,(0.385253)}\frac{1}{x}e^{-\frac{1}{2}\left[(\ln x - 3.83781)^2/0.14842\right]}$$

$$f_X(x) = \frac{1}{\sqrt{2\pi}\,(0.246221)}\frac{1}{x}e^{-\frac{1}{2}\left[(\ln x - 4.57486)^2/0.0606246\right]}$$

where the general form for $f_X(x)$ was given by Equation 4.33.

The CDF of a lognormal random variable is given by Equation 4.47.

$$F_X(x) = P(X \le x) = \int_0^x \frac{1}{\sqrt{2\pi}\,\sigma_Y}\frac{1}{t}e^{-\frac{1}{2}\left[(\ln t - \mu_Y)^2/\sigma_Y^2\right]}\,dt \qquad (4.47)$$

Figure 4.18 presents a family of lognormal CDFs associated with the PDFs in Figure 4.17. The CDF given by Equation 4.47 does not exist in closed form. It can be evaluated by a numerical integration procedure. An alternative to such a procedure involves using a table of values from the standard normal distribution. The following discusses this approach.

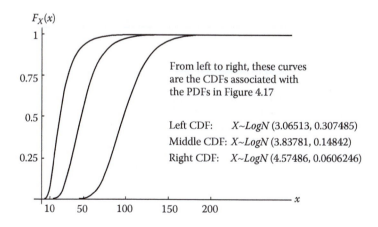

FIGURE 4.18
A family of lognormal CDFs.

If $X \sim LogN(\mu_X, \sigma_X^2)$, then $Y = \ln X \sim N(\mu_Y, \sigma_Y^2)$; therefore,

$$P(X \le x) = P(\ln X \le \ln x) = P\left(\frac{\ln X - \mu_Y}{\sigma_Y} \le \frac{\ln x - \mu_Y}{\sigma_Y}\right) \qquad (4.48)$$

Since $Y = \ln X \sim N(\mu_Y, \sigma_Y^2)$, from Theorem 4.5 it follows that

$$\frac{\ln X - \mu_Y}{\sigma_Y} \sim N(0,1)$$

which is equivalent to the standard normal random variable Z. From this result, Equation 4.48 is equivalent to

$$P(X \le x) = P\left(Z \le \frac{\ln x - \mu_Y}{\sigma_Y}\right) \qquad (4.49)$$

If X has a lognormal distribution, then probabilities associated with various intervals around X can be determined from a table of values of Z, the standard normal distribution.

Example 4.8 *Suppose the uncertainty in a system's cost is described by a lognormal PDF with $E(Cost) = 100$ ($M) and $Var(Cost) = 625$ ($M)^2; this is the right-most PDF in Figure 4.17. Determine*

a. $P(Cost > 2E(Cost))$
b. $P(50 \le Cost \le 150)$

Solution

a. To determine $P(Cost > 2E(Cost))$ recall that

$$P(Cost > 2E(Cost)) = 1 - P(Cost \le 2E(Cost))$$

It is given that $E(Cost) = 100$; therefore,

$$P(Cost > 200) = 1 - P(Cost \le 200)$$

In this example, the random variable *Cost* is given to have a lognormal distribution with $E(Cost) = 100$ and $Var(Cost) = 625$. Thus, the random variable $Y = \ln Cost$ is normally distributed with parameters (determined from Theorem 4.8)

$$\mu_Y = E(\ln Cost) = 4.57486$$

$$\sigma_Y^2 = Var(\ln Cost) = 0.0606246$$

From Equation 4.49

$$P(Cost \le 200) = P\left(Z \le \frac{\ln 200 - 4.57486}{0.246221}\right) = P(Z \le 2.938)$$

From Table A.1, $P(Z \leq 2.938) = 0.998348$, after some interpolation. Therefore,

$$P(Cost > 200) = 1 - P(Z \leq 2.938) = 0.00165$$

This result is consistent with the Markov bound discussion in Section 3.4, as illustrated in Figure 3.19.

b. To determine $P(50 \leq Cost \leq 150)$ note that

$$P(50 \leq Cost \leq 150) = P(\ln 50 \leq \ln(Cost) \leq \ln 150)$$

$$= P\left(\frac{\ln 50 - \mu_Y}{\sigma_Y} \leq \frac{\ln Cost - \mu_Y}{\sigma_Y} \leq \frac{\ln 150 - \mu_Y}{\sigma_Y} \right)$$

$$= P\left(\frac{\ln 50 - \mu_Y}{\sigma_Y} \leq Z \leq \frac{\ln 150 - \mu_Y}{\sigma_Y} \right)$$

$$= P(-2.69 \leq Z \leq 1.77)$$

where

$$Z = \frac{\ln Cost - \mu_Y}{\sigma_Y}$$

$$\mu_Y = E(\ln Cost) = 4.57486 \quad \text{(from Theorem 4.8)}$$

$$\sigma_Y^2 = Var(\ln Cost) = 0.0606246 \quad \text{(from Theorem 4.8)}$$

From Theorem 4.5 we know $Z \sim N(0,1)$, thus,

$$P(50 \leq Cost \leq 150) = P(-2.69 \leq Z \leq 1.77)$$

$$= F_Z(1.77) - F_Z(-2.69)$$

$$= F_Z(1.77) - [1 - F_Z(2.69)]$$

where $F_Z(-2.69) = 1 - F_Z(2.69)$. From Table A.1

$$P(50 \leq Cost \leq 150) = 0.961636 - [1 - 0.9964] = 0.958 \approx 0.96$$

Thus, the system's cost will fall between 50 and 150 million dollars with probability 0.96. This result is also consistent with the discussion presented in Section 3.4, as illustrated in Figure 3.20.

Example 4.9 *In Figure 1.5, the random variable X_2 represented the cost of a system's systems engineering and program management. Furthermore, the point estimate of X_2, denoted by $x_{2PE_{X_2}}$, was equal to 1.26 ($M). If X_2 can be approximated by a lognormal distribution, with $E(X_2) = 1.6875$ ($M) and $Var(X_2) = 0.255677$ ($M)^2, determine*

a. $P(X_2 \leq x_{2PE_{X_2}})$

b. $P(X_2 \leq E(X_2))$

Solution

a. Since the distribution function of X_2 is approximated by a lognormal, from Equation 4.49 we can write

$$P(X_2 \leq x_{2PE_{X_2}}) = P\left(Z \leq \frac{\ln x_{2PE_{X_2}} - \mu_Y}{\sigma_Y}\right)$$

where $Z \sim N(0, 1)$, $\mu_Y = E(\ln X_2)$, and $\sigma_Y^2 = Var(\ln X_2)$. Since $E(X_2) = 1.6875$ and $Var(X_2) = 0.255677$, from Theorem 4.8 $\mu_Y = 0.480258$ and $\sigma_Y^2 = 0.0859804$. Thus,

$$P(X_2 \leq 1.26) = P\left(Z \leq \frac{\ln 1.26 - 0.480258}{0.293224}\right) = P(Z \leq -0.85)$$

From Table A.1

$$P(Z \leq -0.85) = P(Z \geq 0.85) = 1 - P(Z < 0.85)$$
$$= 1 - 0.802 = 0.198$$

thus, $P(X_2 \leq 1.26) = P(Z \leq -0.85) = 0.198$. Therefore, there is nearly a 20% chance the cost of the system's systems engineering and program management will be less than or equal to 1.26 ($M).

b. We are given $E(X_2) = 1.6875$, therefore, $P(X_2 \leq E(X_2)) = P(X_2 \leq 1.6875)$. From Equation 4.49 we can write

$$P(X_2 \leq 1.6875) = P\left(Z \leq \frac{\ln 1.6875 - 0.480258}{0.293224}\right) = P(Z \leq 0.1466)$$

From Table A.1, $P(Z \leq 0.1466) = 0.558$; thus,

$$P(X_2 \leq 1.6875) = P(Z \leq 0.1466) = 0.558$$

Therefore, there is nearly a 56% chance the cost of the system's systems engineering and program management will be less than or equal to 1.6875 ($M). For interest, the PDF and CDF of X_2, for this example, are shown in the Figure 4.19.

This concludes the discussion of the special probability distributions commonly used in cost uncertainty analysis. Chapters 5 through 7 provide further examples of their application to modeling cost uncertainty from a system work breakdown structure perspective. To prepare for that discussion, this chapter concludes with a presentation on how to specify some of these special distributions, when only partial information about them is available.

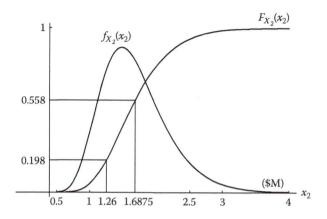

FIGURE 4.19
The PDF and CDF of X_2 in Example 4.9.

4.5 Specifying Continuous Probability Distributions

In systems engineering, probability distributions of variables whose values are uncertain must often be specified by expert technical opinion. This is particularly true in the absence of historical data. In such circumstances, expert opinion can be the only way to quantify a variable's uncertainty. Even when data exists, its quality may be so suspect as to nullify its use altogether. This section discusses strategies for specifying probability distributions when expert subjective assessments are required. This is illustrated in the context of continuous probability distributions.* Before delving into the details of these strategies, we discuss further the concept of subjective probabilities and distribution functions (introduced in Chapter 2).

4.5.1 Subjective Probabilities and Distribution Functions

In systems engineering, probabilities are often used to quantify uncertainties associated with a system's design parameters (e.g., weight), as well as uncertainties in cost and schedule. For reasons mentioned earlier, quantifying this uncertainty is often done in terms of subjective probabilities. Discussed in Chapter 2, subjective probabilities are those assigned to events on the basis of personal judgment. They measure a person's degree-of-belief that

---

* In practice, a continuous distribution is often used to describe the range of possible values for a random variable. This enables subject matter experts to focus on the "shape" that best describes the distribution of probability, rather than assessing individual probabilities associated to each distinct possible value (needed for discrete distributions).

an event will occur. Subjective probabilities are most often associated with one-time, nonrepeatable, events—those whose probabilities cannot be objectively determined from a population of outcomes developed by repeated trials, observations, or experimentation. Subjective probabilities cannot be arbitrary; they *must* adhere to the axioms of probability (refer to Chapter 2). For instance, if an electronics engineer assigns a probability of 0.70 to the event *the number of gates for the new processor chip "will not exceed"* 12,000, it must follow that the chip *will exceed* 12,000 gates with probability 0.30. Subjective probabilities are *conditional* on the state of the person's knowledge, which changes with time. To be credible, subjective probabilities should *only* be assigned to events by subject experts—persons with significant experience with events similar to the one under consideration. In addition, the rationale supporting the assigned probability *must be well documented*.

Instead of assigning a single subjective probability to an event, subject experts often find it easier to describe a function that depicts a subjective distribution of probabilities. Such a distribution is sometimes called a *subjective probability distribution*. Subjective probability distributions are governed by the properties of probability distributions associated with discrete or continuous random variables (refer to Chapter 3). Because of their nature, subjective probability distributions can be thought of as "belief functions"—mathematical representations of a subject expert's best professional judgment in the distribution of probabilities for a particular event.

When formulating subjective probability distributions, subject experts often prefer specifying a range that contains most, but not all, possible values. That is, there is a small nonzero probability that values will occur outside the expert's specified range. One strategy for specifying a subjective probability distribution involves the direct assessment of the distribution's fractiles. Another strategy involves assigning a subjective probability to a subinterval of the range of the distribution function. The following illustrates these strategies. This is done in the context of the distributions presented in this chapter. We begin with the beta distribution.

4.5.2 Specifying a Beta Distribution

The beta distribution has long been the distribution of "choice" for subjective assessments. It can take a wide variety of forms, as seen in Figures 4.9 and 4.10. The following illustrates how the beta distribution can be specified from subjective assessments on the shape parameters α and β and *any* two fractiles.

Case 1 Specify a nonstandard beta distribution for the random variable X given the shape parameters α and β and any two fractiles x_i and x_j, where $(0 \leq i < j \leq 1)$. An illustration of this case is presented in Figure 4.20.

Purposes To determine the minimum and maximum possible values for X, where $X \sim Beta(\alpha, \beta, a, b)$. To compute $E(X)$ and $Var(X)$ from the specified distribution.

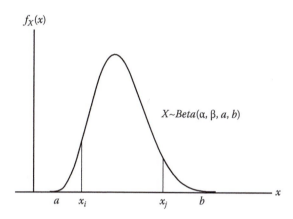

FIGURE 4.20
An illustrative beta distribution—Case 1.

Required information Assessments of α and β and any two fractiles x_i and x_j.

Discussion An assessment of the shape parameters α and β can be facilitated by having a subject expert look at a family of beta distributions, as shown in Figures 4.9 and 4.10. From such a family, an α and β pair can be chosen that reasonably depicts the distribution of probability (e.g., skewed, symmetric) for the variable under consideration. With α and β and any two fractiles x_i and x_j, the minimum and maximum possible values of X are given by Equations 4.50 and 4.51 (refer to Exercise 4.25), respectively.

$$a = \frac{x_i y_j - x_j y_i}{y_j - y_i} \tag{4.50}$$

$$b = \frac{x_j(1 - y_i) - x_i(1 - y_j)}{y_j - y_i} \tag{4.51}$$

In these equations, the terms x_i and x_j are the assessed values of X such that $P(X \le x_i) = i$ and $P(X \le x_j) = j$. The terms y_i and y_j are fractiles computed from the *standard beta distribution* associated with the given (as chosen by the subject expert) α and β. Once a and b have been determined, Theorem 4.4 can be used to compute $E(X)$ and $Var(X)$ associated with the specified distribution.

Example 4.10 *Find the minimum and maximum possible values of X if $X \sim$ Beta$(5, 10, a, b)$, $x_{0.05} = 4.76359$, and $x_{0.95} = 6.70003$. Find $E(X)$ and $Var(X)$.*

Solution Since $X \sim$ Beta$(5, 10, a, b)$, the distribution function of X has shape parameters $\alpha = 5$ and $\beta = 10$. From Equations 4.50 to 4.51 we can write

$$a = \frac{4.76359y_{0.95} - 6.70003y_{0.05}}{y_{0.95} - y_{0.05}} \qquad (4.52)$$

$$b = \frac{6.70003(1 - y_{0.05}) - 4.76359(1 - y_{0.95})}{y_{0.95} - y_{0.05}} \qquad (4.53)$$

Since the random variable Y must have the standard beta distribution $Y \sim Beta(5, 10)$, it can be determined[*] that $y_{0.05} = 0.152718$ and $y_{0.95} = 0.540005$. Substituting these values into Equations 4.52 and 4.53, we have $a = 4$ and $b = 9$, which are the minimum and maximum possible values of X, respectively. The reader should note this example is directly related to Example 4.4 (Section 4.2). Now that values for a and b are determined, the mean and variance of X can be determined directly from Theorem 4.4. It is left to the reader to show that $E(X) = 5.67$ and $Var(X) = 0.347$.

Example 4.11 *Suppose I represents the uncertainty in the number of DSI for a new software application. Suppose a team of software engineers judged 100,000 DSI as a reasonable assessment of the 50th percentile of I and a size of 150,000 DSI as a reasonable assessment of the 95th percentile. Furthermore, suppose the distribution function in Figure 4.21 was considered a good characterization of the uncertainty in the number of DSI. Given this,*

    a. Find the extreme possible values for I.

    b. Compute the mode of I.

    c. Compute E(I) and σ_I.

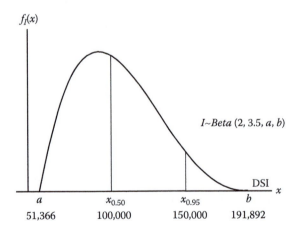

FIGURE 4.21
Beta distribution for Example 4.11.

---

[*] Determined by the *Mathematica* routine Quantile[BetaDistribution[5, 10], k], where k is equal to 0.05 and 0.95.

Solution

a. In Figure 4.21, I is given to be a beta distribution with shape parameters $\alpha = 2$ and $\beta = 3.5$. We are also given two probability assessments for I, specifically, $P(I \leq 100{,}000) = 0.50$ and $P(I \leq 150{,}000) = 0.95$; this is equivalent to the fractiles $x_{0.50} = 100{,}000$ and $x_{0.95} = 150{,}000$ (refer to Figure 4.21). Since $\alpha = 2$ and $\beta = 3.5$, the *standard beta distribution* is $Y \sim Beta(2, 3.5)$. From this we can determine the fractiles $y_{0.50}$ and $y_{0.95}$. Using *Mathematica*, $y_{0.50} = 0.346086$ and $y_{0.95} = 0.70189$ when $\alpha = 2$ and $\beta = 3.5$. Substituting $y_{0.50} = 0.346086$, $y_{0.95} = 0.70189$, $x_{0.50} = 100{,}000$, and $x_{0.95} = 150{,}000$ into Equations 4.50 and 4.51 provides the minimum and maximum possible values for I. These values are denoted by a and b:

$$a = \frac{(100{,}000)0.70189 - (150{,}000)0.346086}{0.70189 - 0.346086} = 51{,}366$$

$$b = \frac{150{,}000(1 - 0.346086) - 100{,}000(1 - 0.70189)}{0.70189 - 0.346086} = 191{,}892$$

b. Since $\alpha > 1$ and $\beta > 1$, from Equation 4.10, the mode of $Y \sim Beta(2, 3.5)$ is

$$y = \frac{1 - \alpha}{2 - \alpha - \beta} = \frac{1 - 2}{2 - 2 - 3.5} = 0.2857$$

By the transformation $y = (x - a)/(b - a)$, where a and b are from part (a), we have $Mode(I) = a + 0.2857(b - a) = 91{,}514$ DSI. Since the beta distribution in this example has a positive skew, the mode of I falls to the left of the 50th percentile of I.

c. From Theorem 4.4 with $\alpha = 2$, $\beta = 3.5$, $a = 51{,}366$ DSI, and $b = 191{,}892$ DSI, we have

$$E(I) = a + (b - a)E(Y) = a + (b - a)\frac{\alpha}{\alpha + \beta}$$

$$= 51{,}366 + (191{,}892 - 51{,}366)\frac{2}{2 + 3.5} = 102{,}466 \text{ DSI}$$

Once again, because the beta distribution in this example has a positive skew, the mean of I falls to the right of the 50th percentile of I. Last, from Equation 4.14 it can be shown that $Var(Y) = 0.0356$. From Equation 4.15 this translates to $Var(I) = 7.03(10)^8 \, DSI^2$; therefore,

$$\sigma_I = \sqrt{Var(I)} = 26{,}514 \text{ DSI} \blacklozenge$$

A nice feature of this approach is its flexibility to fully specify, for a given pair of shape parameters, a *nonstandard beta distribution* from *any two fractiles* of the distribution. This feature has strong practical utility. Subject experts often make "better" judgmental assessments of fractiles that fall near the middle of a distribution (e.g., the $x_{0.40}$ and $x_{0.60}$ fractiles) than out near its tails. Selecting shape parameters that "best" characterize the shape of the distribution has not been considered, in practice, too difficult. Shape parameters can be inferred by asking the expert to visually choose a distribution from a family of beta distributions plotted for various α and β. Representative plots of such a family are shown in Figures 4.9 and 4.10. Visual representations of a variable's uncertainty by distribution functions can be an excellent way to communicate risk to decision-makers.

4.5.3 Specifying Uniform Distributions

The following presents strategies for specifying a uniform distribution, when a subject expert assigns a probability α to a subinterval of the distribution's range. In the following cases, assume the random variable X is *uniformly distributed* over the range $a \leq x \leq b$.

Case 2 Specify a uniform distribution for the random variable X given the subinterval $a \leq x \leq b'$ and α, where a is the minimum possible value of X, $b' < b$, and $\alpha = P(a \leq X \leq b')$. An illustration of this case is presented in Figure 4.22.

Purposes To determine the maximum possible value of X. To compute $E(X)$ and $Var(X)$ from the specified distribution.

Required information Assessments of α and the endpoints of the subinterval $a \leq x \leq b'$.

Discussion In this case, a subject expert defines the subinterval $a \leq x \leq b'$ of the range of possible values for X, given by $a \leq x \leq b$. In addition, an assessment is made on the probability X will fall in this subinterval. If $P(a \leq X \leq b') = \alpha < 1$, then the maximum possible value of X is

$$b = a + \frac{1}{\alpha}(b' - a) \qquad (4.54)$$

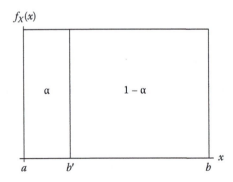

FIGURE 4.22
An illustrative uniform distribution—Case 2.

For example, if $\alpha = 0.25$, $a = 20$, and $b' = 30$, then, from Equation 4.54, the maximum value of X must be $b = 60$. This is illustrated in Figure 4.23.

For an application context, the random variable X might represent the uncertainty in the number of source instructions to develop for a new software application, or in the weight of a new electronic device, or in the number of labor hours to assemble a new widget.

Case 3 Specify a uniform distribution for the random variable X given the subinterval $a' \leq x \leq b'$ and α, where $a < a'$, $b' < b$, and $\alpha = P(a' \leq X \leq b')$. An illustration of this case is presented in Figure 4.24.

Purposes To determine the minimum and maximum possible values of X. To compute $E(X)$ and $Var(X)$ from the specified distribution.

Required information Assessments of α and the endpoints of the subinterval $a' \leq x \leq b'$. Furthermore, assume $a' - a = b - b'$ for this case.

Discussion In this case, a subject expert defines the subinterval $a' \leq x \leq b'$ of the range of possible values for X, given by $a \leq x \leq b$. An assessment

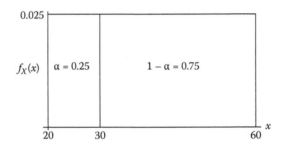

FIGURE 4.23
An illustration of Case 2.

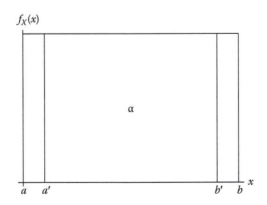

FIGURE 4.24
An illustrative uniform distribution—Case 3.

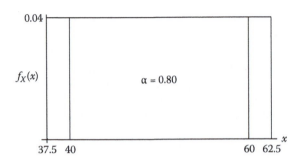

FIGURE 4.25
An illustration of Case 3.

of the probability X will fall in the subinterval $a' \leq x \leq b'$ is also made. If $P(a' \leq X \leq b') = \alpha < 1$, then the minimum and maximum possible values of X are

$$a = a' - \frac{1-\alpha}{2\alpha}(b' - a') \qquad (4.55)$$

$$b = b' + \frac{1-\alpha}{2\alpha}(b' - a') \qquad (4.56)$$

Note that $a' - a = b - b'$. Furthermore, for this case we have

$$P(a \leq X < a') = P(b' < X \leq b) = \frac{1}{2}(1 - \alpha)$$

For example, if $\alpha = 0.80$, $a' = 40$, and $b' = 60$, then, from Equations 4.55 to 4.56, the minimum and maximum possible values of X are $a = 37.5$ and $b = 62.5$. This is illustrated in Figure 4.25. An application context for this case is similar to the previous case.

In this case, it is possible for a to become negative even when a' is positive. In applications where it is sensible that X *be nonnegative* (e.g., if X is the uncertainty in the weight of a new widget), such an occurrence signals a reassessment of a' and α is needed.

4.5.4 Specifying a Triangular Distribution*

The following illustrates one strategy for specifying a triangular distribution when a subject expert assigns a probability α to a subinterval of the distribution's range. In the following case, assume the random variable X has a *triangular distribution* over the range $a \leq x \leq b$.

---

* This case was developed by Dr. Chien-Ching Cho, The MITRE Corporation, Bedford, MA.

Case 4 Specify a triangular distribution for the random variable X given m, the subinterval $a' \leq x \leq b'$, and α, where $a < a'$, $a' < m < b'$, $b' < b$, and $\alpha = P(a' \leq X \leq b')$. An illustration of this case is presented in Figure 4.26.

Purposes To determine the minimum and maximum possible values of X. To compute $E(X)$ and $Var(X)$ from the specified distribution.

Required information Assessments of α and the endpoints of the subinterval $a' \leq x \leq b'$, where $a' < m < b'$. Furthermore, assume for this case

$$\frac{P(X \leq a')}{P(X \geq b')} = \frac{P(X \leq m)}{P(X \geq m)}$$

Discussion In this case, a subject expert defines the subinterval $a' \leq x \leq b'$ of the range of possible values for X, given by $a \leq x \leq b$. An assessment is made of the probability X will fall in the subinterval $a' \leq x \leq b'$.

If $P(a' \leq X \leq b') = \alpha < 1$, then the minimum and maximum possible values of X are respectively

$$a = m - \frac{m - a'}{1 - \sqrt{1 - \alpha}} \tag{4.57}$$

$$b = m + \frac{b' - m}{1 - \sqrt{1 - \alpha}} \tag{4.58}$$

Equations 4.57 and 4.58 originate from the assumption (for this case) that

$$\frac{P(X \leq a')}{P(X \geq b')} = \frac{P(X \leq m)}{P(X \geq m)}$$

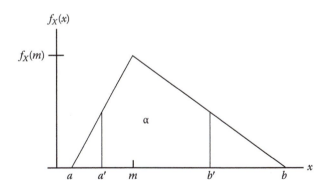

FIGURE 4.26
An illustrative triangular distribution—Case 4.

For example, if $\alpha = 0.75$, $a' = 25$, $m = 35$, and $b' = 60$, then, from Equations 4.57 to 4.58, the minimum and maximum possible values of X are $a = 15$ and $b = 85$. This is illustrated in Figure 4.27.

An application context for this case is similar to the previous cases. It is also possible in this case for a to become negative, even when a' is positive. In applications where it is sensible that X *be nonnegative* (e.g., if X is the uncertainty in the weight of a new widget), such an occurrence signals a reassessment of a' and α is needed.

In summary, Sir Josiah Stamp* once said,

> The government are very keen on amassing statistics. They collect them, raise them to the n-th power, take the cube root, and prepare wonderful diagrams. But one must never forget that every one of these figures comes in the first instance from the village watchman, who puts down what he damn pleases.

Several techniques have been presented for quantifying uncertainty in terms of subjective probabilities and distributions. As discussed, the need to do so is unavoidable on systems engineering projects. An extensive body of social science research exists on techniques for eliciting subjective probabilities and distributions. The book *Uncertainty: A Guide to Dealing With Uncertainty in Quantitative Risk and Policy Analysis* by Morgan and Henrion (1990) provides an excellent summary of this research.

Although the use of expert opinion is sometimes criticized, the basis of the criticism is often traceable to (1) the subject expert was really the "village watchman" or (2) the full scope of the problem being addressed by the expert was poorly described. To lessen the chance of (1) or (2) occurring, it is the prime responsibility of the project's cost and engineering team to collectively do the technical diligence needed to establish credible and defensible assessments.

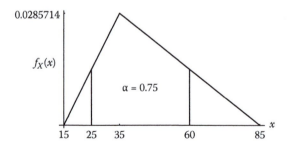

FIGURE 4.27
An illustration of Case 4.

For our purposes, it must be stressed that a key product from subjective assessment efforts must be a well-documented set of assumptions, arguments, and supportive materials. Documentation enables similarly qualified persons (or teams) to conduct independent and objective reviews of the assessments. This alone is an important step toward objectivity and one that would surface the presence of "village watchmen." Credible analyses stem from credible and defensible assessments; credible and defensible assessments stem from credible expertise. Properly conducted and documented assessments, on areas of a project that drive cost, schedule, and technical uncertainties, are among the most important products cost uncertainty analysis drives to produce.

Exercises

4.1 Given the trapezoidal distribution in Example 4.1, show that

    a. $P(25{,}000 \le X \le 28{,}000) = \frac{2}{13}$

    b. $P(25{,}000 \le X \le 35{,}000) = \frac{34}{39}$

4.2 Suppose $X \sim Trap(a, m_1, m_2, b)$ with PDF given in Figure 4.1.

    a. Show that $1 - P(X \le m_1) - P(X > m_2) = 2u_1/u_2$, where $u_1 = m_2 - m_1$ and $u_2 = m_2 + b - a - m_1$.

    b. What region in Figure 4.1 does the probability in Exercise 4.2a represent?

4.3 If $Cost \sim Unif(3, 8)$, then answer the following:

    a. $P(Cost < 5)$

    b. $P(4 < Cost \le 7)$

    c. Find x such that $P(Cost \le x) = 0.80$.

4.4 If $X \sim Unif(a, b)$ show that

    a. $E(X) = \frac{1}{2}(a + b)$

    b. $Var(X) = \frac{1}{12}(b - a)^2$

4.5 For the uniform distributions defined in Case 2 and Case 3, Section 4.5, derive Equations 4.54 (in Case 2), 4.55 (in Case 3), and 4.56 (in Case 3).

4.6 If $X \sim Trng(a, m, b)$, then answer the following:

    a. Verify $f_X(x)$ given by Equation 4.6 is a PDF.

    b. Show $F_X(x)$ changes concavity at $Mode(X)$.

    c. Prove that $E(X) = \frac{1}{3}(a + m + b)$.

4.7 Verify the probabilities in Figure 4.8 by computing the areas under the appropriate regions of each triangle.

4.8 If $X \sim Trng(15, 35, 85)$, then

a. Compute $P(X \leq 60)$

b. Compute $P(X \leq 25)$

c. Show that $P(X \leq 60) - P(X \leq 25) = 0.75$ (as seen in Figure 4.27)

4.9 If $X \sim Trng(0, 1, 1)$ compute

a. $E(5X + 1)$

b. $Var(3X - 1)$

4.10 If $Y \sim Beta(\alpha, \beta)$, verify Equations 4.16 and 4.17 if $E(Y)$ and $Var(Y)$ are known.

4.11 Suppose $Y \sim Beta(\alpha, \beta)$ and $f_Y(y) = 12y^2(1 - y)$, where $0 < y < 1$

a. Find α and β.

b. Compute $E(Y) + \sigma_Y$.

c. Determine $P(0.3 < Y \leq 0.7)$.

4.12 In Example 4.4 (Section 4.2)

a. Determine whether the expected time (mins) to assemble the microcircuit is greater than or less than the most probable time

b. Compute the standard deviation of the assembly time

4.13 If the cost of a system is *normally distributed* with mean 20 ($M) and standard deviation 4 ($M) determine

a. $P(Cost \leq 17)$

b. $P(15 \leq Cost < 22)$

c. $P\left(|Cost - \mu| \geq \frac{1}{2}\right)$

4.14 Suppose the uncertainty in a system's cost is described by a *normal distribution*. Suppose there is a 5% chance the system's cost will not exceed 100 ($M) and an 85% chance its cost will not exceed 200 ($M). From this information determine the mean and standard deviation of the system's cost.

4.15 If $X \sim N(\mu, \sigma^2)$, then show the following is true

a. $f_X(x)$ changes concavity at the points $x = \mu + \sigma$ and $x = \mu - \sigma$.

b. $P(\mu - 2\sigma \leq X \leq \mu + 2\sigma) = 0.9544$.

c. $P(\mu - 3\sigma \leq X \leq \mu + 3\sigma) = 0.9973$.

4.16 If X has a *lognormal distribution*, what does $P(\ln X \leq E(\ln X))$ always equal?

4.17 Compute the mean and variance of $\ln X$ for the three lognormal distributions in Figure 4.17.

4.18 Suppose the uncertainty in a system's cost is described by a *lognormal* PDF with $E(Cost) = 25$ ($M) and $Var(Cost) = 225$ ($M)^2; this is the left-most PDF in Figure 4.17. Determine

a. $P(Cost > E(Cost))$

b. $P(Cost \leq 50)$

4.19 In Figure 1.5, the random variable X_2 represented the cost of a system's systems engineering and program management. The point estimate of X_2, denoted by x_{2PEX_2}, was equal to 1.26 ($M). If X_2 can be approximated by a *lognormal distribution*, with $E(X_2) = 1.6875$ ($M) and $Var(X_2) = 0.255677$ ($M)^2, determine

a. $P(x_{2PEX_2} \leq X_2 < E(X_2))$

b. $P(1 \leq X_2 < 2.5)$

c. $P(X_2 \leq 2.5)$

4.20 If X is a lognormal random variable, show that the maximum value *of its density function* is given by Equation 4.42.

4.21 If X is a lognormal random variable, show that the rth moment of X is given by $E(X^r) = e^{r\mu_Y + \frac{1}{2}\sigma_Y^2 r^2}$.

4.22 Suppose I represents the uncertainty in the number of DSI for a new application. Suppose a team of software engineers judged 35,000 DSI as a reasonable assessment of the 50th percentile of I and a size of 60,000 DSI as a reasonable assessment of the 95th percentile. Furthermore, suppose the distribution function in Figure 4.21 was considered a good characterization of the uncertainty in the number of DSI.

a. Find the extreme possible values for I.

b. Compute the mode of I.

c. Compute $E(I)$ and σ_I.

4.23 Suppose W represents the uncertainty in the weight of a new unmanned spacecraft. Suppose a team of space systems engineers judged 1500 lbs as a reasonable assessment of the minimum possible weight. Furthermore, suppose this team also assessed the chance that W could fall between the minimum possible weight and 2000 lbs to be 80%. If the distribution function for W is *uniform*, determine the expected weight of the spacecraft.

4.24 Suppose I represents the uncertainty in the amount of new code for a software application. Suppose this uncertainty is characterized by the *triangular* PDF in Figure 4.26. If the probability is 0.90 that the amount of code is between 20,000 and 30,000 DSI, with 25,000 DSI as most probable, determine $E(I)$.

4.25 For the beta distribution defined in Case 1, Section 4.5, show that

$$a = \frac{x_i y_j - x_j y_i}{y_j - y_i} \text{ and } b = \frac{x_j(1 - y_i) - x_i(1 - y_j)}{y_j - y_i}$$

Hint: Solve for a and b from a simultaneous equation that involves the transformation $y = (x - a)/(b - a)$. Note that $P(Y \leq y_i) = i = P(X \leq x_i)$ and $P(Y \leq y_j) = j = P(X \leq x_j)$, in the context of Case 1 (Section 4.5).

References

Abramson, R. L. and P. H. Young. 1997 (Spring). FRISKEM–Formal Risk Evaluation Methodology. *The Journal of Cost Analysis*, 14(1), 29–38.

Evans, M., N. Hastings, and B. Peacock. 1993. *Statistical Distributions*, 2nd edn. New York: John Wiley & Sons, Inc.

Garvey, P. R. 1996 (Spring). Modeling cost and schedule uncertainties—A work breakdown structure perspective. *Military Operations Research*, 2(1), 37–43.

Garvey, P. R. and A. E. Taub. 1997 (Spring). A joint probability model for cost and schedule uncertainties. *The Journal of Cost Analysis*, 14(1), 3–27.

Johnson, N. L. and S. Kotz. 1969. *Distributions in Statistics: Discrete Distributions*. 1970. *Continuous Univariate Distributions 1, Continuous Univariate Distributions 2*. 1972. *Continuous Multivariate Distributions*. New York: John Wiley & Sons, Inc.

Morgan, M. G. and M. Henrion. 1990. *Uncertainty: A Guide to Dealing With Uncertainty in Quantitative Risk and Policy Analysis*. New York: Cambridge University Press.

Wolfram, S. 1991. *Mathematica®: A System for Doing Mathematics by Computer*, 2nd edn. Reading, MA: Addison-Wesley Publishing Company, Inc.

Young, D. C. and P. H. Young. 1995. A generalized probability distribution for cost/schedule uncertainty in risk assessment. *Proceedings of the 1995 Western Multi Conference on Business/MIS Simulation Education*, The Society for Computer Simulation. San Diego, CA.

5

Functions of Random Variables and Their Application to Cost Uncertainty Analysis

This chapter presents methods for studying the behavior of *functions of random variables*. Topics include joint probability distributions, linear combinations of random variables, the central limit theorem, and the development of distribution functions specific to a general class of software cost-schedule models.

5.1 Introduction

Functions of random variables occur frequently in cost engineering and analysis problems. For example, the first unit cost UC of an unmanned spacecraft might be derived according to (Lurie and Goldberg 1993)

$$UC = 5.48(SC_{wt})^{0.94}(BOLP)^{0.30}$$

where SC_{wt} is the spacecraft's dry weight (lbs) and $BOLP$ is the beginning-of-life power measured in watts (W). If it is early in a new spacecraft's design, the precise values for SC_{wt} and $BOLP$ might be unknown. The engineering team might better assess ranges of possible values for them instead of single point values. These ranges might be described by probability distributions, such as those presented in Chapter 4. If the first unit cost is a function of the random variables SC_{wt} and $BOLP$, a common question is "What is the probability distribution of UC given probability distributions for SC_{wt} and $BOLP$?" This chapter presents methods to answer this and related questions. First, some mathematical preliminaries.

5.1.1 Joint and Conditional Distributions

When a function is defined by two or more random variables, its probability distribution is called a *joint probability distribution*. Joint probability distributions generalize the concept of univariate distributions to functions of several random variables. Analogous to the univariate case, the *joint cumulative distribution function* (CDF) of random variables X and Y is

$$F_{X,Y}(x,y) = P(X \le x, Y \le y), \quad -\infty < x, y < \infty \qquad (5.1)$$

Discrete random variables: If X and Y are *discrete* random variables, their *joint probability mass function* is defined as follows:

$$p_{X,Y}(x,y) = P(X = x, \ Y = y) \qquad (5.2)$$

Illustrated in Figure 5.1, $p_{X,Y}(x,y)$ is the probability a possible pair of values $p_{X,Y}(x,y)$ will occur. If R is any region in the xy-plane and X and Y are *discrete* random variables, then

$$P((X, Y) \in R) = \sum_{(x,y) \in R} p_{X,Y}(x,y) \qquad (5.3)$$

Equation 5.3 implies the probability of a *random point* falling in a region R is the sum of the heights of the vertical lines that correspond to the points contained in R. The heights of the lines are given by $p_{x,y}(x,y)$. Joint probabilities are defined in terms of R and the joint probability mass function. For example, the probability X is less than Y is represented by the set of all points in the region where $p_{X,Y}(x,y)$. This can be written as follows:

$$P((X, Y) \in \{(x,y) : x < y\}) = \sum_{(x,y): x<y} p_{X,Y}(x,y) \qquad (5.4)$$

If X and Y have a finite number of possible values, it is sometimes convenient to arrange the probabilities associated with these values in a *contingency table*.

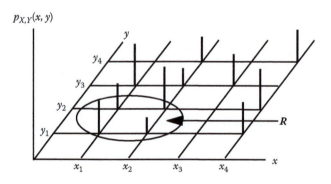

FIGURE 5.1
A joint probability mass function of X and Y.

Table 5.1 illustrates a contingency table for two random variables that each have four possible values. The sum of all $p_{X,Y}(x_i, y_k)$ in a contingency table must equal unity. If X and Y are discrete random variables, then their *marginal probability mass functions* are

$$p_X(x) = P(X = x) = \sum_y p_{X,Y}(x, y) \tag{5.5}$$

$$p_Y(y) = P(Y = y) = \sum_x p_{X,Y}(x, y) \tag{5.6}$$

Equation 5.5 is the marginal probability mass function of X. Equation 5.6 is the marginal probability mass function of Y.

> **Example 5.1** *Suppose the effort (staff months) to modernize a management information system is given by $Eff_{SysEng} = XY$, where X is the number of systems engineering staff needed for Y months. Suppose Table 5.2 is the contingency table for X and Y. Compute*
>
> a. $P(X = 15, Y = 36)$
> b. $P(X = 15)$
> c. $P(Y = 36)$
> d. $P(Eff_{SysEng} < 600)$
>
> *Solution*
>
> a. From Equation 5.2
>
> $$P(X = 15, Y = 36) = p_{X,Y}(15, 36) = 0.25$$

TABLE 5.1

A Contingency Table for X and Y

(X, Y)	y_1	y_2	y_3	y_4
x_1	$p_{X,Y}(x_1, y_1)$	$p_{X,Y}(x_1, y_2)$	$p_{X,Y}(x_1, y_3)$	$p_{X,Y}(x_1, y_4)$
x_2	$p_{X,Y}(x_2, y_1)$	$p_{X,Y}(x_2, y_2)$	$p_{X,Y}(x_2, y_3)$	$p_{X,Y}(x_2, y_4)$
x_3	$p_{X,Y}(x_3, y_1)$	$p_{X,Y}(x_3, y_2)$	$p_{X,Y}(x_3, y_3)$	$p_{X,Y}(x_3, y_4)$
x_4	$p_{X,Y}(x_4, y_1)$	$p_{X,Y}(x_4, y_2)$	$p_{X,Y}(x_4, y_3)$	$p_{X,Y}(x_4, y_4)$

TABLE 5.2

Contingency Table for Example 5.1

	$y_1 = 24$ Months	$y_2 = 36$ Months	Total
$x_1 = 15$ Staff	0.15	0.25	0.40
$x_2 = 25$ Staff	0.20	0.40	0.60
Total	0.35	0.65	1.00

b. $P(X = 15)$ is a marginal probability; from Equation 5.5

$$P(X = 15) = \sum_{k=1}^{2} p_{X,Y}(15, y_k) = 0.15 + 0.25 = 0.40$$

c. $P(Y = 36)$ is a marginal probability; from Equation 5.6

$$P(Y = 36) = \sum_{t=1}^{2} p_{X,Y}(x_t, 36) = 0.25 + 0.40 = 0.65$$

d. From Table 5.2, the region R where the event $\{Eff_{SysEng} < 600\}$ occurs contains only two points; specifically,

$$R = \{(x, y) : xy < 600\} = \{(x_1, y_1), (x_1, y_2)\}$$

where $(x_1, y_1) = (15, 24)$ and $(x_1, y_2) = (15, 36)$. From Equation 5.3

$$P(Eff_{SysEng} < 600) = P(XY < 600)$$

$$= P\left((X, Y) \in \{(x, y) : xy < 600\}\right)$$

$$= \sum_{(x,y):\,xy<600} p_{X,Y}(x, y)$$

$$= p_{X,Y}(x_1, y_1) + p_{X,Y}(x_1, y_2)$$

$$= 0.15 + 0.25 = 0.40.$$

Continuous random variables: If X and Y are *continuous* random variables, the joint probability density function (PDF) of X and Y, denoted by $f(x, y)$, satisfies for any set R in the two-dimensional plane

$$P((X, Y) \in R) = \iint_{(x,y)\in R} f(x, y) dx dy \qquad (5.7)$$

where $f(x, y) \geq 0$ and

$$\int_{-\infty}^{\infty} \int_{-\infty}^{\infty} f(x, y) dx dy = 1$$

The probability associated with a univariate continuous random variable reflects an area under the variable's density function. The probability represented by the double integral in Equation 5.7 is the *volume* over the region R between the xy-plane and the surface determined by $f(x, y)$. In particular,

$$P(a \leq X \leq b \text{ and } c \leq Y \leq d) = \int_{a}^{b} \int_{c}^{d} f(x, y) dy dx \qquad (5.8)$$

With n continuous random variables, $X_1, X_2, X_3, \ldots, X_n$, we have

$$P(a_1 \leq X_1 \leq b_1 \ldots a_n \leq X_n \leq b_n) = \int_{a_1}^{b_1} \cdots \int_{a_n}^{b_n} f(x_1, \ldots, x_n)dx_n \cdots dx_1 \qquad (5.9)$$

The *marginal* PDFs of X and Y are given by

$$f_X(x) = \int_{-\infty}^{\infty} f(x,y)dy, \quad \text{for } -\infty < x < \infty \qquad (5.10)$$

$$f_Y(y) = \int_{-\infty}^{\infty} f(x,y)dx, \quad \text{for } -\infty < y < \infty \qquad (5.11)$$

Example 5.2 *Suppose the effort (staff months) to develop and implement a system's test plans and procedures is given by $Eff_{SysTest} = XY$, where X is the number of test staff needed over Y months. Suppose X and Y are continuous random variables with joint PDF*

$$f(x,y) = \begin{cases} \dfrac{1}{240}, & \text{if } 5 \leq x \leq 15, 12 \leq y \leq 36 \\ 0, & \text{otherwise} \end{cases}$$

This joint PDF has the marginal PDFs in Figure 5.2. Determine

a. *$P(Eff_{SysTest} \leq 120)$*

b. *$P(Eff_{SysTest} \leq 360)$*

c. *$P(Eff_{SysTest} \leq 120)$ given the test staff will not exceed 10 persons*

d. *The probability $Eff_{SysTest}$ is greater than 120 staff months and the test staff and duration will not exceed 10 persons and 24 months, respectively*

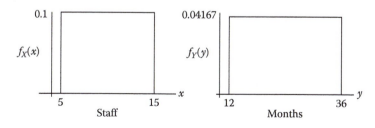

FIGURE 5.2
Marginal distributions for X and Y.

Solution

a. To determine the probability $Eff_{SysTest} \leq 120$, we first sketch the event space. This is shown in Figure 5.3. From Equation 5.8, we have

$$P(Eff_{SysTest} \leq 120) = \iint\limits_{xy \leq 120} f(x,y)dxdy$$

$$= \int_{12}^{24} \int_{5}^{\frac{120}{y}} \frac{1}{240} dxdy = \int_{5}^{10} \int_{12}^{\frac{120}{x}} \frac{1}{240} dydx = 0.09657$$

b. To determine the probability $Eff_{SysTest} \leq 360$, we first sketch the event space. This is shown in Figure 5.4. From Theorem 2.1

$$P(Eff_{SysTest} \leq 360) = 1 - P(Eff_{SysTest} > 360)$$

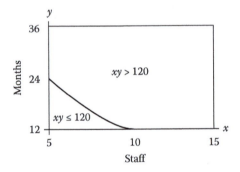

FIGURE 5.3
Event space for $Eff_{SysTest} \leq 120$.

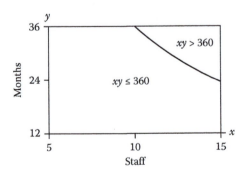

FIGURE 5.4
Event space for $Eff_{SysTest} \leq 360$.

It follows that

$$P(\text{Eff}_{SysTest} \leq 360) = 1 - \iint\limits_{xy>360} f(x,y)dxdy$$

$$= 1 - \int\limits_{24}^{36} \int\limits_{\frac{360}{y}}^{15} \frac{1}{240}dxdy = 1 - \int\limits_{10}^{15} \int\limits_{\frac{360}{x}}^{36} \frac{1}{240}dydx = 0.858$$

c. The probability $\text{Eff}_{SysTest} \leq 120$ staff months *given* the test staff-level will not exceed 10 persons is a conditional probability; specifically, the conditional probability is $P(\text{Eff}_{SysTest} \leq 120 | X \leq 10)$. From Equation 2.2, we can write

$$P(\text{Eff}_{SysTest} \leq 120 | X \leq 10) = \frac{P(\{XY \leq 120\} \cap \{X \leq 10\})}{P(\{X \leq 10\})}$$

In this case,

$$\frac{P(\{XY \leq 120\} \cap \{X \leq 10\})}{P(\{X \leq 10\})} = \frac{\int\limits_{5}^{10} \int\limits_{12}^{\frac{120}{x}} \frac{1}{240}dydx}{\int\limits_{5}^{10} \frac{1}{(15-5)}dx}$$

$$= 2 \int\limits_{5}^{10} \int\limits_{12}^{\frac{120}{x}} \frac{1}{240}dydx = 2(0.09657) = 0.193$$

The conditional probability, in this example, is twice its unconditional probability computed in part (a). Why is this? The unconditional probability is associated with the joint distribution function

$$f(x,y) = (1/240), \quad 5 \leq x \leq 15, \quad 12 \leq y \leq 36$$

If it is given that $X \leq 10$, the joint distribution function essentially becomes

$$f(x,y) = (1/120), \quad 5 \leq x \leq 10, \quad 12 \leq y \leq 36$$

With $f(x,y) = (1/120)$, and $5 \leq x \leq 10$, $12 \leq y \leq 36$, more probability exists in the region where $XY \leq 120$ than in the same region with $f(x,y) = (1/240)$, and $5 \leq x \leq 15$, $12 \leq y \leq 36$.

d. To determine the probability $\text{Eff}_{SysTest} > 120$ staff months and the test staff and duration will not exceed 10 persons and 24 months, define three events A, B, and C as

$$A = \left\{ \text{Eff}_{SysTest} > 120 \right\} = \{XY > 120\}$$

$$B = \{X \leq 10\}$$

$$C = \{Y \leq 24\}$$

Thus, the probability we want to determine is given by

$$P(A \cap B \cap C) = P(\{XY > 120\} \cap \{X \le 10\} \cap \{Y \le 24\})$$

$$= P\left(\left\{\frac{120}{Y} < X\right\} \cap \{X \le 10\} \cap \{Y \le 24\}\right)$$

$$= P\left(\left\{\frac{120}{Y} < X \le 10\right\} \cap \{Y \le 24\}\right)$$

From Equation 5.8

$$P\left(\left\{\frac{120}{Y} < X \le 10\right\} \cap \{Y \le 24\}\right) = \int_{12}^{24} \int_{\frac{120}{y}}^{10} \frac{1}{240} dx dy = 0.1534$$

The probability is just over 0.15 that the effort for system test will exceed 120 staff months, and the test staff-level and duration will not exceed 10 persons and 24 months. This probability is shown by the region R in Figure 5.5.

Example 5.3 *Suppose the effort (staff months) to develop a new software application is given by $Eff_{SW} = X/Y$, where X is the size of a software application (number of delivered source instructions, or DSI) and Y is the development productivity rate (number of DSI per staff month). Suppose X and Y are continuous random variables with joint PDF*

$$f(x, y) = \begin{cases} \dfrac{1}{5(10^6)}, & 50,000 \le x \le 100,000, \quad 100 \le y \le 200 \\ 0, & \text{otherwise} \end{cases}$$

This joint PDF has the marginal PDFs in Figure 5.6. Determine the probability Eff_{SW} will not exceed 300 staff months.

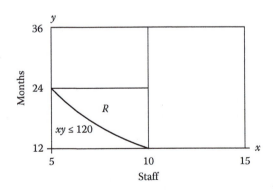

FIGURE 5.5
Region R associated with part (c) of Example 5.2.

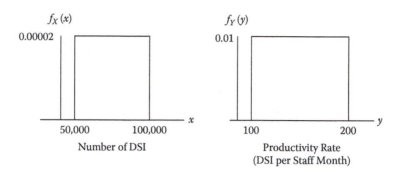

FIGURE 5.6
Marginal distributions for X and Y.

Solution To determine the probability Eff_{SW} will not exceed 300 staff months, we first sketch the event space. This is shown in Figure 5.7. From Equation 5.8, we have

$$P(Eff_{SW} \leq 300) = \iint\limits_{\frac{x}{y} \leq 300} f(x,y)dxdy$$

$$= \int\limits_{166.667}^{200} \int\limits_{50,000}^{300y} \frac{1}{5(10^6)}dxdy$$

$$= \int\limits_{50,000}^{60,000} \int\limits_{\frac{x}{300}}^{200} \frac{1}{5(10^6)}dydx = 0.0333$$

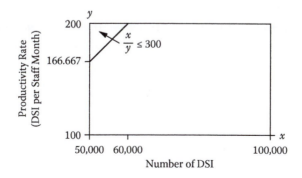

FIGURE 5.7
Event space for $Eff_{SW} \leq 300$.

So far, we have introduced the concept of joint probability distributions for two random variables. Often, it is necessary to know the distribution of one random variable when the other takes a specific value. Such a distribution is known as a conditional probability distribution, which is discussed next in terms of discrete and continuous random variables.

Conditional probability mass function: If two *discrete* random variables X and Y have joint probability mass function $p_{X,Y}(x,y)$, the *conditional probability mass function* of X given $Y = y$ is

$$p_{X|Y=y}(x) = \frac{p_{X,Y}(x,y)}{p_Y(y)} \qquad (5.12)$$

where $p_Y(y) > 0$. Similarly, the *conditional probability mass function* of Y given $X = x$ is

$$p_{Y|X=x}(y) = \frac{p_{X,Y}(x,y)}{p_X(x)} \qquad (5.13)$$

where $p_X(x) > 0$. To illustrate this, return to Example 5.1; suppose we want the probability that the number of systems engineering staff X will be 15 persons, *given* they are needed for 36 months. In this case we want $p_{X|Y=36}(15)$. From Equation 5.12 and Table 5.2 this is

$$p_{X|Y=36}(15) = \frac{p_{X,Y}(15,36)}{p_Y(36)} = \frac{0.25}{0.65} = \frac{5}{13} \approx 0.3846$$

This probability is conditioned on a fixed (or observed) value for Y. It has a value slightly less than the unconditioned probability $P(X = 15)$, which was shown in Example 5.1 to be 0.40.

Conditional probability density function: If two *continuous* random variables X and Y have joint density function $f(x,y)$, then the *conditional* PDF of X, given $Y = y$, is

$$f_{X|Y}(x|y) = \frac{f(x,y)}{f_Y(y)}, \quad f_Y(y) > 0 \qquad (5.14)$$

Similarly, the *conditional PDF* of Y, given $X = x$, is

$$f_{Y|X}(y|x) = \frac{f(x,y)}{f_X(x)}, \quad f_X(x) > 0 \qquad (5.15)$$

Example 5.4 *In Example 5.2, X and Y had the joint PDF*

$$f(x,y) = \begin{cases} \dfrac{1}{240}, & 5 \le x \le 15, 12 \le y \le 36 \\ 0, & \text{otherwise} \end{cases}$$

Find the conditional PDFs of X and Y.

Solution From Equation 5.14, the conditional PDF of X is

$$f_{X|Y}(x|y) = \frac{f(x,y)}{f_Y(y)} = \frac{\frac{1}{240}}{\frac{1}{24}} = \frac{1}{10}, \quad 5 \le x \le 15$$

From Equation 5.15, the conditional PDF of Y is

$$f_{Y|X}(y|x) = \frac{f(x,y)}{f_X(x)} = \frac{\frac{1}{240}}{\frac{1}{10}} = \frac{1}{24}, \quad 12 \le y \le 36$$

Conditional PDFs enable determining the conditional CDF. Specifically,

$$F_{X|Y}(x = a|y) \equiv P(X \le a|Y = y) = \int_{-\infty}^{a} f_{X|Y}(x|y)dx \tag{5.16}$$

$$F_{Y|X}(y = b|x) \equiv P(Y \le b|X = x) = \int_{-\infty}^{b} f_{Y|X}(y|x)dy \tag{5.17}$$

5.1.2 Independent Random Variables

Two random variables X and Y are *independent* if for any two events $\{X \in A\}$ and $\{Y \in B\}$, where A and B are sets of real numbers, we have

$$P(\{X \in A\} \cap \{Y \in B\}) = P(\{X \in A\})P(\{Y \in B\}) \tag{5.18}$$

Equation 5.18 follows if and only if, for any x and y

$$P(\{X \le x\} \cap \{Y \le y\}) \equiv P(X \le x, Y \le y) = P(X \le x)P(Y \le y) \tag{5.19}$$

From Equation 5.19, it follows that

$$F_{X,Y}(x,y) \equiv P(X \le x, Y \le y) = F_X(x)F_Y(y), \quad -\infty < x, y < \infty \tag{5.20}$$

If X and Y are independent *discrete* random variables, Equation 5.18 becomes

$$p_{X,Y}(x,y) = p_X(x)p_Y(y) \tag{5.21}$$

It follows that

$$p_{X|Y=y}(x) = p_X(x) \tag{5.22}$$

$$p_{Y|X=x}(y) = p_Y(y) \tag{5.23}$$

Moreover, if Equation 5.21 holds for two discrete random variables, then the random variables are independent. Similarly, X and Y are independent *continuous* random variables if and only if Equation 5.24 holds for all feasible values of X and Y.

$$f(x,y) = f_X(x)f_Y(y) \tag{5.24}$$

It follows that

$$f_{X|Y}(x|y) = f_X(x) \tag{5.25}$$

$$f_{Y|X}(y|x) = f_Y(y) \tag{5.26}$$

From this, what do you conclude about the random variables X and Y in Examples 5.2 and 5.3? A discussion of this is left as an exercise for the reader. Finally, we say that *dependent* random variables are those that are *not independent*.

5.1.3 Expectation and Correlation

In Chapter 3, the expectation of a random variable was discussed. The expectation of two random variables is stated in the following proposition.

Proposition 5.1 *If X and Y are random variables and $g(x,y)$ is a real-valued function defined for all x and y that are possible values of X and Y, then*

$$E(g(X,Y)) = \begin{cases} \displaystyle\sum_x \sum_y g(x,y) \cdot p_{X,Y}(x,y), & \text{if } X \text{ and } Y \text{ are discrete} \\[2ex] \displaystyle\int_{-\infty}^{\infty} \int_{-\infty}^{\infty} g(x,y) \cdot f(x,y)dxdy, & \text{if } X \text{ and } Y \text{ are continuous} \end{cases} \tag{5.27}$$

where the double summation and double integral must be absolutely convergent.

Example 5.5 *Determine the expectation of $Eff_{SysTest}$ in Example 5.2.*

Solution We need to compute $E(Eff_{SysTest})$. From Example 5.2, the joint distribution of X and Y is given as

$$f(x,y) = \begin{cases} \dfrac{1}{240}, & 5 \le x \le 15, 12 \le y \le 36 \\ 0, & \text{otherwise} \end{cases}$$

In Example 5.2 $Eff_{SysTest}$ is a function of two random variables X and Y; that is, $Eff_{SysTest} = XY = g(X, Y)$. Therefore, in this case, $g(x, y) = xy$. Since X and Y are continuous random variables, from Equation 5.27

$$E(Eff_{SysTest}) = E(XY) = E(g(X, Y)) = \int_5^{15} \int_{12}^{36} xy \cdot \frac{1}{240} dy dx$$

$$= 240 \text{ staff months}$$

It is often of interest to determine where the expected value of a random variable falls along the variable's CDF. Mentioned in Chapters 3 and 4, the expected value of a random variable *is not*, in general, equal to the median of the random variable. This is again illustrated with Example 5.5. It is left as an exercise for the reader to show

$$P(Eff_{SysTest} \le E(Eff_{SysTest})) = P(Eff_{SysTest} \le 240) = 0.56$$

It is often necessary to know the degree to which two random variables associate or vary with each other. In cost analysis, questions such as "How much is the variation in a new satellite's predicted weight attributable to the variation in its cost?" are common. *Covariance* is a measure of how much two random variables covary. Let X and Y be random variables with expected values (means) μ_X and μ_Y, respectively. The covariance of X and Y, denoted by $Cov(X, Y)$, is defined as follows:

$$Cov(X, Y) \equiv \sigma_{XY} = E\{(X - \mu_X)(Y - \mu_Y)\} \tag{5.28}$$

Covariance can be positive, negative, or zero. If X and Y take values simultaneously larger than their respective means, the covariance will be positive. If X and Y take values simultaneously smaller than their respective means, the covariance will also be positive. If one random variable takes a value larger than its mean *and* the other takes a value smaller than its mean, the covariance will be negative. So, when two random variables simultaneously take values on the same sides as their respective means, the covariance will be positive. When two random variables simultaneously take values on opposite sides of their means, the covariance will be negative. The following theorems present useful properties of covariance. Theorem 5.1 presents a way to compute covariance that is easier than using the definition given by Equation 5.28.

Theorem 5.1   *If X and Y are random variables with means μ_X and μ_Y then*

$$Cov(X, Y) = E(XY) - \mu_X \mu_Y$$

Theorem 5.2   *If X and Y are random variables, then*

   a. $Cov(X, Y) = Cov(Y, X)$
   b. $Cov(aX + b, cY + d) = ac\,Cov(X, Y)$ *for any real numbers a, b, c, and d*

Theorem 5.3   *If X and Y are independent random variables, then $Cov(X, Y) = 0$.*

Covariance as a measure of the degree two random variables covary can be hard to interpret. Suppose X_1 and Y_1 are random variables such that $X_2 = 2X_1$ and $Y_2 = 2Y_1$. From Theorem 5.2, $Cov(X_2, Y_2) = 4Cov(X_1, Y_1)$. Although X_1 and Y_1 and X_2 and Y_2 behave in precisely the same way with respect to each other, the random variables X_2 and Y_2 have a covariance four times greater than the covariance of X_1 and Y_1. A more convenient measure is one where the relationship between pairs of random variables could be interpreted along a common scale. Consider the following.

Suppose we have two standard random variables Z_X and Z_Y, where

$$Z_X = \frac{X - \mu_X}{\sigma_X} \text{ and } Z_Y = \frac{Y - \mu_Y}{\sigma_Y}$$

Using Theorem 5.2, the covariance of Z_X and Z_Y reduces to

$$Cov(Z_X, Z_Y) = Cov\left(\frac{X - \mu_X}{\sigma_X}, \frac{Y - \mu_Y}{\sigma_Y}\right)$$

$$= \frac{1}{\sigma_X} \cdot \frac{1}{\sigma_Y} Cov\,(X - \mu_X, Y - \mu_Y)$$

$$= \frac{1}{\sigma_X} \cdot \frac{1}{\sigma_Y} Cov(X, Y)$$

$$= \rho_{X,Y}$$

The term $\rho_{X,Y}$ is known as the Pearson correlation coefficient. It is the traditional statistic to measure the degree to which two random variables linearly correlate (or covary). Formally, the *Pearson correlation coefficient* between two random variables X and Y is

$$Corr(X, Y) \equiv \rho_{X,Y} = \frac{Cov(X, Y)}{\sigma_X \sigma_Y} \qquad (5.29)$$

provided $\sigma_X > 0$ and $\sigma_Y > 0$. From Theorem 5.1, Equation 5.29 simplifies to

$$Corr(X, Y) \equiv \rho_{X,Y} = \frac{E(XY) - \mu_X \mu_Y}{\sigma_X \sigma_Y} \qquad (5.30)$$

The correlation coefficient is dimensionless. Pearson's correlation coefficient measures the *strength of the linear relationship* between two random variables. It is bounded by the interval $-1 \le \rho_{X,Y} \le 1$. If $Y = aX + b$, where a and b are real numbers and $a > 0$, then $\rho_{X,Y} = 1$; if $a < 0$ then $\rho_{X,Y} = -1$. When $\rho_{X,Y} = 0$, we say X and Y are *uncorrelated*. There is a complete absence of linearity between them. Figure 5.8 illustrates the types of correlation that can exist between random variables.

Example 5.6 *If $Y = X^2$ and $X \sim Unif(-1, 1)$, show that $\rho_{X,Y} = 0$.*

Solution From Equation 5.30

$$Corr(X, Y) \equiv Corr(X, X^2) \equiv \rho_{X,X^2} = \frac{E(XX^2) - \mu_X \mu_{X^2}}{\sigma_X \sigma_{X^2}}$$

Since $X \sim Unif(-1, 1)$, we have $f_X(x) = \frac{1}{2}$ on $-1 \le x \le 1$; therefore,

$$E(XX^2) = E(X^3) = \int_{-1}^{1} x^3 f_X(x)\, dx = \int_{-1}^{1} x^3 \frac{1}{2} dx = 0$$

$$\mu_X = E(X) = \int_{-1}^{1} x f_X(x)\, dx = \int_{-1}^{1} x \frac{1}{2} dx = 0$$

$$\mu_{X^2} = E(X^2) = \int_{-1}^{1} x^2 \frac{1}{2} dx = \frac{1}{3}$$

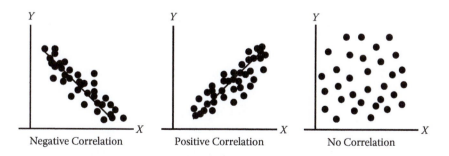

| Negative Correlation | Positive Correlation | No Correlation |

FIGURE 5.8
Correlation between random variables X and Y.

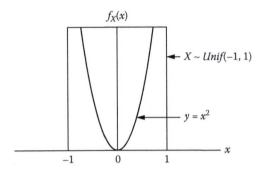

FIGURE 5.9
Graph of $Y = X^2$ and $X \sim Unif(-1, 1)$.

Therefore,

$$Corr(X, X^2) \equiv \rho_{X, X^2} = \frac{0 - 0 \cdot \frac{1}{3}}{\sigma_X \sigma_{X^2}} = 0$$

In this example, we conclude there is a complete absence of linearity between X and Y. This can be seen in Figure 5.9.

Theorem 5.4   *If X and Y are independent random variables, then $\rho_{X,Y} = 0$.*

Proof.   This follows from Theorem 5.3 and Equation 5.29. Since X and Y are independent random variables, from Theorem 5.3 we have $Cov(X, Y) = 0$. From Equation 5.29, if $Cov(X, Y) = 0$, it immediately follows that $\rho_{X,Y} = 0$.

The converse of Theorem 5.4 is not true. If $\rho_{X,Y} = 0$ then X and Y are *uncorrelated*. However, it *does not* follow that X and Y are independent. Again, if X is uniformly distributed in $-1 \leq x \leq 1$ and $Y = X^2$, then $\rho_{X,Y} = 0$; however, Y is dependent on X in this case. Theorem 5.4 gives rise to the following:

Theorem 5.5   *If X and Y are independent random variables, then*

$$E(XY) = E(X)E(Y) \tag{5.31}$$

Proof.   Since X and Y are independent random variables, from Theorem 5.3 we have $Cov(X, Y) = 0$. From Theorem 5.1, this is equivalent to $E(XY) - \mu_X \mu_Y = 0$; thus, $E(XY) = E(X)E(Y)$.

Theorem 5.6 *If a_1 and a_2 are either both positive or both negative, and a_1, a_2, b_1, and b_2 are real numbers, then $Corr(a_1 X + b_1, a_2 Y + b_2) = Corr(X, Y)$.*

Proof. Let $Z = a_1 X + b_1$ and $W = a_2 Y + b_2$. We need to show

$$Corr(Z, W) = \frac{E(ZW) - \mu_Z \mu_W}{\sigma_Z \sigma_W} = Corr(X, Y) \tag{5.32}$$

From Theorem 3.9

$$
\begin{aligned}
E(ZW) &= E((a_1 X + b_1)(a_2 Y + b_2)) \\
&= E(a_1 a_2 XY + a_1 b_2 X + a_2 b_1 Y + b_1 b_2) \\
&= a_1 a_2 E(XY) + a_1 b_2 E(X) + a_2 b_1 E(Y) + b_1 b_2
\end{aligned}
$$

From Theorem 3.9

$$\mu_Z \equiv E(Z) = a_1 E(X) + b_1 \quad \text{and} \quad \mu_W \equiv E(W) = a_2 E(Y) + b_2$$

From Theorem 3.11

$$\sigma_Z^2 = a_1^2 \sigma_X^2 \quad \text{and} \quad \sigma_W^2 = a_2^2 \sigma_Y^2$$

Combining these

$$E(ZW) - \mu_Z \mu_W = a_1 a_2 E(XY) - a_1 a_2 E(X)E(Y) = a_1 a_2 (E(XY) - E(X)E(Y))$$

and

$$\sigma_Z = |a_1| \sigma_X, \quad \sigma_W = |a_2| \sigma_Y$$

Substituting into Equation 5.32 yields

$$Corr(Z, W) = \frac{a_1 a_2 (E(XY) - E(X)E(Y))}{a_1 a_2 \sigma_X \sigma_Y} = Corr(X, Y)$$

This theorem states that the correlation between two random variables is unaffected by a linear change in either X or Y.

Example 5.7 *Suppose X denotes the number of engineering staff required to test a new rocket propulsion system. Suppose X is uniformly distributed in the interval $5 \le x \le 15$. If the number of months Y required to design, conduct, and analyze the test is given by $Y = 2X + 3$, compute the expected test effort, measured in staff months.*

Solution We are given $X \sim Unif(5, 15)$ and $Y = 2X + 3$. The test effort, in staff months, is the product XY. To determine the *expected test effort*, we need to compute $E(XY)$. From Equation 5.30

$$E(XY) = \rho_{X,Y}\sigma_X\sigma_Y + \mu_X\mu_Y$$

Since Y is a linear function of X, we have $\rho_{X,Y} = 1$; thus,

$$E(XY) = \sigma_X\sigma_Y + \mu_X\mu_Y$$

Since $X \sim Unif(5, 15)$, the mean and variance of X (Theorem 4.2) is

$$\mu_X \equiv E(X) = \frac{1}{2}(5 + 15) = 10$$

$$\sigma_X^2 \equiv Var(X) = \frac{1}{12}(15 - 5)^2 = \frac{100}{12} \text{ and } \sigma_X = \frac{10}{\sqrt{12}}$$

Since $Y = 2X + 3$, the mean and variance of Y (Theorems 3.9 and 3.11) is

$$\mu_Y \equiv E(Y) = E(2X + 3) = 2E(X) + 3 = 2 \cdot 10 + 3 = 23$$

$$\sigma_Y^2 \equiv Var(Y) = Var(2X + 3) = 2^2 Var(X) = 4\sigma_X^2 = 4 \cdot \frac{100}{12} = \frac{100}{3}$$

$$\sigma_Y = \frac{10}{\sqrt{3}}$$

Substituting these values into $E(XY)$ we have

$$E(XY) = \sigma_X\sigma_Y + \mu_X\mu_Y = \frac{10}{\sqrt{12}} \cdot \frac{10}{\sqrt{3}} + 10 \cdot 23 = 246.7 \text{ staff months}$$

Thus, the expected effort to test the new rocket's propulsion system is nearly 247 staff months.

Example 5.8 *Suppose the effort Eff_{SW} (staff months) to develop software is given by $Eff_{SW} = 2.8I^{1.2}$, where I is thousands of DSI to be developed. If $I \sim Unif(20, 60)$ determine $\rho_{Eff_{SW}, I}$.*

Solution From Equation 5.30

$$Corr(Eff_{SW}, I) \equiv \rho_{Eff_{SW}, I} = \frac{E(Eff_{SW}I) - \mu_{Eff_{SW}}\mu_I}{\sigma_{Eff_{SW}}\sigma_I} \tag{5.33}$$

Computation of $E(Eff_{SW}I)$

$$E(Eff_{SW}I) = E(2.8I^{1.2}I) = E(2.8I^{2.2})$$

From Proposition 3.1,

$$E(2.8I^{2.2}) = \int_{20}^{60} 2.8x^{2.2}f_I(x)dx$$

Since $I \sim Unif(20, 60)$, the PDF of I is

$$f_I(x) = \frac{1}{40}, \quad 20 \leq x \leq 60$$

Therefore,

$$E(Eff_{SW}I) = E(2.8I^{2.2}) = \int_{20}^{60} 2.8x^{2.2} \frac{1}{40} dx = 10{,}397.385$$

Computation of $\mu_{Eff_{SW}}$

$$\mu_{Eff_{SW}} \equiv E(Eff_{SW}) = E(2.8I^{1.2}) = \int_{20}^{60} 2.8x^{1.2} \frac{1}{40} dx = 236.6106$$

Computation of $\sigma_{Eff_{SW}}$

$$\sigma_{Eff_{SW}} = \sqrt{Var(Eff_{SW})} = \sqrt{E((Eff_{SW})^2) - (\mu_{Eff_{SW}})^2} = 80.8256$$

It is left for the reader to show $E((Eff_{SW})^2) = 62{,}517.36251$.

Computation of μ_I and σ_I
Since $I \sim Unif(20, 60)$, it follows immediately from Theorem 4.2

$$\mu_I \equiv E(I) = \frac{1}{2}(20 + 60) = 40$$

$$\sigma_I \equiv \sqrt{Var(I)} = \sqrt{\frac{1}{12}(60 - 20)^2} = 11.547$$

Computation of $\rho_{Eff_{SW}, I}$
Substituting the preceeding computations into Equation 5.33 yields

$$Corr(Eff_{SW}, I) \equiv \rho_{Eff_{SW}, I} = \frac{E(Eff_{SW}I) - \mu_{Eff_{SW}} \mu_I}{\sigma_{Eff_{SW}} \sigma_I} = \frac{932.961}{933.293} \approx 0.9996$$

Although the relationship between Eff_{SW} and I is nonlinear, a Pearson correlation coefficient of this magnitude suggests, in this case, the relationship is not distinguishably different from linear.

Rank correlation: In 1904, statistician C. Spearman developed a correlation coefficient that uses the ranks of values observed for n-pairs of random variables. The coefficient is known as *Spearman's rank correlation coefficient*. Let

$$(U_1, V_1), (U_2, V_2), (U_3, V_3), \ldots, (U_n, V_n)$$

be n-pairs of random samples (a set of independent random variables from the same PDF) from a continuous bivariate distribution. To determine the rank correlation between the pairs of random variables

$$(U_1, V_1), (U_2, V_2), (U_3, V_3), \ldots, (U_n, V_n)$$

the values of $U_1, U_2, U_3, \ldots, U_n$ and $V_1, V_2, V_3, \ldots, V_n$ are ranked among themselves. For instance, the values of $U_1, U_2, U_3, \ldots, U_n$ would be ranked in increasing order, with the smallest value receiving a rank of one. Likewise, the values of $V_1, V_2, V_3, \ldots, V_n$ would also be ranked in increasing order, with the smallest value receiving a rank of one. The difference between these rankings is the basis behind Spearman's coefficient. Specifically, Spearman's rank correlation coefficient, denoted by r_s, is given by*

$$r_s = 1 - \frac{6}{n^3 - n} \sum_{i=1}^{n} d_i^2 \tag{5.34}$$

where d_i is the difference in the ranks between U_i and V_i.

Where Pearson's correlation coefficient determines the degree of linearity between two random variables, Spearman's rank correlation coefficient measures their monotonicity. Like Pearson's correlation coefficient, Spearman's rank correlation coefficient is bounded by the interval $-1 \leq r_s \leq 1$. If r_s is close to 1, then larger values of U tend to be paired (or associated) with larger values of V. If r_s is close to -1, then larger values of U tend to be paired (or associated) with smaller values of V. An r_s near zero is seen when the ranks reflect a random arrangement.

In Example 5.8, recall Eff_{SW} and I have a Pearson correlation of 0.9996 in the interval $20 \leq I \leq 60$. Mentioned in that example, this suggests the two random variables have a strong linear relationship in the indicated interval. Furthermore, since Eff_{SW} (given in Example 5.8) is a strictly monotonically increasing function of I, the rank correlation between Eff_{SW} and I would be unity ($r_s = 1$). However, Pearson's correlation coefficient and Spearman's rank correlation coefficient can be very different. This is seen in Figure 5.10. In Figure 5.10 we have $Y = X^{100}$ and $X \sim Unif(0, 1)$. Pearson's correlation coefficient between X and Y is 0.24 (showing this is left as an exercise for the reader), while their rank correlation is unity. Looking at Figure 5.10, why (in this case) are these correlation coefficients so different?

Correlation is *not* causation. A strong positive correlation between two random variables does not necessarily imply large values for one *causes* large values for the other. Correlation close to unity *only* means two random variables are strongly associated and the hypothesis of a linear

---

* Keeping, E. S. 1962. *Introduction to Statistical Inference*. Princeton, NJ: D. Van Nostrand Company, Inc.

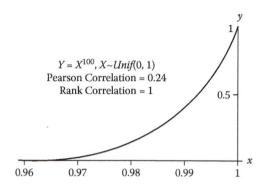

FIGURE 5.10
Illustrative correlation coefficients.

association (for Pearson's correlation coefficient) or a monotonic association (for Spearman's rank correlation coefficient) cannot be rejected.

5.2 Linear Combinations of Random Variables

It is often necessary to work with sums of random variables. Sums of random variables arise frequently in cost analysis. For instance, in Figure 1.3, a system's total cost can be expressed as follows:

$$Cost = X_1 + X_2 + X_3 + \cdots + X_n \tag{5.35}$$

where $X_1, X_2, X_3, \ldots, X_n$ are random variables that represent the cost of the system's work breakdown structure (WBS) cost elements. From this, we can often think of $Cost$ as a linear combination of the random variables $X_1, X_2, X_3, \ldots, X_n$. In general, given a collection of n-random variables $X_1, X_2, X_3, \ldots, X_n$ and constants $a_1, a_2, a_3, \ldots, a_n$ the random variable

$$Y = a_1 X_1 + a_2 X_2 + a_3 X_3 + \cdots + a_n X_n \tag{5.36}$$

is called a *linear combination* of $X_1, X_2, X_3, \ldots, X_n$.

Theorem 5.7 *If $Y = a_1 X_1 + a_2 X_2 + a_3 X_3 + \cdots + a_n X_n$ then*

$$E(Y) = a_1 E(X_1) + a_2 E(X_2) + a_3 E(X_3) + \cdots + a_n E(X_n) \tag{5.37}$$

Theorem 5.7 is an extension of Theorem 3.9. It states the expected value of a sum of random variables is the sum of the expected values of the individual random variables. Theorem 5.7 is valid whether or not the random variables $X_1, X_2, X_3, \ldots, X_n$ are independent.

Theorem 5.8   *If* $Y = a_1 X_1 + a_2 X_2 + a_3 X_3 + \cdots + a_n X_n$ *then*

$$Var(Y) = \sum_{i=1}^{n} a_i^2 Var(X_i) + 2 \sum_{i=1}^{n-1} \sum_{j=i+1}^{n} a_i a_j \rho_{X_i,X_j} \sigma_{X_i} \sigma_{X_j} \qquad (5.38)$$

Theorem 5.8 is an extension of Theorem 3.11. It states the variance of a sum of random variables is the sum of the variances of the individual random variables, plus the sum of the covariances between them. If the random variables $X_1, X_2, X_3, \ldots, X_n$ are *independent* then

$$Var(Y) = a_1^2 Var(X_1) + a_2^2 Var(X_2) + a_3^2 Var(X_3) + \cdots + a_n^2 Var(X_n) \qquad (5.39)$$

> **Example 5.9** *Suppose the total cost of a system is given by* Cost = $X_1 + X_2 + X_3$. *Let* X_1 *denote the cost of the system's prime mission product (PMP)*. Let* X_2 *denote the cost of the system's systems engineering, program management, and system test. Suppose* X_1 *and* X_2 *are dependent random variables and* $X_2 = \frac{1}{2}X_1$. *Let* X_3 *denote the cost of the system's data, spare parts, and support equipment. Suppose* X_1 *and* X_3 *are independent random variables with distribution functions given in Figure 5.11. Compute* E(Cost) *and* Var(Cost).
>
> *Solution* Since $X_1 \sim N(30,100)$, we have from Theorem 4.6
>
> $$E(X_1) = 30, \quad Var(X_1) = 100, \quad \sigma_{X_1} = \sqrt{Var(X_1)} = 10$$
>
> From Theorems 3.9 and 3.11 we have
>
> $$E(X_2) = E\left(\frac{1}{2}X_1\right) = \frac{1}{2}E(X_1) = 15$$
>
> $$Var(X_2) = Var\left(\frac{1}{2}X_1\right) = \frac{1}{4}Var(X_1) = 25, \sigma_{X_2} = \sqrt{Var(X_2)} = 5$$
>
> Since $X_3 \sim Unif(5,8)$, we have from Theorem 4.2
>
> $$E(X_3) = \frac{1}{2}(5+8) = 6.5, \ Var(X_3) = \frac{1}{12}(8-5)^2 = 0.75$$

---

* In systems cost analysis, PMP cost typically refers to the total cost of the system's hardware, software, and hardware-software integration. Chapter 6 provides a detailed discussion.

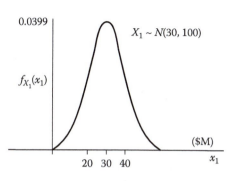
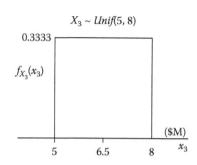

FIGURE 5.11
Density functions for Example 5.9.

Computation of E(*Cost*)
From Theorem 5.7 (for $i = 1, 2, 3$)

$$E(Cost) = E(X_1) + E(X_2) + E(X_3) = 30 + 15 + 6.5 = 51.5 \ (\$M)$$

Computation of Var(*Cost*)
From Theorem 5.8 (for $i = 1, 2, 3$)

$$Var(Cost) = Var(X_1) + Var(X_2) + Var(X_3)$$
$$+ 2\left[\rho_{X_1,X_2}\sigma_{X_1}\sigma_{X_2} + \rho_{X_1,X_3}\sigma_{X_1}\sigma_{X_3} + \rho_{X_2,X_3}\sigma_{X_2}\sigma_{X_3}\right]$$

Since X_1 and X_2 are *linearly related*, in this example, we know $\rho_{X_1,X_2} = 1$. Since X_1 and X_3 were given to be independent random variables, from Theorem 5.4 we know $\rho_{X_1,X_3} = 0$. With Theorem 5.6, this means

$$\rho_{X_2,X_3} = \rho_{\frac{1}{2}X_1,X_3} = \rho_{X_1,X_3} = 0$$

Substituting these values into *Var(Cost)* we have

$$Var(Cost) = 100 + 25 + 0.75$$
$$+ 2\left[1(10)(5) + 0(10)(\sqrt{0.75}) + 0(5)(\sqrt{0.75})\right] = 225.75 \ (\$M)^2$$

The units of variance $(\$M)^2$ have little meaning; it is better to think of the range of dollars in terms of the standard deviation; that is,

$$\sigma_{Cost} = \sqrt{Var(Cost)} = 15.02 \ (\$M)$$

5.2.1 Cost Considerations on Correlation

In Example 5.9, X_1 and X_2 are *dependent* random variables. As discussed, the nature of their dependency was such that $\rho_{X_1,X_2} = 1$. Suppose X_1 and X_2 were *independent* random variables with $X_1 \sim N(30, 100)$ and $X_2 \sim N(15, 25)$. How would this impact $E(Cost)$ and $Var(Cost)$, as computed in Example 5.9? The value of $E(Cost)$ would remain the same. Why? However, if X_1 and X_2 are *independent* random variables then $\rho_{X_1, X_2} = 0$; the value of $Var(Cost)$ reduces in magnitude; specifically,

$$Var(Cost) = Var(X_1) + Var(X_2) + Var(X_3) = 125.75 \ (\$M)^2$$

In Example 5.9, the *dependency* between X_1 and X_2 results in a value for $Var(Cost)$ nearly 80% greater than its value would be if X_1 and X_2 were *independent*. Seen in Example 5.9, dependencies between random variables can significantly affect the variance of their sum. Since a system's total cost is essentially a sum of n-work breakdown structure cost element costs, such as,

$$Cost = X_1 + X_2 + X_3 + \cdots + X_n$$

it is *critically* important for cost analysts to capture dependencies among $X_1, X_2, X_3, \ldots, X_n$, particularly those with nonnegative correlations. Not doing so can significantly misstate the true variability (uncertainty) in a system's total cost. Figure 5.12, Theorem 5.9, and Chapter 9 address how nonnegative correlation can affect the variance of a sum of n-random variables. Shown is how the variance increases dramatically with the number of random variables being summed and the extent that ρ approaches unity.

Theorem 5.9   *Let* $Cost = X_1 + X_2 + X_3 + \cdots + X_n$ *where* $X_1, X_2, X_3, \ldots, X_n$ *are random variables that represent a system's WBS cost element costs. If each*

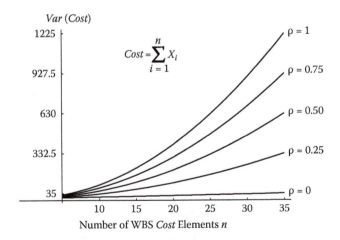

FIGURE 5.12
Theorem 5.9 with $\sigma^2 = 1$.

pair of $X_1, X_2, X_3, \ldots, X_n$ have common variance σ^2 and common nonnegative correlation ρ, then Var(Cost) $= \sigma^2[n + n(n-1)\rho]$.

Proof. *From Theorem 5.8, we have*

$$Var(Cost) = \sum_{i=1}^{n} Var(X_i) + 2\sum_{i=1}^{n-1}\sum_{j=i+1}^{n} \rho_{X_i, X_j} \sigma_{X_i} \sigma_{X_j}$$

Each pair of $X_1, X_2, X_3, \ldots, X_n$ is given to have common variance σ^2 and common nonnegative correlation ρ; therefore,

$$Var(Cost) = \sum_{i=1}^{n} \sigma^2 + 2\sum_{i=1}^{n-1}\sum_{j=i+1}^{n} \rho\sigma^2$$

$$= n\sigma^2 + n(n-1)\rho\sigma^2 = \sigma^2[n + n(n-1)\rho]$$

Some interesting results follow from this theorem; in particular,

$$Var(Cost) = \begin{cases} n\sigma^2, & \rho = 0 \\ [n + n(n-1)\rho]\sigma^2, & 0 < \rho < 1 \\ n^2\sigma^2, & \rho = 1 \end{cases}$$

Figure 5.12 illustrates this theorem with $\sigma^2 = 1$.

5.3 Central Limit Theorem and a Cost Perspective

This section describes one of the most important theorems in probability theory, the *central limit theorem*. It states that, under certain conditions, the distribution function of a sum of independent random variables approaches the normal distribution. From a cost analysis perspective, this theorem has great practical importance. As mentioned previously, a system's total cost is a summation of WBS cost element costs $X_1, X_2, X_3, \ldots, X_n$. Because of this, the distribution function of a system's total cost will often be approximately normal. We will see many examples of this in the chapters that follow.

Theorem 5.10: The Central Limit Theorem (CLT) *Suppose $X_1, X_2, X_3, \ldots, X_n$ is a sequence of n independent random variables with $E(X_i) = \mu_i$ and $Var(X_i) = \sigma_i^2$ (each finite). If*

$$Y = X_1 + X_2 + X_3 + \cdots + X_n$$

then, under certain conditions,* as $n \to \infty$ the random variable $Z = (Y - \mu)/\sigma$ approaches the standard normal, where

$$\mu = \sum_{i=1}^{n} \mu_i \text{ and } \sigma = \sqrt{\sum_{i=1}^{n} \sigma_i^2}$$

Theorem 5.10 places no restriction on the types of distribution functions that characterize the random variables $X_1, X_2, X_3, \ldots, X_n$. However, for a given n, the "rate" that the distribution function of Y approaches the normal distribution is affected by the shapes of the distribution functions for $X_1, X_2, X_3, \ldots, X_n$. If these distributions are approximately "bell-shaped," then the distribution function of Y may approach the normal for small n. If they are asymmetric, then n may need to be large for Y to approach the normal distribution.

The central limit theorem is frequently cited to explain why the distribution function of a system's total cost is often approximately normal. This is illustrated in the following case discussion.

Case Discussion 5.1: The electronic components of a 20 W solid-state amplifier (SSA) for a satellite communication workstation are listed in Table 5.3. Let the total component-level cost of the SSA be given by

$$Cost_{SSA} = X_1 + X_2 + X_3 + \cdots + X_{12} \tag{5.40}$$

Suppose $X_1, X_2, X_3, \ldots, X_{12}$ are independent random variables representing the costs of the SSA's components. Suppose the distribution function of each component is triangular, with parameters given in Table 5.3. Furthermore, suppose the mode of X_i represents its point estimate, that is,

$$x_{iPEX_i} = Mode(X_i) \quad i = 1, 2, \ldots, 12$$

---

* Informally, the individual random variables $X_1, X_2, X_3, \ldots, X_n$ that constitute Y should make only a small contribution to Y. In addition, none of the random variables $X_1, X_2, X_3, \ldots, X_n$ should dominate in standard deviation. For a further discussion of these conditions, as well as other forms of the central limit theorem, refer to Feller, W. 1968. *An Introduction to Probability Theory and Its Applications*, vol. 2, 3rd rev. edn. New York: John Wiley & Sons, Inc.

TABLE 5.3

20-Watt (W) SSA Component Cost

	Components	Cost ($K)			Mean ($K)	Variance ($K)^2
		Min	Mode	Max		
X_1	Transmitter Synthesizer	12.8	16.9	22.4	17.37	3.87
X_2	Receiver Synthesizer	12.8	16.9	22.4	17.37	3.87
X_3	Reference Generator	15.5	18.3	21.1	18.30	1.31
X_4	Receiver Loopback	7.4	9.2	11.1	9.23	0.57
X_5	BITE Control CCA	6.4	9.1	13.6	9.70	2.21
X_6	Power Supply	17.8	25.1	32.4	25.10	8.88
X_7	IMPATT Modules	36.4	66.5	100.5	67.80	171.41
X_8	Combiner Plate	15.2	18.7	22.7	18.87	2.35
X_9	SHF upconverter	12.1	16.6	24.6	17.77	6.68
X_{10}	Chassis	21.1	29.6	44.8	31.83	24.03
X_{11}	Backplane	3.3	4.8	6.1	4.73	0.33
X_{12}	Wave Guide Components	4.8	6.7	8.7	6.73	0.63
Component Cost		165.6	238.4	330.4	244.8	226.13

Note: The sum of the modes is not the mode of the distribution function of $Cost_{SSA}$.

From this, determine the mean and variance of $Cost_{SSA}$, as well as an approximation to its underlying distribution function.

Since distribution function of each X_i is given to be triangular, Theorem 4.3 can be applied to determine the mean and variance of each component's cost. For instance,

$$E(X_1) = \frac{1}{3}(a_1 + m_1 + b_1) = \frac{1}{3}(12.8 + 16.9 + 22.4) = 17.37 \ (\$K)$$

$$Var(X_1) = \frac{1}{18}\left[(m_1 - a_1)(m_1 - b_1) + (b_1 - a_1)^2\right] = 3.87 \ (\$K)^2$$

where a_1 is the minimum value of X_1, m_1 is the mode of X_1, and b_1 is the maximum value of X_1. Similar notation assumptions and calculations apply to the other components in Table 5.3. From Theorems 5.7 and 5.8, the mean and variance of the total component-level cost of the SSA are

$$E(Cost_{SSA}) = \mu_{Cost_{SSA}} = E\left(\sum_{i=1}^{12} X_i\right) = \sum_{i=1}^{12} E(X_i) = 244.8 \ (\$K) \qquad (5.41)$$

$$Var(Cost_{SSA}) = \sigma^2_{Cost_{SSA}} = Var\left(\sum_{i=1}^{12} X_i\right) = \sum_{i=1}^{12} Var(X_i) = 226.13 \ (\$K)^2 \quad (5.42)$$

Since $X_1, X_2, X_3, \ldots, X_{12}$ are independent random variables (with finite means and variances), from the central limit theorem (Theorem 5.10)

$$Z = \frac{Cost_{SSA} - E(Cost_{SSA})}{\sqrt{Var(Cost_{SSA})}} = \frac{Cost_{SSA} - 244.8}{\sqrt{226.13}} \tag{5.43}$$

is approximately $N(0, 1)$. This is equivalent to saying

$$Cost_{SSA} \sim N\left(\mu_{Cost_{SSA}}, \sigma^2_{Cost_{SSA}}\right) \tag{5.44}$$

We will next assess the applicability of this theorem that suggests the distribution function for $Cost_{SSA}$ is approximately normal with parameters given by (Equation 5.44). Monte Carlo simulation is one way to make this assessment.

In the context of Case Discussion 5.1, the Monte Carlo approach involves taking a random sample from each $X_1, X_2, X_3, \ldots, X_{12}$ and summing these sampled values according to Equation 5.40. This produces one random sample for $Cost_{SSA}$. This sampling process is repeated many thousands of times to produce an empirical frequency distribution of $Cost_{SSA}$. From the frequency distribution, an empirical CDF of $Cost_{SSA}$ is established. In Figure 5.13, the curve implied by the "points" is the empirical CDF of $Cost_{SSA}$. The curve given by the *solid line* is an assumed normal distribution, with parameters given by (Equation 5.44). Observe how closely the "points" fall along the solid line. On the basis of this empirical evidence, the central limit theorem appears applicable in this case.

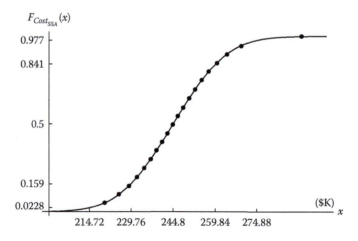

FIGURE 5.13
Cumulative distribution function of $Cost_{SSA}$.

The analysis summarized in Figure 5.13 provides empirical evidence *only* that the normal distribution is a reasonable form for the distribution function of $Cost_{SSA}$. It might next be asked "Could the underlying distribution function for $Cost_{SSA}$ *be normal*?" To answer this, a procedure known as the Kolmogorov–Smirnov (K-S) test can be used. The K-S test (Law and Kelton 1991) applies *only* to continuous distribution functions. It is a formal statistical procedure for testing whether a sample of observations (such as samples generated by a Monte Carlo simulation) could come from a hypothesized theoretical distribution. The following illustrates the K-S test in the context of Case Discussion 5.1.

The Kolmogorov-Smirnov test:

- Let $\hat{F}_{Cost_{SSA}}(x)$ represent the observed CDF of $Cost_{SSA}$ (Equation 5.40) generated from a Monte Carlo sample of $n = 5000$ observations. This CDF is shown in Table 5.4.
- Let $F_{Cost_{SSA}}(x)$ represent a hypothesized CDF. Suppose $F_{Cost_{SSA}}(x)$ is normal with mean 244.8 ($K) and variance 226.13 ($K)^2. Since the

TABLE 5.4

Kolmogorov–Smirnov Test for Case Discussion 5.1

	Values in the Left-Most Column are in Dollars Thousand		
x	$\hat{F}_{Cost_{SSA}}(x)$	$F_{Cost_{SSA}}(x)$	$|F_{Cost_{SSA}}(x) - \hat{F}_{Cost_{SSA}}(x)|$
220.19	0.05	0.0509	0.0009
225.15	0.10	0.0957	0.0043
228.64	0.15	0.1413	0.0087
231.80	0.20	0.1937	0.0063
234.35	0.25	0.2436	0.0064
236.72	0.30	0.2955	0.0045
238.89	0.35	0.3472	0.0028
240.66	0.40	0.3915	0.0085
242.50	0.45	0.4392	0.0108
244.34	0.50	0.4878	0.0122
246.21	0.55	0.5374	0.0126
248.11	**0.60**	**0.5871**	**0.0129**
250.28	0.65	0.6422	0.0078
252.53	0.70	0.6964	0.0036
254.99	0.75	0.7510	0.0010
257.49	0.80	0.8006	0.0006
260.49	0.85	0.8516	0.0016
263.80	0.90	0.8968	0.0032
269.22	0.95	0.9478	0.0022

hypothesized distribution for $Cost_{SSA}$ is normal, values for $F_{Cost_{SSA}}(x)$ in Table 5.4 reflect

$$P\left(Z \le \frac{x - 244.8}{\sqrt{226.13}}\right)$$

The mean and variance of the hypothesized CDF were *not* derived from the observations generated by the Monte Carlo samples.

- Compute the statistic $D = \underset{x}{\text{Max}} \left| F_{Cost_{SSA}}(x) - \hat{F}_{Cost_{SSA}}(x) \right|$. From Table 5.4, it is seen by the bold numbers that $D = 0.0129$.

- Suppose we wish to test the claim that the observed values summarized in Table 5.4 come from the hypothesized distribution. Let α be the probability of rejecting the claim when it is actually true. Suppose we let $\alpha = 0.01$.

- Referring to Table A.2, if

$$\left(\sqrt{n} + 0.12 + \frac{0.11}{\sqrt{n}}\right) D > c_{1-\alpha}$$

reject the claim; otherwise, accept it. Since α was chosen to be 0.01 for this test, from Table A.2 we have $c_{1-\alpha} = c_{0.99} = 1.628$. With $n = 5000$ and $D = 0.0129$ we have $(70.8322)(0.0129) = 0.9137 < c_{0.99} = 1.628$; thus, we accept the claim.

In a strict sense, accepting the claim that the distribution function for $Cost_{SSA}$ is normal *only* means it is a statistically plausible mathematical model of the underlying distribution. Acceptance does not mean the normal is the "best" or "unique" model form. Other hypothesized distributions might be accepted by the K-S test. It can be shown, in this case, the test also accepts the lognormal distribution as another statistically plausible model of the underlying distribution of $Cost_{SSA}$. Showing this is left as an exercise for the reader.

In cost analysis, the "precise" mathematical form of distribution functions, such as those for $Cost_{SSA}$, are rarely known. A credible analysis must provide decision-makers defensible analytical evidence that the form of a distribution function is mathematically plausible. Looking into whether central limit theorem applies, plotting hypothesized versus simulated distribution functions (e.g., Figure 5.13) and conducting statistical tests (i.e., the K-S test) are among the ways such evidence is established.

5.3.1 Further Considerations

As mentioned previously, the cost of a system can be expressed as

$$Cost = X_1 + X_2 + X_3 + \cdots + X_n \tag{5.45}$$

where $X_1, X_2, X_3, \ldots, X_n$ are random variables representing the costs of n WBS elements that constitute the system. From the preceding case discussion, we saw a circumstance where $F_{Cost}(x)$ could be approximated by a normal distribution. This is sometimes viewed as a paradox. Since a system's cost historically exceeds the value anticipated, or planned, why is its distribution function not positively skewed? The normal distribution is symmetric about its mean; it has no skew.

There are many reasons why the cost of a system exceeds the value anticipated, or planned. A prime reason is a system's cost is often based *only* on its point estimate. From Equation 1.2 the point estimate of the cost of a system is given by

$$x_{PE_{Cost}} = x_{1PE_{X_1}} + x_{2PE_{X_2}} + x_{3PE_{X_3}} + \cdots + x_{nPE_{X_n}}$$

where $x_{iPE_{X_i}}$ are the point estimates of each X_i $(i = 1, \ldots, n)$. Recall $x_{PE_{Cost}}$ is a value for *Cost* that traditionally contains *no reserve dollars* for uncertainties in a system's technical definition or cost estimation approaches. Because of this, $x_{PE_{Cost}}$ often falls below the 50th percentile of *Cost*; that is, $x_{PE_{Cost}}$ can have a high probability of being exceeded. This is illustrated by considering further Case Discussion 5.1. In this case discussion, $X_1, X_2, X_3, \ldots, X_{12}$ are *independent* random variables representing the costs of the SSA's twelve components. Suppose the point estimates of these components are the modes of X_i $(i = 1, \ldots, 12)$, given in Table 5.3. The point estimate of the cost of the SSA, denoted by $x_{PE_{Cost_{SSA}}}$ is

$$x_{PE_{Cost_{SSA}}} = Mode(X_1) + Mode(X_2) + Mode(X_3) + \cdots + Mode(X_{12}). \quad (5.46)$$

From Table 5.3, $x_{PE_{Cost_{SSA}}} = 238.4$ ($K). Since the distribution function of $Cost_{SSA}$ is approximately normal, in this case, we have

$$P(Cost_{SSA} > x_{PE_{Cost_{SSA}}} = 238.4) = P\left(Z > \frac{238.4 - 244.8}{\sqrt{226.13}}\right) = 0.665.$$

The normal PDF of $Cost_{SSA}$ is shown in Figure 5.14. Note that more probability exists to the right of $x_{PE_{Cost_{SSA}}}$ than to its left. If the cost of the SSA was anticipated, or planned, as the value given by $x_{PE_{Cost_{SSA}}}$, then there *is* a high probability (nearly 67%) it will be exceeded. This is true despite the distribution function of $Cost_{SSA}$ being approximately normal.

What drives this probability is the degree to which the distribution functions of each X_i $(i = 1, \ldots, 12)$ are skewed. The greater the positive skew, the greater the probability that $x_{PE_{Cost_{SSA}}}$ (defined by Equation 5.46) *will* be exceeded. The greater the negative skew, the greater the probability that $x_{PE_{Cost_{SSA}}}$ *will not* be exceeded. In either circumstance, the distribution function of the sum of these X_i's will, because of the central limit theorem,

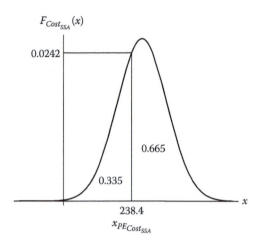

FIGURE 5.14
PDF for $Cost_{SSA}$.

frequently approach a normal. This may seem nonintuitive; nonetheless, the sum of many random variables characterized by skewed distributions *can* result in a distribution function that has no skew at all.

Last, since $Cost_{SSA}$ is considered to have a normal distribution, in this case, the mode of $Cost_{SSA}$ is equal to its mean—244.8 ($K). The sum of the modes of each X_i ($i = 1, \dots, 12$), seen in Table 5.3, is 238.4 ($K). In general, the sum of the modes of n-random variables *will not equal* the mode of the distribution function of the sum of these variables. If $Cost = X_1 + X_2 + X_3 + \dots + X_n$ and

$$x_{PE_{Cost}} = Mode(X_1) + Mode(X_2) + Mode(X_3) + \dots + Mode(X_n)$$

then, $x_{PE_{Cost}} \neq Mode(Cost)$. If the distribution function of each X_i is normal, then, $x_{PE_{Cost}} = Mode(Cost)$; in general, if the distribution function of each X_i is normal, then, $x_{PE_{Cost}} = Mode(Cost) = E(Cost) = Med(Cost)$.

5.4 Transformations of Random Variables

It is often necessary to determine the distribution function of a random variable that is a function (or transformation) of one or more random variables. For instance, the direct engineering hours to design a communication satellite may be a function of the satellite's weight W (lbs). Such a function might be given by Equation 5.47.

$$Hours = 4 + 2\sqrt{W} \tag{5.47}$$

If W is a random variable, then *Hours* is a function (or transformation) of the random variable W. In software cost analysis, the effort Eff_{SW} (staff months) to develop software can be a function of the number of source instructions to develop. A general form of this function is

$$Eff_{SW} = c_1 I^{c_2} \tag{5.48}$$

where c_1 and c_2 are positive constants and I is the number of thousands of delivered source instructions (KSDI) to be developed.* If I is a random variable, then Eff_{SW} is a function of the random variable I. A question that might be asked is "What is the 50th percentile of Eff_{SW} if the uncertainty in the number of delivered source instructions to develop is characterized by a uniform distribution in the interval $30 \leq x \leq 80$ KDSI?" To answer this question we need the distribution function of Eff_{SW} given the distribution function for I. In the preceding section, we discussed a possible distribution function for the random variable *Cost*, where

$$Cost = X_1 + X_2 + X_3 + \cdots + X_n \tag{5.49}$$

and the X_i's $(i = 1, \ldots, n)$ were random variables representing the costs of a system's n WBS cost elements. In Equation 5.49, *Cost* is a function of n random variables. From the central limit theorem, we saw the distribution function of *Cost* can, under certain conditions, be approximately normal. What if the central limit theorem does not apply? How, then, is the distribution function determined for a random variable that is a function of other random variables? The following presents methods to address this question.

5.4.1 Functions of a Single Random Variable

This section presents how to determine the distribution function of a random variable that is a function of another random variable. This is presented in the context of continuous random variables.† Consider the following example.

Example 5.10 *Suppose the direct engineering hours to design a new communication satellite is given by*

$$Hours = 4 + 2\sqrt{W} \tag{5.50}$$

where W is the satellite's weight, in lbs. Suppose the uncertainty in the satellite's weight is captured by a uniform distribution whose range of possible values is

---

* Section 5.4.2 presents a detailed discussion of the function given by Equation 5.48.
† Refer to Case Discussion 3.1 for a view of this discussion from the perspective of discrete random variables.

*given by $1000 \leq w \leq 2000$. Suppose the satellite design team assessed 1500 lbs
to be the point estimate for weight; that is, $w_{PE} = 1500$.*[*]

a. *Determine the CDF of Hours.*

b. *Compute $P(Hours \leq h_{PE})$, where $h_{PE} = 4 + 2\sqrt{w_{PE}}$.*

c. *Determine the PDF of Hours.*

Solution

a. We are given $W \sim Unif(1000, 2000)$. From Equation 4.4

$$f_W(w) = \frac{1}{1000} \quad 1000 \leq w \leq 2000$$

The CDF of *Hours* is $F_{Hours}(h) = P(Hours \leq h)$, where h denotes
the possible values of *Hours*. Since $Hours = 4 + 2\sqrt{W}$, the interval
$1000 \leq w \leq 2000$ is mapped onto the interval $67.2456 \leq h \leq 93.4427$.
Thus, for h this interval

$$F_{Hours}(h) = P(Hours \leq h) = P(4 + 2\sqrt{W} \leq h) = P\left(W \leq \left(\frac{h-4}{2}\right)^2\right)$$

$$= \int_{1000}^{[(h-4)/2]^2} f_W(w)\, dw \quad = \frac{1}{1000}\left(\frac{h-4}{2}\right)^2 - 1$$

Thus, the CDF of *Hours*, presented in Figure 5.15, is

$$F_{Hours}(h) = P(Hours \leq h)$$

$$= \begin{cases} 0, & \text{if } h < 67.2456 \\ \dfrac{1}{1000}\left(\dfrac{h-4}{2}\right)^2 - 1, & \text{if } 67.2456 \leq h \leq 93.4427 \\ 1, & \text{if } h > 93.4427 \end{cases} \quad (5.51)$$

b. From Equation 5.50 we have $h_{PE} = 4 + 2\sqrt{w_{PE}}$; thus, $h_{PE} = 81.46$
when $w_{PE} = 1500$. Therefore, $P(Hours \leq h_{PE}) = P(Hours \leq 81.46)$.
From Equation 5.51 this probability is

$$P(Hours \leq h_{PE}) = P(Hours \leq 81.46) = \frac{1}{1000}\left(\frac{81.46-4}{2}\right)^2 - 1 = 0.50$$

c. To compute the PDF of *Hours*, we can differentiate $F_{Hours}(h)$ with
respect to h. From Chapter 3, recall that

$$f_{Hours}(h) = \frac{d}{dh}(F_{Hours}(h))$$

It follows that $f_{Hours}(h) = \frac{1}{2000}(h - 4)$, for $67.2456 \leq h \leq 93.4427$.

---

[*] Instead of using w_{PE_W} to denote the point estimate of the random variable W, we simplify the
notation and let w_{PE} represent this value.

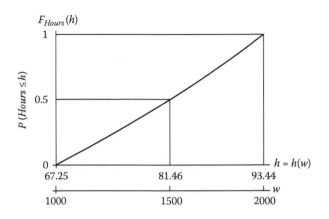

FIGURE 5.15
The CDF of *Hours*.

In Example 5.10, the procedures to develop $F_{Hours}(h)$ and $f_{Hours}(h)$ are generalized by the following theorem.

Theorem 5.11   *Suppose X is a continuous random variable with PDF $f_X(x) > 0$ for $a \leq x \leq b$. Consider the random variable $Y = g(X)$ where $y = g(x)$ is a strictly increasing or decreasing differentiable function of x. Let the inverse of $y = g(x)$ be given by $x = v(y)$, then $Y = g(X)$ has the PDF*

$$f_Y(y) = \begin{cases} f_X(v(y)) \cdot \left| \dfrac{d[v(y)]}{dy} \right|, & g(a) \leq y \leq g(b), \quad \text{if } g(x) \text{ increasing} \\[3ex] f_X(v(y)) \cdot \left| \dfrac{d[v(y)]}{dy} \right|, & g(b) \leq y \leq g(a), \quad \text{if } g(x) \text{ decreasing} \end{cases} \tag{5.52}$$

If $y = g(x)$ is strictly increasing

$$F_Y(y) = P(Y \leq y) = P(g(X) \leq y) = P(X \leq v(y)) = F_X(v(y)) = F_X(x) \quad (5.53)$$

If $y = g(x)$ is strictly decreasing

$$F_Y(y) = P(Y \leq y) = P(g(X) \leq y) = P(X > v(y))$$

$$= 1 - F_X(v(y)) = 1 - F_X(x) \tag{5.54}$$

Discussion of Theorem 5.11: Applying Theorem 5.11 to Example 5.10 yields the following:

$$f_{Hours}(h) = f_W(v(h)) \cdot \left| \frac{d[v(h)]}{dh} \right|, \quad g(1000) \le h \le g(2000) \tag{5.55}$$

where $h = g(w) = 4 + 2\sqrt{w}$ and $w = v(h) = \left(\frac{h-4}{2} \right)^2$. Since

$$f_W(w) = \frac{1}{1000}, \quad 1000 \le w \le 2000$$

we have $f_W(v(h)) = f_W\left(\left(\frac{h-4}{2} \right)^2 \right) = \frac{1}{1000}$ and $\left| \frac{d[v(h)]}{dh} \right| = \frac{h-4}{2}$. Substituting this into Equation 5.55 yields

$$f_{Hours}(h) = \frac{1}{1000} \cdot \frac{h-4}{2}, \quad 67.2456 \le h \le 93.4427 \tag{5.56}$$

which is the same as the PDF in part (c) of Example 5.10.

Theorem 5.11 also provides insight into the fractiles of a distribution function. In Example 5.10, $h = g(w) = 4 + 2\sqrt{w}$ is a *strictly increasing* differentiable function of w. From Theorem 5.11, this implies

$$F_{Hours}(h) = F_W(w)$$

Thus, the value of h associated with the α-fractile of W will also be the α-fractile of *Hours*. For example, in Figure 5.15 observe that

$$F_W(1500) = 0.50 = F_{Hours}(81.46)$$

Here, the value of h associated with the 0.50-fractile of W is the 0.50-fractile of *Hours*. Specifically,

$$w_{0.50} = 1500 \text{ and } P(W \le w_{0.50}) = 0.50$$
$$h_{0.50} = 81.46 = 4 + 2\sqrt{w_{0.50}} \text{ and } P(Hours \le h_{0.50}) = 0.50$$

Similarly, it can be shown (left as an exercise for the reader) that

$$F_W(1750) = 0.75 = F_{Hours}(87.67)$$

The practical value of this aspect of Theorem 5.11 is high, because cost-related equations (e.g., Equation 5.50) are often simple increasing or decreasing differentiable functions of one variable. When $Y = g(X)$ and Theorem 5.11

applies, the CDF of Y is *not needed* to determine its fractiles. The α-fractiles of Y are, in fact, completely determined from the α-fractiles of X. In practice, not having to determine the CDF of Y, either analytically or through Monte Carlo simulation, can save a great deal of mathematical effort. When possible, cost analysts should readily take advantage of this aspect of Theorem 5.11.

Example 5.11 *From the information in Example 5.10 compute*
(a) $E(Hours)$ (b) σ_{Hours}

Solution

a. Two approaches are shown.

Approach 1: From Equation 3.22, we can write

$$E(Hours) = \int_{67.2456}^{93.4427} h \cdot f_{Hours}(h)dh = \int_{67.2456}^{93.4427} h \cdot \frac{1}{2000}(h-4)dh = 81.09 \text{ hours}$$

Approach 2: Since $Hours = g(W) = 4 + 2\sqrt{W}$, it follows from Proposition 3.1

$$E(Hours) = E(g(W)) = \int_{1000}^{2000} g(w) \cdot f_W(w)dw$$

$$= \int_{1000}^{2000} (4 + 2\sqrt{w}) \cdot \frac{1}{1000} dw = 81.09 \text{ hours}$$

b. To determine σ_{Hours}, from Theorem 3.10 we have

$$Var(Hours) = E(Hours^2) - [E(Hours)]^2$$

Since

$$E(Hours^2) = \int_{1000}^{2000} [g(w)]^2 \cdot f_W(w)\,dw$$

$$= \int_{1000}^{2000} [(4 + 2\sqrt{w})]^2 \cdot \frac{1}{1000} dw = 6632.75 \text{ (hours)}^2$$

we have

$$Var(Hours) = E(Hours^2) - [E(Hours)]^2 = 6632.75 - (81.09)^2$$
$$= 57.1619 \text{ (hours)}^2$$

therefore

$$\sigma_{Hours} = \sqrt{Var(Hours)} = 7.56 \text{ hours}$$

The reader should also verify that $E(Hours^2)$ can be computed by

$$E(Hours^2) = \int_{67.2456}^{93.4427} h^2 \cdot f_{Hours}(h)dh = \int_{67.2456}^{93.4427} h^2 \cdot \frac{1}{2000}(h-4)dh$$

5.4.2 Applications to Software Cost-Schedule Models

This section presents a further discussion on functions of a single random variable as they apply to software cost-schedule models. These models are often used in cost analysis to determine the effort (staff months), cost (dollars), and schedule (months) of a software development project. The general forms of these models are as follows:

$$Eff_{SW} = c_1 I^{c_2} \tag{5.57}$$

$$Cost_{SW} = \ell_r Eff_{SW} \tag{5.58}$$

$$T_{SW} = k_1 (Eff_{SW})^{k_2} \tag{5.59}$$

In Equation 5.57, Eff_{SW} is a random variable representing the software project's development effort (staff months), c_1 and c_2 are positive constants, and I is a random variable representing the number of KDSI to be developed.* In Equation 5.58, $Cost_{SW}$ is a random variable representing the software project's development cost (dollars) and ℓ_r is a constant† representing a labor rate (dollars per staff month). Notice $Cost_{SW}$ can also be expressed as a function of I, that is,

$$Cost_{SW} = \ell_r(c_1 I^{c_2}) \tag{5.60}$$

In Equation 5.59, T_{SW} is a random variable representing the software project's development schedule (months) and k_1 and k_2 are positive constants. Notice T_{SW} can also be expressed as a function of I, that is,

$$T_{SW} = k_1(c_1 I^{c_2})^{k_2} \tag{5.61}$$

Equations 5.57 through 5.61 represent one approach (Boehm 1981) for determining a software development project's effort, cost, and schedule; there are others. For instance, Eff_{SW} might be determined as the ratio of two random variables I and P_r as shown by Equation 5.62. Here, P_r is the software project's development productivity rate (e.g., the number of DSI per staff month).

$$Eff_{SW} = \frac{I}{P_r} \tag{5.62}$$

---

* Throughout this book, when I appears in the formula given by Equation 5.57 it is assumed that I is always in KDSI. It is further assumed that I is always greater than zero.
† In this section, we treat ℓ_r as a constant to keep the discussion focused on functions of a single random variable; however, in practice, ℓ_r is often treated as a random variable.

Equation 5.62 is an example of a function of two random variables. Working with such functions is discussed in Section 5.4.3.

Case Discussion 5.2: If the development effort Eff_{SW} for a software project is defined by $Eff_{SW} = c_1 I^{c_2}$, and $I \sim Unif(a, b)$, determine $F_{Eff_{SW}}(s)$, $f_{Eff_{SW}}(s)$, $E(Eff_{SW})$, and $Var(Eff_{SW})$.

Determination of $F_{Eff_{SW}}(s)$

We want the distribution function of Eff_{SW} given the distribution function for I is *uniform*, in the interval $a \le x \le b$. From Equation 4.4 we know

$$f_I(x) = \frac{1}{b-a}, \quad a \le x \le b$$

where a and b represent the minimum and maximum possible values of I. By definition

$$F_{Eff_{SW}}(s) = P(Eff_{SW} \le s) = P(c_1 I^{c_2} \le s)$$

$$= P\left(I \le \left(\frac{s}{c_1}\right)^{\frac{1}{c_2}}\right) = \int_a^{\left(\frac{s}{c_1}\right)^{\frac{1}{c_2}}} f_I(x)\, dx = \frac{1}{b-a}\left[\left(\frac{s}{c_1}\right)^{\frac{1}{c_2}} - a\right]$$

where $a \le \left(\frac{s}{c_1}\right)^{\frac{1}{c_2}} \le b$; therefore,

$$F_{Eff_{SW}}(s) = P(Eff_{SW} \le s) = \frac{1}{b-a}\left[\left(\frac{s}{c_1}\right)^{\frac{1}{c_2}} - a\right], \quad a \le \left(\frac{s}{c_1}\right)^{\frac{1}{c_2}} \le b \quad (5.63)$$

Determination of $f_{Eff_{SW}}(s)$

Given $Eff_{SW} = g(I) = c_1 I^{c_2}$ we can write $s = g(x) = c_1 x^{c_2}$, which is a strictly increasing differentiable function of x. Let the inverse of x be given by

$$x = v(s) = \left(\frac{s}{c_1}\right)^{\frac{1}{c_2}}$$

From Theorem 5.11, we have

$$f_{Eff_{SW}}(s) = f_I(v(s)) \cdot \frac{d[v(s)]}{ds}, \quad g(a) \le s \le g(b)$$

Therefore,

$$f_{Eff_{SW}}(s) = \frac{1}{b-a} \cdot \frac{1}{c_1 c_2} \cdot \left(\frac{s}{c_1}\right)^{\frac{1}{c_2}-1}, \quad c_1 a^{c_2} \le s \le c_1 b^{c_2} \tag{5.64}$$

which is also the derivative of $F_{Eff_{SW}}(s)$ with respect to s. It is left to the reader to verify Equation 5.64 is a probability density function.

Determination of $E(Eff_{SW})$

From Proposition 3.1, the expected software development effort is

$$E(Eff_{SW}) = E(g(I)) = \int_a^b g(x) f_I(x)\, dx$$

$$= \int_a^b c_1 x^{c_2} \cdot \frac{1}{b-a}\, dx = c_1 \cdot \frac{1}{b-a} \int_a^b x^{c_2}\, dx$$

Therefore

$$E(Eff_{SW}) = \frac{c_1}{c_2 + 1} \cdot \frac{1}{b-a} \left[b^{c_2+1} - a^{c_2+1} \right] \tag{5.65}$$

Alternatively, Equation 5.65 could have been derived as follows:

$$E(Eff_{SW}) = \int_{c_1 a^{c_2}}^{c_1 b^{c_2}} s \cdot f_{Eff_{SW}}(s)\, ds = \int_{c_1 a^{c_2}}^{c_1 b^{c_2}} s \cdot \frac{1}{b-a} \cdot \frac{1}{c_1 c_2} \cdot \left(\frac{s}{c_1}\right)^{\frac{1}{c_2}-1}\, ds$$

Determination of $Var(Eff_{SW})$

From Theorem 3.10, we know

$$Var(Eff_{SW}) = E(Eff_{SW}^2) - \left[E(Eff_{SW}) \right]^2$$

Now

$$E(Eff_{SW}^2) = E(g(I)^2) = \int_a^b g(x)^2 f_I(x)\, dx$$

$$= \int_a^b (c_1 x^{c_2})^2 \frac{1}{b-a}\, dx = \frac{1}{b-a} \cdot \frac{c_1^2}{2 c_2 + 1} \left[b^{2 c_2+1} - a^{2 c_2+1} \right]$$

Therefore,

$$Var(Eff_{SW}) = \frac{1}{b-a} \cdot \frac{c_1^2}{2c_2+1} \left[b^{2c_2+1} - a^{2c_2+1} \right] - \left[E(Eff_{SW}) \right]^2 \qquad (5.66)$$

where

$$E(Eff_{SW}) = \frac{c_1}{c_2+1} \cdot \frac{1}{b-a} \left[b^{c_2+1} - a^{c_2+1} \right]$$

This concludes Case Discussion 5.2. The following illustrates how these results can be applied to a software development project.

Example 5.12 *Suppose the effort (staff months) to develop software for a new system is given by $Eff_{SW} = 2.8\, I^{1.2}$. Suppose the uncertainty in the number of KDSI is represented by the distribution $I \sim Unif(30, 80)$. Determine*

a. $P(Eff_{SW} \leq 300)$

b. $P(Eff_{SW} \leq E(Eff_{SW}))$

c. $\sigma_{Eff_{SW}}$

d. $P(Cost_{SW} \leq 4,500,000)$ *given $\ell_r = 15,000$ dollars per staff month.*

Solution

a. Given $Eff_{SW} = 2.8\, I^{1.2}$, we know from Equation 5.57 that $c_1 = 2.8$, $c_2 = 1.2$. Since $I \sim Unif(30, 80)$, from Equation 5.63

$$P(Eff_{SW} \leq 300) = \frac{1}{80-30} \left[\left(\frac{300}{2.8} \right)^{\frac{1}{1.2}} - 30 \right] \qquad 30 \leq 49.16 \leq 80$$

$$= 0.383$$

Figure 5.16 shows this region of probability for Eff_{SW}, as well as the PDF of Eff_{SW}. The PDF comes from Equation 5.64 (in Case Discussion 5.2).

b. From Equation 5.65

$$E(Eff_{SW}) = \frac{2.8}{1.2+1} \cdot \frac{1}{50} \left[80^{1.2+1} - 30^{1.2+1} \right] = 346.12 \text{ staff months}$$

From Equation 5.63

$$P(Eff_{SW} \leq E(Eff_{SW})) = P(Eff_{SW} \leq 346.12) = F_{Eff_{SW}}(346.12) = 0.508$$

c. From Equation 5.66

$$Var(Eff_{SW}) = \frac{1}{50} \cdot \frac{(2.8)^2}{2(1.2)+1} \left[80^{2(1.2)+1} - 30^{1.2+1} \right] - [346.12]^2$$

$$= 11,608.65 \text{ (staff months)}^2$$

Therefore, $\sigma_{Eff_{SW}} = \sqrt{Var(Eff_{SW})} = 107.7$ staff months.

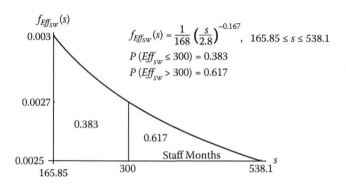

FIGURE 5.16
The PDF of Eff_{SW} in Example 5.12.

  d. Given $\ell_r = 15{,}000$ dollars per staff month, we have

$$P(Cost_{SW} \leq 4{,}500{,}000) = P(\ell_r \cdot Eff_{SW} \leq 4{,}500{,}000)$$

$$= P\left(Eff_{SW} \leq \frac{4{,}500{,}000}{\ell_r}\right)$$

$$= P\left(Eff_{SW} \leq \frac{4{,}500{,}000}{15{,}000}\right)$$

$$= P(Eff_{SW} \leq 300) = 0.383$$

Example 5.13 *Once again, suppose the effort (staff months) to develop software for a new system is determined by $Eff_{SW} = 2.8\,I^{1.2}$, where $I \sim Unif(30, 80)$. If the software development schedule (months) is given by $T_{SW} = 2.5(Eff_{SW})^{0.32}$, determine the schedule that has a 95% chance of not being exceeded.*

Solution Three solution approaches are presented.

Approach 1: This approach operates from the CDF of I. From the information given in this example, we have

$$T_{SW} = 2.5(Eff_{SW})^{0.32}$$

$$= 2.5(2.8\,I^{1.2})^{0.32} = 3.48\,I^{0.384} \tag{5.67}$$

Since $T_{SW} = g(I) = 3.48\,I^{0.384}$ and $I > 0$, we can write $t = g(x) = 3.48\,x^{0.384}$ where t and x are the values possible for T_{SW} and I, respectively. Since t

is a strictly increasing differentiable function of x, in this example, from Theorem 5.11

$$F_{T_{SW}}(t) = F_I(x) \tag{5.68}$$

The value of t associated with the 0.95-fractile of I will equal the 0.95-fractile of T_{SW}. From Equation 4.5, we know

$$F_I(x) = \frac{x - 30}{80 - 30} = \frac{x - 30}{50}, \quad 30 \le x \le 80$$

The 0.95-fractile of I is $x_{0.95}$ such that $F(x_{0.95}) = P(I \le x_{0.95}) = 0.95$; that is, $x_{0.95}$ is the solution to

$$\frac{x_{0.95} - 30}{50} = 0.95 \tag{5.69}$$

Solving Equation 5.69 for $x_{0.95}$ yields $x_{0.95} = 77.5$ KDSI; thus

$$x_{0.95} = 77.5 \quad \text{and} \quad P(I \le x_{0.95}) = 0.95$$
$$t_{0.95} = 18.5 = 3.48 x_{0.95}^{0.384} \quad \text{and} \quad P(T_{SW} \le t_{0.95}) = 0.95$$

This is equivalent to

$$F_I(77.5) = F_{T_{SW}}(18.5) = 0.95$$

Therefore, 18.5 months is the software development schedule that has a 95% chance of not being exceeded.

Approach 2: This approach operates from the CDF of Eff_{SW}. Since $T_{SW} = g(Eff_{SW}) = 2.5(Eff_{SW})^{0.32}$, we can write $t = g(s) = 2.5 s^{0.32}$ where t and s are the values possible for T_{SW} and Eff_{SW}, respectively. Since t is a strictly increasing differentiable function of s, from Theorem 5.11

$$F_{T_{SW}}(t) = F_{Eff_{SW}}(s)$$

Thus, the value of t associated with the 0.95-fractile of Eff_{SW} will equal the 0.95-fractile of T_{SW}. From Case Discussion 5.2, the general formula for $F_{Eff_{SW}}(s)$ is given by Equation 5.63. It is left as an exercise for the reader to show, for this example, that $F_{Eff_{SW}}(518) = F_{T_{SW}}(18.5) = 0.95$.

Approach 3: This approach involves explicitly determining the functional form of $F_{T_{SW}}(t)$ and then solving the expression $F_{T_{SW}}(t_{0.95}) = 0.95$ for $t_{0.95}$. It is left as an exercise for the reader to show, for this example,

$$F_{T_{SW}}(t) = \frac{1}{50}\left[\left(\frac{t}{3.48}\right)^{\frac{1}{0.384}} - 30\right], \quad 12.8 \le t < 18.7^*$$

From this expression it follows, after rounding, that $F_{T_{SW}}(18.5) = 0.95$.

Example 5.14 *If $T_{SW} = 2.5(Eff_{SW})^{0.32}$ and $Eff_{SW} = 2.8I^{1.2}$ then write a general formula for $P(T_{SW} \le t)$, if $I \sim Trng(30, 50, 80)$.*

Solution Notice T_{SW} can be written as $T_{SW} = 2.5(2.8I^{1.2})^{0.32} = 3.48I^{0.384}$. This implies $t = g(x) = 3.48x^{0.384}$, where t and x are possible values of T_{SW} and I, respectively. Note that t is a strictly increasing differentiable function of x; therefore, from Theorem 5.11.

$$F_{T_{SW}}(t) = P(T_{SW} \le t) = P(g(I) \le t) = P(I \le v(t)) = F_I(v(t)) = F_I(x)$$

where

$$x = v(t) = \left(\frac{t}{3.48}\right)^{\frac{1}{0.384}}$$

In this equation, x is the inverse of $t = g(x) = 3.48x^{0.384}$. Since I is given to have a triangular distribution function, from Equation 4.7

$$F_I(x) = \begin{cases} 0, & \text{if } x < a \\ \dfrac{(x-a)^2}{(b-a)(m-a)}, & \text{if } a \le x < m \\ 1 - \dfrac{(b-x)^2}{(b-a)(b-m)}, & \text{if } m \le x < b \\ 1, & \text{if } x \ge b \end{cases}$$

thus,

$$P(T_{SW} \le t) = \begin{cases} 0, & \text{if } \left(\dfrac{t}{3.48}\right)^{\frac{1}{0.384}} < 30 \\ \dfrac{1}{50}\cdot\dfrac{1}{20}\left(\left(\dfrac{t}{3.48}\right)^{\frac{1}{0.384}} - 30\right)^2, & \text{if } 30 \le \left(\dfrac{t}{3.48}\right)^{\frac{1}{0.384}} < 50 \\ 1 - \dfrac{1}{50}\cdot\dfrac{1}{30}\left(80 - \left(\dfrac{t}{3.48}\right)^{\frac{1}{0.384}}\right)^2, & \text{if } 50 \le \left(\dfrac{t}{3.48}\right)^{\frac{1}{0.384}} < 80 \\ 1, & \text{if } \left(\dfrac{t}{3.48}\right)^{\frac{1}{0.384}} \ge 80 \end{cases}$$

---

* These endpoints are rounded from the interval $12.8467 \le t < 18.7226$.

Equations 5.70 through 5.85 present general probability formulas for the software effort and software schedule models defined by Equation 5.57 and Equation 5.59, respectively (Garvey and Powell 1989).

Given $Eff_{SW} = c_1 I^{c_2}$, if $I \sim Unif(a, b)$ then

- Cumulative Distribution Function

$$F_{Eff_{SW}}(s) = P(Eff_{SW} \leq s) = \begin{cases} 0, & \left(\dfrac{s}{c_1}\right)^{\frac{1}{c_2}} < a \\[3mm] \dfrac{1}{b-a}\left[\left(\dfrac{s}{c_1}\right)^{\frac{1}{c_2}} - a\right], & a \leq \left(\dfrac{s}{c_1}\right)^{\frac{1}{c_2}} < b \\[3mm] 1, & \left(\dfrac{s}{c_1}\right)^{\frac{1}{c_2}} \geq b \end{cases}$$

$$(5.70)$$

- Probability Density Function

$$f_{Eff_{SW}}(s) = \frac{1}{b-a} \cdot \frac{1}{c_1 c_2} \cdot \left(\frac{s}{c_1}\right)^{\frac{1}{c_2}-1}, \quad a \leq \left(\frac{s}{c_1}\right)^{\frac{1}{c_2}} \leq b \qquad (5.71)$$

- Mean (staff months)

$$E(Eff_{SW}) = \frac{c_1}{c_2 + 1} \cdot \frac{1}{b - a}\left[b^{c_2+1} - a^{c_2+1}\right] \qquad (5.72)$$

- Variance (staff months)^2

$$Var(Eff_{SW}) = \frac{1}{b-a} \cdot \frac{c_1^2}{2c_2 + 1}\left[b^{2c_2+1} - a^{2c_2+1}\right] - [E(Eff_{SW})]^2 \qquad (5.73)$$

Given $Eff_{SW} = c_1 I^{c_2}$, if $I \sim Trng(a, m, b)$ then

- Cumulative Distribution Function

$$F_{Eff_{SW}}(s) = P(Eff_{SW} \leq s)$$

$$= \begin{cases} 0, & \left(\dfrac{s}{c_1}\right)^{\frac{1}{c_2}} < a \\[3mm] \dfrac{1}{b-a} \cdot \dfrac{1}{m-a} \left[\left(\dfrac{s}{c_1}\right)^{\frac{1}{c_2}} - a\right]^2, & a \leq \left(\dfrac{s}{c_1}\right)^{\frac{1}{c_2}} < m \\[3mm] 1 - \dfrac{1}{b-a} \cdot \dfrac{1}{b-m} \left[b - \left(\dfrac{s}{c_1}\right)^{\frac{1}{c_2}}\right]^2, & m \leq \left(\dfrac{s}{c_1}\right)^{\frac{1}{c_2}} < b \\[3mm] 1, & \left(\dfrac{s}{c_1}\right)^{\frac{1}{c_2}} \geq b \end{cases}$$

$$\tag{5.74}$$

- Probability Density Function

$$f_{Eff_{SW}}(s) =$$

$$= \begin{cases} \dfrac{2}{b-a} \cdot \dfrac{1}{m-a} \cdot \dfrac{1}{c_2} \cdot \dfrac{1}{c_1} \left(\dfrac{s}{c_1}\right)^{\frac{1}{c_2}-1} \\[3mm] \qquad \times \left[\left(\dfrac{s}{c_1}\right)^{\frac{1}{c_2}} - a\right], & a \leq \left(\dfrac{s}{c_1}\right)^{\frac{1}{c_2}} < m \\[3mm] \dfrac{2}{b-a} \cdot \dfrac{1}{b-m} \cdot \dfrac{1}{c_2} \cdot \dfrac{1}{c_1} \left(\dfrac{s}{c_1}\right)^{\frac{1}{c_2}-1} \\[3mm] \qquad \times \left[b - \left(\dfrac{s}{c_1}\right)^{\frac{1}{c_2}}\right], & m \leq \left(\dfrac{s}{c_1}\right)^{\frac{1}{c_2}} \leq b \end{cases}$$

$$\tag{5.75}$$

- Mean (staff months)

$$E(Eff_{SW}) = c_1 \frac{2}{b-a} \cdot \frac{1}{m-a} \left[\frac{m^{c_2+2} - a^{c_2+2}}{c_2+2} + \frac{a^{c_2+2} - am^{c_2+1}}{c_2+1}\right]$$

$$+ c_1 \frac{2}{b-a} \cdot \frac{1}{m-b} \left[\frac{b^{c_2+2} - m^{c_2+2}}{c_2+2} + \frac{bm^{c_2+1} - b^{c_2+2}}{c_2+1}\right]$$

$$\tag{5.76}$$

- Variance (staff months)^2

$$Var(Eff_{SW}) = c_1^2 \frac{2}{b-a} \cdot \frac{1}{m-a} \left[\frac{m^{2c_2+2} - a^{2c_2+2}}{2c_2 + 2} + \frac{a^{2c_2+2} - am^{2c_2+1}}{2c_2 + 1} \right]$$

$$+ c_1^2 \frac{2}{b-a} \cdot \frac{1}{m-b} \left[\frac{b^{2c_2+2} - m^{2c_2+2}}{2c_2 + 2} + \frac{bm^{2c_2+1} - b^{2c_2+2}}{2c_2 + 1} \right]$$

$$- [E(Eff_{SW})]^2 \tag{5.77}$$

Given $T_{SW} = k_1(Eff_{SW})^{k_2} \equiv T_{SW} = k_1(c_1 I^{c_2})^{k_2}$, if $I \sim Unif(a,b)$ then

- Cumulative Distribution Function

$$F_{T_{SW}}(t) = P(T_{SW} \leq t)$$

$$= \begin{cases} 0, & \left(\dfrac{t}{k_1 c_1^{k_2}}\right)^{\frac{1}{c_2 k_2}} < a \\[3mm] \dfrac{1}{b-a}\left[\left(\dfrac{t}{k_1 c_1^{k_2}}\right)^{\frac{1}{c_2 k_2}} - a\right], & a \leq \left(\dfrac{t}{k_1 c_1^{k_2}}\right)^{\frac{1}{c_2 k_2}} < b \quad (5.78) \\[3mm] 1, & \left(\dfrac{t}{k_1 c_1^{k_2}}\right)^{\frac{1}{c_2 k_2}} \geq b \end{cases}$$

- Probability Density Function

$$f_{T_{SW}}(t) = \frac{1}{b-a} \cdot \frac{1}{k_1 k_2} \cdot \frac{1}{c_2 c_1^{k_2}} \cdot \left(\frac{t}{k_1 c_1^{k_2}}\right)^{\frac{1}{c_2 k_2}-1}, \quad a \leq \left(\frac{t}{k_1 c_1^{k_2}}\right)^{\frac{1}{c_2 k_2}} \leq b$$

$$\tag{5.79}$$

- Mean (staff months)

$$E(T_{SW}) = \frac{k_1 c_1^{k_2}}{c_2 k_2 + 1} \cdot \frac{1}{b-a}\left[b^{c_2 k_2 + 1} - a^{c_2 k_2 + 1}\right] \tag{5.80}$$

- Variance (staff months)^2

$$Var(T_{SW}) = \frac{k_1^2 c_1^{2k_2}}{2c_2k_2 + 1} \cdot \frac{1}{b - a} \left[b^{2c_2k_2+1} - a^{2c_2k_2+1} \right] - [E(T_{SW})]^2 \quad (5.81)$$

Given $T_{SW} = k_1(\mathit{Eff}_{SW})^{k_2} \equiv T_{SW} = k_1(c_1 I^{c_2})^{k_2}$, if $I \sim Trng(a, m, b)$ then

- Cumulative Distribution Function

$$F_{T_{SW}}(t) = P(T_{SW} \le t)$$

$$
= \begin{cases}
0, & \left(\dfrac{t}{k_1 c_1^{k_2}} \right)^{\frac{1}{c_2 k_2}} < a \\[2ex]
\dfrac{1}{b - a} \cdot \dfrac{1}{m - a} \\
\quad \times \left[\left(\dfrac{t}{k_1 c_1^{k_2}} \right)^{\frac{1}{c_2 k_2}} - a \right]^2, & a \le \left(\dfrac{t}{k_1 c_1^{k_2}} \right)^{\frac{1}{c_2 k_2}} < m \\[2ex]
1 - \dfrac{1}{b - a} \cdot \dfrac{1}{b - m} \\
\quad \times \left[b - \left(\dfrac{t}{k_1 c_1^{k_2}} \right)^{\frac{1}{c_2 k_2}} \right]^2, & m \le \left(\dfrac{t}{k_1 c_1^{k_2}} \right)^{\frac{1}{c_2 k_2}} < b \\[2ex]
1, & \left(\dfrac{t}{k_1 c_1^{k_2}} \right)^{\frac{1}{c_2 k_2}} \ge b
\end{cases}
\quad (5.82)
$$

- Probability Density Function

$$
f_{T_{SW}}(t) = \begin{cases}
\dfrac{2}{b - a} \cdot \dfrac{1}{m - a} \cdot \dfrac{1}{c_2 k_2} \cdot \dfrac{1}{k_1 c_1^{k_2}} \left(\dfrac{t}{k_1 c_1^{k_2}} \right)^{\frac{1}{c_2 k_2} - 1} \\
\quad \times \left[\left(\dfrac{t}{k_1 c_1^{k_2}} \right)^{\frac{1}{c_2 k_2}} - a \right], & a \le \left(\dfrac{t}{k_1 c_1^{k_2}} \right)^{\frac{1}{c_2 k_2}} < m \\[3ex]
\dfrac{2}{b - a} \cdot \dfrac{1}{b - m} \cdot \dfrac{1}{c_2 k_2} \cdot \dfrac{1}{k_1 c_1^{k_2}} \left(\dfrac{t}{k_1 c_1^{k_2}} \right)^{\frac{1}{c_2 k_2} - 1} \\
\quad \times \left[b - \left(\dfrac{t}{k_1 c_1^{k_2}} \right)^{\frac{1}{c_2 k_2}} \right], & m \le \left(\dfrac{t}{k_1 c_1^{k_2}} \right)^{\frac{1}{c_2 k_2}} \le b
\end{cases}
$$

$$(5.83)$$

- Mean (staff months)

$$E(T_{SW}) = k_1 c_1^{k_2} \frac{2}{b-a} \cdot \frac{1}{m-a} \left[\frac{m^{c_2 k_2 + 2} - a^{c_2 k_2 + 2}}{c_2 k_2 + 2} + \frac{a^{c_2 k_2 + 2} - a m^{c_2 k_2 + 1}}{c_2 k_2 + 1} \right]$$

$$+ k_1 c_1^{k_2} \frac{2}{b-a} \cdot \frac{1}{m-b} \left[\frac{b^{c_2 k_2 + 2} - m^{c_2 k_2 + 2}}{c_2 k_2 + 2} + \frac{b m^{c_2 k_2 + 1} - b^{c_2 k_2 + 2}}{c_2 k_2 + 1} \right]$$

$$(5.84)$$

- Variance (staff months)^2

$$Var(T_{SW}) = (k_1 c_1^{k_2})^2 \frac{2}{b-a} \cdot \frac{1}{m-a}$$

$$\times \left[\frac{m^{2 c_2 k_2 + 2} - a^{2 c_2 k_2 + 2}}{2 c_2 k_2 + 2} + \frac{a^{2 c_2 k_2 + 2} - a m^{2 c_2 k_2 + 1}}{2 c_2 k_2 + 1} \right]$$

$$+ (k_1 c_1^{k_2})^2 \frac{2}{b-a} \cdot \frac{1}{m-b}$$

$$\times \left[\frac{b^{2 c_2 k_2 + 2} - m^{2 c_2 k_2 + 2}}{2 c_2 k_2 + 2} + \frac{b m^{2 c_2 k_2 + 1} - b^{2 c_2 k_2 + 2}}{2 c_2 k_2 + 1} \right]$$

$$- [E(T_{SW})]^2 \qquad (5.85)$$

Example 5.15 *Suppose the effort and schedule of a software project are given by $Eff_{SW} = c_1 I^{c_2}$ and $T_{SW} = k_1 (Eff_{SW})^{k_2}$.*

a. *Develop the correlation formula between Eff_{SW} and T_{SW}, if $I \sim Unif(a, b)$.*

b. *Compute this correlation if $c_1 = 2.8$, $c_2 = 1.2$, $k_1 = 2.5$, $k_2 = 0.32$, and $I \sim Unif(30, 80)$.*

c. *Discuss what the correlation implies about Eff_{SW} and T_{SW}.*

Solution

a. From Equation 5.30, the correlation between Eff_{SW} and T_{SW} is

$$\rho_{Eff_{SW}, T_{SW}} = \frac{E(Eff_{SW} T_{SW}) - E(Eff_{SW}) E(T_{SW})}{\sigma_{Eff_{SW}} \sigma_{T_{SW}}} \qquad (5.86)$$

The first term in the numerator can be written as

$$E(Eff_{SW} T_{SW}) = E(c_1 I^{c_2} \cdot k_1 (c_1 I^{c_2})^{k_2}) = k_1 c_1^{k_2 + 1} E(I^{c_2 (k_2 + 1)})$$

Since

$$E(I^{c_2 (k_2 + 1)}) = \int_a^b t^{c_2 (k_2 + 1)} f_I(t) \, dt = \frac{1}{b-a} \cdot \frac{1}{c_2 (k_2 + 1) + 1}$$

$$\times \left[b^{c_2 (k_2 + 1) + 1} - a^{c_2 (k_2 + 1) + 1} \right]$$

we have

$$E(\mathit{Eff}_{SW}T_{SW}) = k_1 c_1^{k_2+1} \frac{1}{b-a} \cdot \frac{1}{c_2(k_2+1)+1} \left[b^{c_2(k_2+1)+1} - a^{c_2(k_2+1)+1} \right]$$

From Equation 5.65, we have

$$E(\mathit{Eff}_{SW}) = \frac{c_1}{c_2+1} \cdot \frac{1}{b-a} \left[b^{c_2+1} - a^{c_2+1} \right] \tag{5.87}$$

From Equation 5.66, we have

$$\sigma_{\mathit{Eff}_{SW}} = \sqrt{ \frac{1}{b-a} \cdot \frac{c_1^2}{2c_2+1} \left[b^{2c_2+1} - a^{2c_2+1} \right] - \left[E(\mathit{Eff}_{SW}) \right]^2 } \tag{5.88}$$

It is left as an exercise for the reader to show that

$$E(T_{SW}) = \frac{k_1 c_1^{k_2}}{c_2 k_2 + 1} \cdot \frac{1}{b-a} \left[b^{c_2 k_2+1} - a^{c_2 k_2+1} \right] \tag{5.89}$$

$$\sigma_{T_{SW}} = \sqrt{ \frac{k_1^2 c_1^{2k_2}}{2c_2 k_2 + 1} \cdot \frac{1}{b-a} \left[b^{2c_2 k_2+1} - a^{2c_2 k_2+1} \right] - \left[E(T_{SW}) \right]^2 } \tag{5.90}$$

Thus, if $I \sim \mathit{Unif}(a,b)$, then the general formula for the correlation between Eff_{SW} and T_{SW} is given by Equation 5.91.

$$\rho_{\mathit{Eff}_{SW},T_{SW}} =$$

$$\frac{ k_1 c_1^{k_2+1} \cdot \frac{1}{b-a} \cdot \frac{1}{c_2(k_2+1)+1} \left[b^{c_2(k_2+1)+1} - a^{c_2(k_2+1)+1} \right] - E(\mathit{Eff}_{SW})E(T_{SW}) }{ \sqrt{ \frac{c_1^2}{2c_2+1} \cdot \frac{1}{b-a} \left[b^{2c_2+1} - a^{2c_2+1} \right] - [E(\mathit{Eff}_{SW})]^2 } \sqrt{ \frac{k_1^2 c_1^{2k_2}}{2c_2 k_2 + 1} \cdot \frac{1}{b-a} \left[b^{2c_2 k_2+1} - a^{2c_2 k_2+1} \right] - [E(T_{SW})]^2 } }$$

$$\tag{5.91}$$

b. Substituting $c_1 = 2.8$, $c_2 = 1.2$, $k_1 = 2.5$, $k_2 = 0.32$, $a = 30$, and $b = 80$ into these expressions yields

$$E(\mathit{Eff}_{SW}T_{SW}) = 5736.2323$$

$$E(\mathit{Eff}_{SW}) = 346.12, \quad Var(\mathit{Eff}_{SW}) = 11{,}610.31$$

$$E(T_{SW}) = 16.055, \quad Var(T_{SW}) = 2.798$$

Therefore, from Equation 5.91, the correlation between Eff_{SW} and T_{SW} is $\rho_{\mathit{Eff}_{SW},T_{SW}} = 0.995$.

c. Although the true relationship between Eff_{SW} and T_{SW} is nonlinear, a correlation coefficient this close to unity indicates the relationship is not statistically significantly different from linear in the region $165.85 \leq s \leq 538.10$. This is illustrated in Figure 5.17.

Example 5.16 *Suppose a new radar system requires developing 14 software functions listed in Table 5.5. Let the uncertainties in the amount of code to develop*

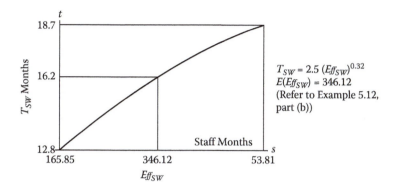

FIGURE 5.17
A plot of T_{SW} versus Eff_{SW} from Example 5.15.

TABLE 5.5

Radar Software Functions and Size Uncertainty Assessments

		Min (KDSI)	Mode (KDSI)	Max (KDSI)
Post Processor				
Radar Report Processor	I_1	3.6	4.0	4.8
Radar Control Processor	I_2	5.4	6.0	7.2
Seco Processor	I_3	1.8	2.0	2.4
Auto Monitoring	I_4	4.5	5.0	6.0
Network Interfacing	I_5	1.8	2.0	2.4
System Control Processor				
Mode Control	I_6	10.8	12.0	14.4
Display Console	I_7	13.5	15.0	18.0
Missile Impact Prediction				
OS and Utilities	I_8	12.6	14.0	16.8
Operational Program	I_9	27.0	30.0	36.0
Satellite Test Program	I_{10}	12.6	14.0	16.8
Library	I_{11}	10.8	12.0	14.4
Data Reduction	I_{12}	29.7	33.0	39.6
Seco Support	I_{13}	14.4	16.0	19.2
Communications	I_{14}	6.3	7.0	8.4
Total	I_{Total}	154.8	172.0	206.4

Note: The sum of the modes is not the mode of the distribution function of I_{Total}.

be represented by the random variables $I_1, I_2, I_3, \ldots, I_{14}$, where each I is in KDSI. Assume each I is characterized by a triangular distribution function. Suppose $I_1, I_2, I_3, \ldots, I_{14}$ are independent random variables and

$$I_{Total} = I_1 + I_2 + I_3 + \cdots + I_{14}.$$

a. *What is the mean and variance of I_{Total}?*
b. *What distribution function approximates the distribution of I_{Total}?*
c. *Determine the 0.50-fractile of $Eff_{SW} = 2.8(I_{Total})^{1.2}$.*

Solution

a. We are given the distribution function for each I is triangular, that is,

$$I_1 \sim Trng(3.6, 4.0, 4.8), \quad I_2 \sim Trng(5.4, 6.0, 7.2),$$
$$I_3 \sim Trng(1.8, 2.0, 2.4), \ldots, I_{14} \sim Trng(6.3, 7.0, 8.4)$$

From Theorem 5.7 (Equation 5.37)

$$E(I_{Total}) = E(I_1) + E(I_2) + E(I_3) + \cdots + E(I_{14}) \qquad (5.92)$$

Since each I has a triangular distribution, from Theorem 4.3

$$E(I_1) = \frac{1}{3}(3.6 + 4.0 + 4.8) = 4.13$$

$$E(I_2) = \frac{1}{3}(5.4 + 6.0 + 7.2) = 6.2$$

and so forth. Substituting these values into Equation 5.92 yields

$$E(I_{Total}) = 4.13 + 6.2 + 2.067 + \cdots + 7.23 = 177.73 \ \text{KDSI}$$

Since $I_1, I_2, I_3, \ldots, I_{14}$ are *independent,*[*] from Equation 5.39

$$Var(I_{Total}) = Var(I_1) + Var(I_2) + Var(I_3) + \cdots + Var(I_{14})$$

From Theorem 4.3

$$Var(I_1) = \frac{1}{18}\left\{(4.0 - 3.6)(4.0 - 4.8) + (4.8 - 3.6)^2\right\} = 0.0622$$

Following a similar set of calculations for $I_2, I_3, \ldots, I_{14}$, it can be shown that

$$Var(I_{Total}) = 12.77 \ \text{KDSI}^2$$

b. Since $I_1, I_2, I_3, \ldots, I_{14}$ are given to be independent random variables, the total size of the radar software I_{Total} is the sum of 14 independent random variables. By the central limit theorem (Theorem

---

[*] From Theorem 5.4, since $I_1, I_2, I_3, \ldots, I_{14}$ are independent random variables the correlation between each pair of $I_1, I_2, I_3, \ldots, I_{14}$ is zero.

5.10), it is reasonable to assume the distribution function of I_{Total} will be approximately normal. From part (a) this means $I_{Total} \sim N(E(I_{Total}), Var(I_{Total})) = N(177.73, 12.77)$.

c. In this example, we are given $Eff_{SW} = 2.8(I_{Total})^{1.2}$. If x and s are the values possible for I_{Total} and Eff_{SW}, respectively, then $s = 2.8x^{1.2}$ is a strictly increasing differentiable function of x. From Theorem 5.11, this implies

$$F_{I_{Total}}(x) = F_{Eff_{SW}}(s) \qquad (5.93)$$

From part (b) we know that $F_{I_{Total}}(x) = 0.50$ when $x = 177.73$ KDSI; therefore, $x_{0.50} = 177.73$, which is the 0.50-fractile of I_{Total}. From Equation 5.93

$$F_{I_{Total}}(177.73) = 0.50 = F_{Eff_{SW}}(s)$$

Since $s = 2.8x^{1.2}$, when $x = x_{0.50} = 177.73$ we have $s = 1402.4$; thus

$$F_{I_{Total}}(177.73) = 0.50 = F_{Eff_{SW}}(1402.4)$$

In summary, the 0.50-fractile of Eff_{SW} is 1402.4 staff months. This is the same as saying $Med(Eff_{SW}) = 1402.4$ staff months. It is left as an exercise to determine the 0.25 and 0.75 fractiles of Eff_{SW}.

5.4.3 Functions of Two Random Variables

Thus far, we have focused on deriving the probability distribution function for a function of a single random variable. Functions of two or more random variables commonly occur in cost uncertainty analysis. For instance, if the unit cost of an unmanned spacecraft is determined by

$$UC = 5.48(SC_{wt})^{0.94}(BOLP)^{0.30}$$

then UC is a function of two random variables—spacecraft dry weight SC_{wt} and beginning-of-life power $BOLP$. Likewise, if the software development effort for a project is determined by

$$Eff_{SW} = \frac{I}{P_r} \qquad (5.94)$$

then Eff_{SW} is a function of two random variables—the amount of code to develop I (in DSI) and the development productivity P_r (in DSI per staff month). The following theorem provides a set of general integral formulas for determining the density functions of sums, differences, products, and quotients of two random variables. We shall see that determining this density function, in closed form, can be computationally challenging. In many cases a closed form is not even possible. In such circumstances, computer-based methods (e.g., Monte Carlo simulation) are often used to approximate the density function.

Theorem 5.12 (Mood et al. 1974) *Let X and Y be continuous random variables with joint density $f(x,y)$. If U is a function of X and Y with density function $g(u)$, then*

$$U = X + Y \text{ has density } g(u) = \int_{-\infty}^{\infty} f(x, u - x)\, dx = \int_{-\infty}^{\infty} f(u - y, y)\, dy$$

$$U = X - Y \text{ has density } g(u) = \int_{-\infty}^{\infty} f(x, x - u)\, dx = \int_{-\infty}^{\infty} f(u + y, y)\, dy$$

$$U = XY \text{ has density } g(u) = \int_{-\infty}^{\infty} \frac{1}{|x|} f\left(x, \frac{u}{x}\right) dx = \int_{-\infty}^{\infty} \frac{1}{|y|} f\left(\frac{u}{y}, y\right) dy$$

$$U = X/Y \text{ has density } g(u) = \int_{-\infty}^{\infty} |x| f(ux, x)\, dx = \int_{-\infty}^{\infty} |y| f(uy, y)\, dy$$

The reader is directed to Mood et al. (1974) for a proof of this theorem. Theorem 5.12 provides a number of interesting results. For instance, suppose U_1, U_2, and U_3 are *independent* random variables with $U_1 \sim Unif(0, 1)$, $U_2 \sim Unif(0, 1)$, and $U_3 \sim Unif(0, 1)$. If $U = U_1 + U_2$, then the density function for U can be shown to be triangular (Cramer 1966). Furthermore, if $U = U_1 + U_2 + U_3$, then the density function for U is "bell-shaped"—but not yet normally distributed. Figure 5.18 (Cramer 1966) illustrates these results.

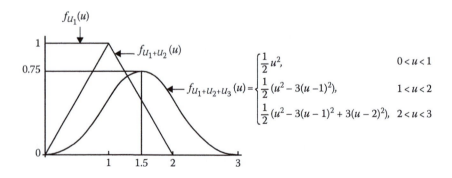

FIGURE 5.18
Sums of independent *Unif* (0, 1) random variables.

In continuation, suppose the random variable U is defined by

$$U = U_1 + U_2 + U_3 + \cdots + U_n$$

where $U_1, U_2, U_3, \ldots, U_n$ are independent random variables and $U_i \sim$ $Unif(0,1)$ for $i = 1,\ldots,n$. By the central limit theorem, as n increases the distribution function of U will rapidly approach a normal distribution. This remarkable result is further discussed and illustrated in Appendix A.

The following presents an application of Theorem 5.12. A PDF for software development effort, defined by Equation 5.94, is derived.

Example 5.17 *In Example 5.3, the effort Eff_{SW} to develop a new software application was given by*

$$Eff_{SW} = \frac{X}{Y}$$

where $X = I$ is the amount of code to develop (in DSI) and $Y = P_r$ is the development productivity (in DSI per staff month). Suppose X and Y are continuous random variables with joint PDF

$$f(x,y) = \begin{cases} \dfrac{1}{5(10^6)}, & 50,000 \le x \le 100,000, \quad 100 \le y \le 200 \\ 0, & \text{otherwise} \end{cases}$$

a. *Use Theorem 5.12 to find the PDF of Eff_{SW}.*
b. *Verify $P(Eff_{SW} \le 300) = 0.0333$ and $P(Eff_{SW} \le 610) \approx 0.75$.*
c. *From part (a), determine $E(Eff_{SW})$.*

Solution

a. Since Eff_{SW} is a ratio of two random variables, from Theorem 5.12 Eff_{SW} has the PDF $g(u)$, where

$$g(u) = \int_{-\infty}^{\infty} |y| f(uy, y)\, dy$$

In this equation, u represents feasible values of the random variable Eff_{SW} (staff months). To use the integral given by $g(u)$, it is necessary to define the regions of integration specific to this example. These regions are shown in Figure 5.19. From Figure 5.19, we see that

$$g(u) = \begin{cases} \displaystyle\int_{\frac{50,000}{u}}^{200} \dfrac{1}{5(10^6)}\, y\, dy, & 250 \le u \le 500 \\[4mm] \displaystyle\int_{100}^{\frac{100,000}{u}} \dfrac{1}{5(10^6)}\, y\, dy, & 500 \le u \le 1000 \end{cases}$$

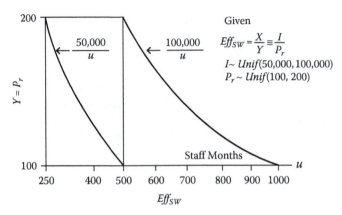

FIGURE 5.19
Regions of integration for $g(u)$ in Example 5.17.

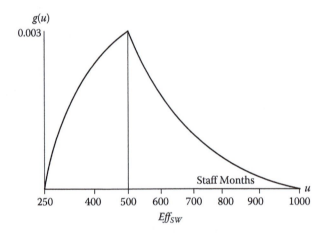

FIGURE 5.20
Probability density function for Eff_{SW}.

The PDF of Eff_{SW} is, therefore, given by Equation 5.95.

$$g(u) = \begin{cases} \dfrac{1}{2} \cdot \dfrac{1}{5(10^6)} \left\{ (200)^2 - \left(\dfrac{50,000}{u} \right)^2 \right\}, & 250 \le u \le 500 \\[4mm] \dfrac{1}{2} \cdot \dfrac{1}{5(10^6)} \left\{ \left(\dfrac{100,000}{u} \right)^2 - (100)^2 \right\}, & 500 \le u \le 1000 \end{cases}$$

(5.95)

A plot of this PDF is shown in Figure 5.20.

b. Using Equation 5.95, probabilities associated with various values of Eff_{SW} can be computed. For instance, the probability that $Eff_{SW} \le 300$ staff months is

$$P(Eff_{SW} \leq 300) = \int_{250}^{300} \frac{1}{2} \cdot \frac{1}{5(10^6)} \left\{ (200)^2 - \left(\frac{50,000}{u} \right)^2 \right\} du = 0.0333$$

This result is consistent with Example 5.3. The probability $Eff_{SW} \leq$ 610 staff months is

$$P(Eff_{SW} \leq 610) = \int_{250}^{500} \frac{1}{2} \cdot \frac{1}{5(10^6)} \left\{ (200)^2 - \left(\frac{50,000}{u} \right)^2 \right\} du$$

$$+ \int_{500}^{610} \frac{1}{2} \cdot \frac{1}{5(10^6)} \left\{ \left(\frac{100,000}{u} \right)^2 - (100)^2 \right\} du$$

$$= 0.50 + 0.250656 \approx 0.75$$

A family of boundary curves for Eff_{SW} is presented in Figure 5.21. Shown are values of Eff_{SW} for various combinations of the number of DSI to develop $X = I$ and the development productivity rate $Y = P_r$ (DSI per staff month).

c. Last, from Equation 5.95 the expected effort can be computed; specifically,

$$E(Eff_{SW}) = \int_{250}^{500} u \cdot \frac{1}{2} \cdot \frac{1}{5(10^6)} \left\{ (200)^2 - \left(\frac{50,000}{u} \right)^2 \right\} du$$

$$+ \int_{500}^{1000} u \cdot \frac{1}{2} \cdot \frac{1}{5(10^6)} \left\{ \left(\frac{100,000}{u} \right)^2 - (100)^2 \right\} du$$

$$= 519.86 \text{ staff months}$$

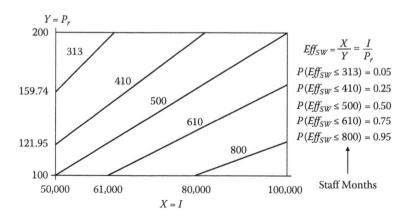

FIGURE 5.21
Boundary curves for Eff_{SW} and associated probabilities.

Theorem 5.12 provides a way to determine the PDF of sums, differences, products, and quotients of two random variables. The integrals in Theorem 5.12 are classically known as *convolution integrals*. In many applied problems, these integrals are hard to determine. In cost uncertainty analysis, conditions often prevail that enable analysts to approximate the form of a PDF. If an approximation can be found (or theoretically claimed), then it is unnecessary to compute a convolution integral. For instance, we know (from the central limit theorem) the sum of a sufficiently large number of independent random variables will approach the normal distribution. Similarly, from the central limit theorem, we know the product of a sufficiently large number of independent random variables will approach the lognormal distribution.

The last topic discussed in this chapter is the Mellin transform. The Mellin transform is a useful technique for computing the moments of products and quotients of many random variables. The application of the Mellin transform to cost functions comprised of two or more random variables is emphasized.

5.5 Mellin Transform and Its Application to Cost Functions

This section presents a little known technique for determining moments of products and quotients of random variables. Known as the Mellin transform (Epstein 1948, Giffin 1975), it works on random variables that are continuous, independent, and nonnegative.* The Mellin transform is well suited to cost functions since *Cost* is essentially a nonnegative random variable. The following defines the Mellin transform. Examples are provided to illustrate its use from a cost perspective.

Definition 5.1 *If X is a nonnegative random variable, $0 < x < \infty$, the Mellin transform of its PDF $f_X(x)$ is*

$$M_X(s) = \int_0^\infty x^{s-1} f_X(x)\, dx \tag{5.96}$$

for all s for which the integral exists.

From Equation 5.96 it can be seen that

$$M_X(1) = \int_0^\infty f_X(x)\, dx = 1 \tag{5.97}$$

---

* An extension of the Mellin transform technique to random variables that are not everywhere positive is discussed by Epstein (1948).

$$M_X(2) = \int_0^\infty x f_X(x)\, dx = E(X) \tag{5.98}$$

$$M_X(3) = \int_0^\infty x^2 f_X(x)\, dx = E(X^2) \tag{5.99}$$

From these equations it follows from Equation 3.33 that

$$M_X(s) = E(X^{s-1}) \tag{5.100}$$

It also immediately follows that

$$Var(X) = M_X(3) - [M_X(2)]^2 \tag{5.101}$$

The Mellin transform is very useful when dealing with random variables raised to a power. For example, if for any real a we have $Y = X^a$, then

$$M_Y(s) = E\left(Y^{s-1}\right) = E\left((X^a)^{s-1}\right) = E\left((X^{as-a})\right)$$
$$= E\left(\left(X^{(as-a+1)-1}\right)\right) = M_X(as - a + 1) \tag{5.102}$$

As an illustration, consider the Mellin transform of $Eff_{SW} = 2.8I^{1.2}$. This yields

$$M_{Eff_{SW}}(s) = E\left(Eff_{SW}^{s-1}\right) = E\left(\left(2.8I^{1.2}\right)^{s-1}\right) = E\left(\left(2.8^{s-1}I^{1.2s-1.2}\right)\right)$$
$$= 2.8^{s-1}E((I^{(1.2s-1.2+1)-1})) = 2.8^{s-1}M_I(1.2s - 1.2 + 1) \tag{5.103}$$

therefore,

$$M_{Eff_{SW}}(s) = 2.8^{s-1}M_I(1.2s - 1.2 + 1) \tag{5.104}$$

Equation 5.104 provides a way to generate moments of the random variable Eff_{SW}. For instance, the expected effort $E(Eff_{SW})$ can be written in terms of Equation 5.104 as follows:

$$E(Eff_{SW}) = M_{Eff_{SW}}(2) = 2.8M_I(2.2)$$

For example, if $I \sim Unif(30, 80)$ then from Equation 5.96

$$M_I(s) = \int_0^\infty t^{s-1} f_I(t)\, dt = \int_{30}^{80} t^{s-1} \frac{1}{50}\, dt = \frac{1}{50}\left[\frac{80^s - 30^s}{s}\right]$$

where $s \neq 0$. Therefore,

$$E(\mathit{Eff}_{SW}) = M_{\mathit{Eff}_{SW}}(2) = 2.8 M_I(2.2)$$

$$= (2.8)\frac{1}{50}\left[\frac{80^{2.2} - 30^{2.2}}{2.2}\right] = 346.12 \text{ staff months} \qquad (5.105)$$

This value agrees with the value of $E(\mathit{Eff}_{SW})$ computed by Equation 5.65 in Example 5.12. Furthermore, note that Equation 5.105 is a specific application of the general formula for $E(\mathit{Eff}_{SW})$ given by Equation 5.65. The following presents an important convolution property of the Mellin transform.

Theorem 5.13 (Giffin 1975) *Let X, Y, and W be independent random variables with PDFs $f_X(x)$, $f_Y(y)$, and $f_W(w)$, respectively. If α, β_1, β_2, β_3 are constants and*

$$Z = \alpha X^{\beta_1} Y^{\beta_2} W^{\beta_3}$$

then

$$M_Z(s) = \alpha^{s-1} M_X(\beta_1 s - \beta_1 + 1) M_Y(\beta_2 s - \beta_2 + 1) M_W(\beta_3 s - \beta_3 + 1) \blacklozenge$$

From Theorem 5.13, if $Z = XY$, then

$$M_Z(s) = M_X(s) M_Y(s) \qquad (5.106)$$

Similarly, from Theorem 5.13, if $Z = X/Y$, then

$$M_Z(s) = M_X(s) M_Y(2 - s) \qquad (5.107)$$

Equations 5.108 through 5.111 present Mellin transforms for selected distribution functions often used in cost uncertainty analysis. The distribution of X and its Mellin transform $M_X(s)$, $s \neq 0, -1$, are given.

- $X \sim \mathit{Unif}(a, b)$

$$M_X(s) = \frac{1}{s(b - a)}(b^s - a^s) \qquad (5.108)$$

- $X \sim \mathit{Trng}(a, m, b)$

$$M_X(s) = \frac{2}{s(s + 1)(b - a)}\left[\frac{b(b^s - m^s)}{b - m} - \frac{a(m^s - a^s)}{m - a}\right] \qquad (5.109)$$

- $X \sim Trap(a, m_1, m_2, b)$

$$M_X(s) = L_1 L_3 \frac{\left[m_1^s (sm_1 - (s+1)a) + a^{s+1}\right]}{s(s+1)} + L_1 \frac{(m_2^s - m_1^s)}{s}$$

$$+ L_1 L_2 \frac{\left[m_2^s (sm_2 - (s+1)b) + b^{s+1}\right]}{s(s+1)} \qquad (5.110)$$

where $L_1 = \frac{2}{(m_2 + b - a - m_1)}$ $L_2 = \frac{1}{(b - m_2)}$ $L_3 = \frac{2}{(m_1 - a)}$

- $X \sim LogN(\mu_Y, \sigma_Y^2)$

$$M_X(s) = e^{\mu_Y(s-1) + \frac{1}{2}\sigma_Y^2(s-1)^2} \qquad (5.111)$$

Example 5.18* *Let the unit cost UC of an unmanned spacecraft be given by*

$$UC = 5.48(SC_{wt})^{0.94}(BOLP)^{0.30}$$

where UC is a function of SC_{wt} (the spacecraft's weight in lbs) and BOLP (the spacecraft's beginning-of-life power in watts). Suppose point estimates for weight and power are 6500 lbs and 2000 watts; that is,

$$wPE_{SC_{wt}} = 6500 \text{ and } jPE_{BOLP} = 2000$$

where possible values for SC_{wt} and BOLP are given by w and j, respectively. If the uncertainties around these point estimates are described by the PDFs in Figure 5.22, use the Mellin transform to compute the expected unit cost E(UC).

Solution To simplify notation, let $X = SC_{wt}$, $Y = BOLP$, and $Z = UC$. We then need to compute $E(Z)$, where $Z = 5.48X^{0.94}Y^{0.30}$. From Theorem 5.13, the Mellin transform of Z is

$$M_Z(s) = 5.48^{s-1}M_X(0.94s - 0.94 + 1)M_Y(0.30s - 0.30 + 1)$$

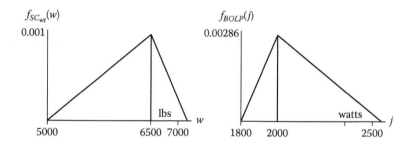

FIGURE 5.22
PDFs for SC_{wt} and BOLP.

---

* This example is an adaptation from Lurie, P. M., and M. S. Goldberg. 1993. *A Handbook of Cost Risk Analysis Methods*, P-2734. Alexandria, VA: The Institute for Defense Analyses.

From Equation 5.100

$$E(UC) = E(Z) = M_Z(2) = 5.48 M_X(1.94) M_Y(1.30)$$

Since the PDFs for weight and power are triangular, from Equation 5.109

$$M_X(1.94) = \frac{2}{1.94(2.94)(2000)}$$

$$\times \left[\frac{7000(7000^{1.94} - 6500^{1.94})}{7000 - 6500} - \frac{5000(6500^{1.94} - 5000^{1.94})}{6500 - 5000} \right]$$

$$= 3652.486$$

$$M_Y(1.30) = \frac{2}{1.30(2.30)(700)}$$

$$\times \left[\frac{2500(2500^{1.30} - 2000^{1.30})}{2500 - 2000} - \frac{1800(2000^{1.30} - 1800^{1.30})}{2000 - 1800} \right]$$

$$= 9.918$$

Therefore

$$E(UC) = E(Z) = M_Z(2) = 198.5 \ (\$K)$$

Let us discuss this example further. If the point estimates for SC_{wt} and $BOLP$ were substituted into UC, then

$$UC_{PE} = 5.48(6500)^{0.94}(2000)^{0.30} = 205.7 \ (\$K)$$

In this example, why is $E(UC) < UC_{PE}$? As seen in Figure 5.22 the skew of SC_{wt} is negative. There is far more probability the spacecraft's weight will fall to the left of 6500 lbs than to the right of 6500 lbs. Furthermore, the variance of SC_{wt} is significantly greater than the variance of $BOLP$; showing this is left for the reader. For these reasons, we have an expected cost that is less than the point estimate of the unit cost.

Example 5.19 *A new software application is to be developed. Suppose the application consists of a mixture of new code I_{New} and reused code I_{Reused}. Let the effort associated with developing the application be a function of the equivalent size I_{Equiv}, where (from [Conte et al. 1986])*

$$I_{Equiv} = I_{New} + I_{Reused}^{0.857} \tag{5.112}$$

Suppose values for I_{New} and I_{Reused} are uncertain. If I_{New} and I_{Reused} are independent random variables with PDFs given in Figure 5.23, use the Mellin transform to compute $E(I_{Equiv})$ and $\sigma_{I_{Equiv}}$.

Solution We are given $I_{Equiv} = I_{New} + I_{Reused}^{0.857}$. From Theorems 5.7 and 5.8

$$E(I_{Equiv}) = E(I_{New}) + E(I_{Reused}^{0.857})$$

$$Var(I_{Equiv}) = Var(I_{New}) + Var(I_{Reused}^{0.857})$$

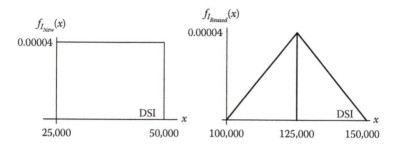

FIGURE 5.23
Probability density functions for I_{New} and I_{Reused}.

Computing $E(I_{Equiv})$

We have $E(I_{Equiv}) = E(I_{New}) + E(I_{Reused}^{0.857})$. From Equation 5.98 $E(I_{New}) = M_{I_{New}}(2)$. Suppose we let $Z = I_{Reused}^{0.857}$, then from Theorem 5.13

$$M_Z(s) = M_{I_{Reused}}(0.857(s-1)+1)$$

$$E\left(I_{Reused}^{0.857}\right) = M_Z(2) = M_{I_{Reused}}(0.857(2-1)+1) = M_{I_{Reused}}(1.857)$$

From this, we have

$$E(I_{Equiv}) = M_{I_{New}}(2) + M_{I_{Reused}}(1.857)$$

Since $I_{New} \sim Unif(25{,}000, 50{,}000)$, from Equation 5.108 $M_{I_{New}}(2) = 37{,}500$. Similarly, since $I_{Reused} \sim Trng(100{,}000, 125{,}000, 150{,}000)$, from Equation 5.109 $M_{I_{Reused}}(1.857) = 23{,}327.8$; therefore,

$$E(I_{Equiv}) = 37{,}500 + 23{,}327.8 = 60{,}827.8 \text{ DSI} \approx 61 \text{ KDSI}$$

Computing $\sigma_{I_{Equiv}}$

To compute $\sigma_{I_{Equiv}}$, we begin by computing $Var(I_{Equiv})$. Since I_{New} and I_{Reused} are independent random variables

$$Var(I_{Equiv}) = Var(I_{New}) + Var\left(I_{Reused}^{0.857}\right)$$

From Equation 5.101

$$Var(I_{New}) = M_{I_{New}}(3) - \left(M_{I_{New}}(2)\right)^2$$

We can write

$$Var\left(I_{Reused}^{0.857}\right) = E\left(\left(I_{Reused}^{0.857}\right)^2\right) - \left(E\left(I_{Reused}^{0.857}\right)\right)^2$$

$$= E\left(I_{Reused}^{1.714}\right) - \left(M_{I_{Reused}}(1.857)\right)^2$$

Suppose we let $W = I_{Reused}^{1.714}$, then from Theorem 5.13

$$M_W(s) = M_{I_{Reused}}(1.714(s-1)+1)$$

$$E\left(I_{Reused}^{1.714}\right) = M_W(2) = M_{I_{Reused}}(1.714(2-1)+1) = M_{I_{Reused}}(2.714)$$

Therefore, $Var\left(I_{Reused}^{0.857}\right) = M_{I_{Reused}}(2.714) - (M_{I_{Reused}}(1.857))^2$, from which

$$Var(I_{Equiv}) = M_{I_{New}}(3) - (M_{I_{New}}(2))^2 + M_{I_{Reused}}(2.714) - (M_{I_{Reused}}(1.857))^2$$

From Equation 5.108, $M_{I_{New}}(3) = 1.45833(10)^9$ and $M_{I_{New}}(2) = 37{,}500$. From Equation 5.109, $M_{I_{Reused}}(2.714) = 5.46856(10)^8$ and $M_{I_{Reused}}(1.857) = 23{,}327.8$. Substituting these values into

$$Var(I_{Equiv}) = M_{I_{New}}(3) - (M_{I_{New}}(2))^2 + M_{I_{Reused}}(2.714) - (M_{I_{Reused}}(1.857))^2$$

$$\text{produces } \sigma_{I_{Equiv}} = \sqrt{Var(I_{Equiv})} = 7{,}399.49 \text{ DSI} \approx 7.4 \text{ KDSI} \blacklozenge$$

Case Discussion 5.3: In Example 5.2, the effort for system test was given by the $Eff_{SysTest} = XY$, where X is staff-level and Y is the number of months. Suppose X and Y are independent random variables with distribution functions shown in Figure 5.24.*

a. Use a convolution integral in Theorem 5.12 to develop a general formula for the PDF of $Eff_{SysTest}$. Plot the density function.
b. Using the PDF of $Eff_{SysTest}$ compute the mean of $Eff_{SysTest}$, $P(Eff_{SysTest} \leq E(Eff_{SysTest}))$, and $P(Eff_{SysTest} \leq 173)$.
c. Use the *Mellin transform* to compute the mean and variance of $Eff_{SysTest}$.

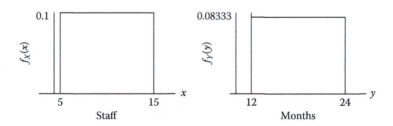

FIGURE 5.24
Marginal distribution for X (staff) and Y (months).

---

* This is a slight variation from Example 5.2, where the range of possible values for Y was given as 12–36 months. It is left to the reader to study how the problem solution presented in Case Discussion 5.3 changes, if Y varies from 12–36 months instead of 12–24 months.

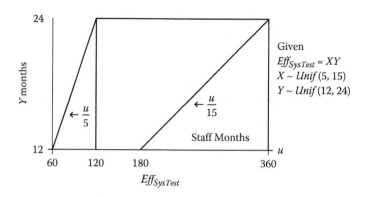

FIGURE 5.25
Region of integration for $g(u)$.

Discussion:

a. Since X and Y are independent, their joint distribution function is

$$f(x,y) = \frac{1}{10} \cdot \frac{1}{12} = \frac{1}{120}, \quad 5 \le x \le 15, \; 12 \le y \le 24 \qquad (5.113)$$

Let $Eff_{SysTest} = U = XY$. Let $g(u)$ represent the PDF of $Eff_{SysTest}$. Since $Eff_{SysTest}$ is a product of two random variables, from Theorem 5.12

$$g(u) = \int_{-\infty}^{\infty} \frac{1}{|y|} f\left(\frac{u}{y}, y\right) dy \qquad (5.114)$$

The regions of integration for $g(u)$ are shown in Figure 5.25.
From Figure 5.25, and Equation 5.114, the PDF of $Eff_{SysTest}$ is given by the three integrals over the following regions:

$$g(u) = \begin{cases} \displaystyle\int_{12}^{\frac{u}{5}} \frac{1}{y} \cdot \frac{1}{120} \, dy = \frac{1}{120} \ln\left(\frac{u}{60}\right), & 60 \le u \le 120 \\[2ex] \displaystyle\int_{12}^{24} \frac{1}{y} \cdot \frac{1}{120} \, dy = \frac{1}{120} \ln(2), & 120 \le u \le 180 \\[2ex] \displaystyle\int_{\frac{u}{15}}^{24} \frac{1}{y} \cdot \frac{1}{120} \, dy = \frac{1}{120} \ln\left(\frac{360}{u}\right), & 180 \le u \le 360 \end{cases} \qquad (5.115)$$

Equation 5.115 is the PDF of $Eff_{SysTest}$. It is left to the reader to check that $g(u)$ has a unit area over the interval $60 \le u \le 360$. Figure 5.26 shows a plot of this density function.

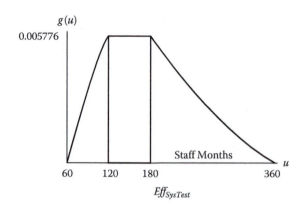

FIGURE 5.26
PDF of $Eff_{SysTest}$.

b. From the density function we can compute the mean effort for system test, as well as various probabilities. These computations are as follows:

$$E(Eff_{SysTest}) = \int_{60}^{360} u\,g(u)\,du = \int_{60}^{120} u\,\frac{1}{120}\ln\left(\frac{u}{60}\right)du + \int_{120}^{180} u\,\frac{1}{120}\ln(2)\,du$$

$$+ \int_{180}^{360} u\,\frac{1}{120}\ln\left(\frac{360}{u}\right)du = 180$$

Knowledge of the density function facilitates computing various probabilities of interest. From Equation 5.115

$$P(Eff_{SysTest} \le E(Eff_{SysTest})) = P(Eff_{SysTest} \le 180)$$

$$= \int_{60}^{120} \frac{1}{120}\ln\left(\frac{u}{60}\right)du + \int_{120}^{180} \frac{1}{120}\ln(2)\,du$$

$$= 0.19315 + 0.34657 = 0.54$$

Similarly,

$$P(Eff_{SysTest} \le 173) = \int_{60}^{120} \frac{1}{120}\ln\left(\frac{u}{60}\right)du + \int_{120}^{173} \frac{1}{120}\ln(2)\,du$$

$$= 0.19315 + 0.30613 \approx 0.50$$

In this case, the median test effort is approximately 173 staff months. Shown in Figure 5.27 are curves of constant effort for various pairs of x (staff) and y (months). A probability associated with each effort is also shown.

As discussed, developing a general formula for the PDF of $Eff_{SysTest}$ involves some tricky mathematics. A slight alteration in the problem statement can further complicate the mathematics. If, for instance, the distribution function of X was triangular instead of uniform, it would be quite difficult to develop an analytical form of $g(u)$.

c. The following illustrates how the Mellin transform applies to this case discussion. The first two moments, which lead to the mean and variance of the test effort, are developed. It is given that

$$Eff_{SysTest} = U = XY, \quad 0 < x < \infty, \quad 0 < y < \infty$$

From Theorem 5.13

$$M_{Eff_{SysTest}}(s) = M_U(s) = M_X(s)M_Y(s) \tag{5.116}$$

From Equation 5.98

$$E\left(Eff_{SysTest}\right) = E(U) = M_U(2) = M_X(2)M_Y(2) \tag{5.117}$$

From Equation 5.101

$$Var\left(Eff_{SysTest}\right) = Var(U) = M_U(3) - [M_U(2)]^2$$
$$= M_X(3)M_Y(3) - [M_X(2)M_Y(2)]^2 \tag{5.118}$$

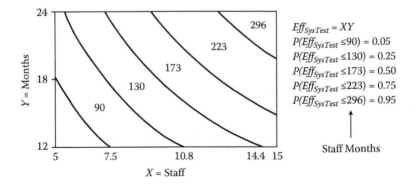

FIGURE 5.27
Boundary curves for $Eff_{SysTest}$.

Since the distribution functions for X and Y are uniform with parameters shown in Figure 5.24, from Equation 5.108 it follows that $M_X(2) = 10$, $M_Y(2) = 18$, $M_X(3) = 108.333$, and $M_Y(3) = 336$. Substituting these values into Equations 5.117 and 5.118 yields

$$E\left(\text{Eff}_{\text{SysTest}}\right) = E(U) = 180 \text{ staff months}$$

$$\text{Var}\left(\text{Eff}_{\text{SysTest}}\right) = \text{Var}(U) = 4000 \text{ (staff months)}^2$$

$$\sigma_{\text{Eff}_{\text{SysTest}}} = \sqrt{\text{Var}\left(\text{Eff}_{\text{SysTest}}\right)} = 63.25 \text{ staff months}$$

The Mellin transform is clearly a convenient way to compute the moments of $\text{Eff}_{\text{SysTest}}$ without the need for its density function.

Next, a final case discussion is presented. It will show how concepts throughout this chapter combine to produce useful results. Specifically, formulas for the mean and variance of a ratio of two uniformly distributed random variables and two beta distributed random variables are developed. Seen in previous examples, ratios of random variables can arise frequently in cost uncertainty analysis.

Case Discussion 5.4: Suppose I and P_r are independent random variables. Develop general formulas for $E(\text{Eff}_{\text{SW}})$ and $\text{Var}(\text{Eff}_{\text{SW}})$ if $\text{Eff}_{\text{SW}} = I/P_r$ and

a. $I \sim \text{Unif}(a_1, b_1)$ and $P_r \sim \text{Unif}(a_2, b_2)$
b. $I \sim \text{Beta}(\alpha_1, \beta_1, a_1, b_1)$ and $P_r \sim \text{Beta}(\alpha_2, \beta_2, a_2, b_2)$

Discussion Since I and P_r are independent, from Theorem 5.5

$$E\left(\text{Eff}_{\text{SW}}\right) = E\left(\frac{I}{P_r}\right) = E(I)E\left(\frac{1}{P_r}\right) = \mu_I E\left(\frac{1}{P_r}\right) \qquad (5.119)$$

By definition

$$\text{Var}\left(\text{Eff}_{\text{SW}}\right) = E\left(\text{Eff}_{\text{SW}}^2\right) - \left[E\left(\text{Eff}_{\text{SW}}\right)\right]^2$$

$$= E\left(I^2 \frac{1}{P_r^2}\right) - \mu_I^2 \left[E\left(\frac{1}{P_r}\right)\right]^2 = E\left(I^2\right)E\left(\frac{1}{P_r^2}\right) - \mu_I^2 \left[E\left(\frac{1}{P_r}\right)\right]^2$$

By definition $\text{Var}(I) = E(I^2) - [E(I)]^2$. This is equivalent to

$$E(I^2) = \sigma_I^2 + \mu_I^2$$

Substituting into *Var* (Eff_{SW}) yields

$$Var\left(Eff_{SW}\right) = \left(\sigma_I^2 + \mu_I^2\right) E\left(\frac{1}{P_r^2}\right) - \mu_I^2 \left[E\left(\frac{1}{P_r}\right)\right]^2 \qquad (5.120)$$

a. We are interested in using these equations to develop general formulas for the mean and variance of Eff_{SW}, when I and P_r are uniformly distributed random variables. It has just been shown that

$$E\left(Eff_{SW}\right) = E\left(\frac{I}{P_r}\right) = E(I)E\left(\frac{1}{P_r}\right) = \mu_I E\left(\frac{1}{P_r}\right)$$

Since $I \sim Unif(a_1, b_1)$, we know $\mu_I = \frac{1}{2}(a_1 + b_1)$; therefore,

$$E\left(Eff_{SW}\right) = \frac{1}{2}(a_1 + b_1) E\left(\frac{1}{P_r}\right)$$

To produce a general formula for $E(Eff_{SW})$, it remains to determine

$$E\left(\frac{1}{P_r}\right) \equiv E\left((P_r)^{-1}\right)$$

Determining $E((P_r)^{-1})$ will be accomplished from the PDF of $(P_r)^{-1}$. Let $Z = (P_r)^{-1}$; therefore,

$$Z = g(P_r) \Rightarrow z = g(y) = \frac{1}{y} \Rightarrow y = v(z) = \frac{1}{z}$$

Since $g(y)$ is a strictly decreasing differentiable function of y, from Theorem 5.11

$$f_Z(z) = f_{P_r}(v(z)) \cdot \left|\frac{d[v(z)]}{dz}\right|, \quad g(b_2) \le z \le g(a_2)$$

$$f_Z(z) = \frac{1}{b_2 - a_2} \cdot \frac{1}{z^2}, \quad \frac{1}{b_2} \le z \le \frac{1}{a_2}$$

A picture of this density function is shown in Figure 5.28. From the PDF we know that

$$E\left(\frac{1}{P_r}\right) = E(Z) = \int_{\frac{1}{b_2}}^{\frac{1}{a_2}} z f_Z(z)\, dz = \int_{\frac{1}{b_2}}^{\frac{1}{a_2}} z \cdot \frac{1}{b_2 - a_2} \cdot \frac{1}{z^2}\, dz = \frac{1}{b_2 - a_2} \ln\left(\frac{b_2}{a_2}\right)$$

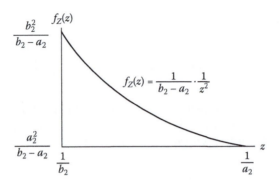

FIGURE 5.28
PDF of $Z = 1/P_r$.

Therefore,

$$E\left(Eff_{SW}\right) = \frac{1}{2} \cdot \frac{a_1 + b_1}{b_2 - a_2} \ln\left(\frac{b_2}{a_2}\right) \qquad (5.121)$$

Next, we will develop a formula for the variance of Eff_{SW}. By definition

$$Var(Eff_{SW}) = E(Eff_{SW}^2) - \left[E(Eff_{SW})\right]^2$$

From Equation 5.121

$$Var(Eff_{SW}) = E(Eff_{SW}^2) - \left[\frac{1}{2} \cdot \frac{a_1 + b_1}{b_2 - a_2} \ln\left(\frac{b_2}{a_2}\right)\right]^2$$

It remains, then, to determine $E(Eff_{SW}^2)$; this will be done by the Mellin transform technique. Let

$$Q = Eff_{SW}^2 = \frac{I^2}{P_r^2} \Rightarrow E(Q) = E(Eff_{SW}^2) = E\left(\frac{I^2}{P_r^2}\right)$$

From Theorem 5.13

$$M_Q(s) = M_I(2s - 1)M_{P_r}(3 - 2s) \qquad (5.122)$$

$$E(Q) = M_Q(2) = M_I(3)M_{P_r}(-1) \qquad (5.123)$$

Since I and P_r are uniformly distributed random variables, from Equation 5.108

$$M_I(3) = \frac{1}{3(b_1 - a_1)} \left(b_1^3 - a_1^3\right) \text{ and}$$

$$M_{P_r}(-1) = \frac{1}{(-1)(b_2 - a_2)} \left(b_2^{-1} - a_2^{-1}\right)$$

Following some algebraic manipulation, we have

$$E(Q) = \frac{1}{3} \cdot \frac{1}{b_2 a_2} \left(b_1^2 + b_1 a_1 + a_1^2\right)$$

Therefore,

$$Var(Eff_{SW}) = \frac{1}{3} \cdot \frac{1}{b_2 a_2} \left(b_1^2 + b_1 a_1 + a_1^2\right) - \left[\frac{1}{2} \cdot \frac{a_1 + b_1}{b_2 - a_2} \ln \left(\frac{b_2}{a_2}\right)\right]^2 \tag{5.124}$$

Equations 5.121 and 5.124 are general formulas for the mean and variance of Eff_{SW}, if I and P_r are independent uniformly distributed random variables. Suppose we apply these formulas to Example 5.17; this implies $a_1 = 50,000$, $b_1 = 100,000$, $a_2 = 100$, and $b_2 = 200$. Substituting these values into Equations 5.121 and 5.124 yields

$$E(Eff_{SW}) = 519.86 \text{ staff months}$$

$$Var(Eff_{SW}) = 21,411.8 \text{ (staff months)}^2$$

$$\sigma_{Eff_{SW}} = \sqrt{Var(Eff_{SW})} = 146.328 \text{ staff months}$$

b. In this part, formulas are developed for the mean and variance of Eff_{SW} if I and P_r are *each beta-distributed*. From Equation 4.8 a random variable X is beta-distributed with shape parameters α and β ($\alpha > 0$ and $\beta > 0$) if its PDF is

$$f_X(x \mid \alpha, \beta) = \begin{cases} \dfrac{1}{b-a} \cdot \dfrac{\Gamma(\alpha + \beta)}{\Gamma(\alpha)\Gamma(\beta)} \left(\dfrac{x-a}{b-a}\right)^{\alpha-1} \left(\dfrac{b-x}{b-a}\right)^{\beta-1}, & a < x < b \\ 0, & \text{otherwise} \end{cases}$$

Continuing with this case discussion, let

$$Z = Eff_{SW} = \frac{I}{P_r} \Rightarrow M_Z(s) = M_I(s)M_{P_r}(2 - s)$$

Therefore,

$$E(Z) = E(Eff_{SW}) = M_Z(2) = M_I(2)M_{P_r}(0) \qquad (5.125)$$

The Mellin transform of X is, in this case is,

$$
\begin{aligned}
M_X(s) &= \int_a^b x^{s-1} f_X(x \,|\, \alpha, \beta)\, dx \\
&= \frac{1}{b-a} \cdot \frac{\Gamma(\alpha+\beta)}{\Gamma(\alpha)\Gamma(\beta)} \int_a^b x^{s-1} \left(\frac{x-a}{b-a}\right)^{\alpha-1} \left(\frac{b-x}{b-a}\right)^{\beta-1} dx \\
&= \frac{1}{(b-a)^{\alpha+\beta-1}} \cdot \frac{\Gamma(\alpha+\beta)}{\Gamma(\alpha)\Gamma(\beta)} \int_a^b x^{s-1}(x-a)^{\alpha-1}(b-x)^{\beta-1}\, dx
\end{aligned}
$$

$$(5.126)$$

We are given $I \sim Beta(\alpha_1, \beta_1, a_1, b_1)$. From Theorem 4.4, we know that

$$M_I(2) = E(I) = a_1 + (b_1 - a_1)\frac{\alpha_1}{\alpha_1 + \beta_1} \qquad (5.127)$$

Given $P_r \sim Beta(\alpha_2, \beta_2, a_2, b_2)$, from Equation 5.126

$$
\begin{aligned}
M_{P_r}(0) = \xi &= \frac{1}{(b_2 - a_2)^{\alpha_2+\beta_2-1}} \cdot \frac{\Gamma(\alpha_2+\beta_2)}{\Gamma(\alpha_2)\Gamma(\beta_2)} \\
&\times \int_{a_2}^{b_2} y^{-1}(y-a_2)^{\alpha_2-1}(b_2-y)^{\beta_2-1}\, dy
\end{aligned}
$$

$$(5.128)$$

Substituting Equations 5.127 and 5.128 into Equation 5.125, we have

$$E(Z) = E(Eff_{SW}) = \xi \left(a_1 + (b_1 - a_1)\frac{\alpha_1}{\alpha_1 + \beta_1} \right) \qquad (5.129)$$

As an illustration, consider the case where $I \sim Beta(5, 10, 50(10)^3, 100(10)^3)$ and $P_r \sim Beta(5, 5, 100, 200)$. The expected effort $E(Eff_{SW})$ is

$$E(Eff_{SW}) = \xi \left(50(10)^3 + (100(10)^3 - 50(10)^3)\frac{5}{5 + 10} \right)$$

$$= (66,666.67)\,\xi$$

where, from numerical integration

$$\xi = \frac{1}{(100)^9} \cdot \frac{\Gamma(10)}{\Gamma(5)\Gamma(5)} \int_{100}^{200} y^{-1}(y - 100)^4(200 - y)^4 \, dy = 0.0067358$$

Therefore,

$$E(Eff_{SW}) = (66,666.67)\,(0.0067358) = 449.053 \text{ staff months}$$

A determination of $Var(Eff_{SW})$ completes this discussion. By definition

$$Var(Eff_{SW}) = E(Eff_{SW}^2) - [E(Eff_{SW})]^2$$

From Equation 5.123

$$E(Eff_{SW}^2) = M_I(3)M_{P_r}(-1)$$

where

$$M_I(3) = \frac{1}{(b_1 - a_1)^{\alpha_1 + \beta_1 - 1}} \cdot \frac{\Gamma(\alpha_1 + \beta_1)}{\Gamma(\alpha_1)\Gamma(\beta_1)}$$

$$\times \int_{a_1}^{b_1} t^2(t - a_1)^{\alpha_1 - 1}(b_1 - t)^{\beta_1 - 1} \, dt$$

$$M_{P_r}(-1) = \frac{1}{(b_2 - a_2)^{\alpha_2 + \beta_2 - 1}} \cdot \frac{\Gamma(\alpha_2 + \beta_2)}{\Gamma(\alpha_2)\Gamma(\beta_2)}$$

$$\times \int_{a_2}^{b_2} y^{-2}(y - a_2)^{\alpha_2 - 1}(b_2 - y)^{\beta_2 - 1} \, dy$$

If $I \sim Beta(5, 10, 50(10)^3, 100(10)^3)$ and $P_r \sim Beta(5, 5, 100, 200)$, then a numerical integration of the two integrals yields

$$M_I(3) = 4.47917(10)^9 \quad \text{and} \quad M_{P_r}(-1) = 0.000045852$$

therefore,

$$E(Eff_{SW}^2) = (4.47917(10)^9)(0.000045852)$$
$$= 205,378.9028 \text{ (staff months)}^2$$

so,

$$Var(Eff_{SW}) = 3730.3 \text{ (staff months)}^2$$

$$\sigma_{Eff_{SW}} = \sqrt{Var(Eff_{SW})} = 61.07 \text{ staff months}$$

In summary, the effort mean and standard deviation (rounded) is

$$E(Eff_{SW}) = 449 \text{ staff months}$$

$$\sigma_{Eff_{SW}} = 61 \text{ staff months} \blacklozenge$$

This chapter concludes with series of transformation formulas associated with various algebraic operations on random variables. A cost analysis context is offered for each case.

- Multiplication of a random variable X by a constant a. Denote this operation by $U = aX$. There are many types of cost analysis applications. For example a could represent a labor rate (dollars per staff month) and X could represent an effort (staff months).

  Transformation Formulas

$$E(U) = aE(X) \tag{5.130}$$

$$Var(U) = a^2 Var(X) \tag{5.131}$$

- Addition of a constant a to a random variable X. Denote this operation by $U = a + X$. There are many types of cost analysis applications. For example a could represent a fixed cost, while X could represent a variable cost (whose precise value is uncertain).

  Transformation Formulas

$$E(U) = a + E(X) \tag{5.132}$$

$$Var(U) = Var(X) \tag{5.133}$$

- Sum of two independent uniform random variables X_1 and X_2. Denote this operation by $U = X_1 + X_2$, where $X_1 \sim Unif\,(a_1, b_1)$ and $X_2 \sim Unif\,(a_2, b_2)$. Exercise 5.8 and the discussion pertaining to Figure 5.18 provide a cost analysis application.

  Transformation Formulas

$$U \sim Trap\,((a_1 + a_2), (a_2 + b_1), (a_1 + b_2), (b_1 + b_2)),$$

$$\text{if } b_1 - a_1 < b_2 - a_2 \tag{5.134}$$

$$U \sim Trng\,((a_1 + a_2), m, (b_1 + b_2)), \quad \text{if } b_1 - a_1 = b_2 - a_2 \tag{5.135}$$

where $m = \frac{1}{2}[(a_1 + a_2) + (b_1 + b_2)]$.

- Sum of two independent normal random variables X_1 and X_2. Denote this operation by $U = X_1 + X_2$, where $X_1 \sim N(a_1, b_1)$ and $X_2 \sim N(a_2, b_2)$. Refer to Section 6.2.2 for a cost analysis application.

Transformation Formula

$$U \sim N(a_1 + a_2, \; b_1 + b_2) \tag{5.136}$$

- Sum of n independent normal random variables. Denote this operation by $U = X_1 + X_2 + \cdots + X_n$. The most common cost analysis application is summing cost element costs across a system's work breakdown structure (WBS). In this context, X_i might represent the cost of the ith cost element in the WBS (refer to Section 6.2.2 for a cost analysis application).

Transformation Formulas

In accordance with Theorem 5.10 (Central Limit Theorem), as n becomes increasingly large, the random variable U approaches a normal probability distribution with mean and variance

$$E(U) = E(X_1) + E(X_2) + \cdots + E(X_n) \tag{5.137}$$

$$Var(U) = Var(X_1) + Var(X_2) + \cdots + Var(X_n) \tag{5.138}$$

- Ratio of a uniformly distributed random variable. Denote this operation by $U = 1/X$ where $X \sim Unif(a, b)$. A cost analysis context is provided in Case Discussion 5.4 and in Exercise 5.19.

Transformation Formulas

$$f_U(u) = \frac{1}{b-a} \cdot \frac{1}{u^2}, \quad \frac{1}{b} \le u \le \frac{1}{a} \tag{5.139}$$

$$E\left(\frac{1}{X}\right) = \frac{1}{b-a} \ln\left(\frac{b}{a}\right) \tag{5.140}$$

$$Var\left(\frac{1}{X}\right) = \frac{1}{ba} - \left(\frac{1}{b-a} \ln\left(\frac{b}{a}\right)\right)^2 \tag{5.141}$$

- Product of two independent random variables. Denote this operation by $U = X_1 X_2$. Case Discussion 5.3 provides a cost analysis context.

Transformation Formulas

$$E(U) = \mu_1 \mu_2 \tag{5.142}$$

$$Var(U) = \left(\sigma_1^2 + \mu_1^2\right)\left(\sigma_2^2 + \mu_2^2\right) - (\mu_1 \mu_2)^2 \tag{5.143}$$

where $E(X_1) = \mu_1$, $E(X_2) = \mu_2$, $Var(X_1) = \sigma_1^2$, and $Var(X_2) = \sigma_2^2$.

- Product of n independent random variables. Denote this operation by $U = X_1 X_2 \cdots X_n$. Example 4.9 provides a cost analysis context.

Transformation Formulas
The random variable $U = X_1 X_2 \cdots X_n$ approaches the lognormal distribution as $n \to \infty$.

$$E(U) = \mu_1 \mu_2 \cdots \mu_{n-1} \mu_n \qquad (5.144)$$

$$Var(U) = \prod_{i=1}^{n} (\sigma_i^2 + \mu_i^2) - \prod_{i=1}^{n} \mu_i^2 \qquad (5.145)$$

where $E(X_i) = \mu_i$, $Var(X_i) = \sigma_i^2$.

Exercises

5.1 In Example 5.2, $Eff_{SysTest} = XY$ and X and Y have a joint PDF.

$$f(x,y) = \begin{cases} \dfrac{1}{240}, & 5 \le x \le 15,\ 12 \le y \le 36 \\ 0, & \text{otherwise} \end{cases}$$

a. Sketch the event spaces associated with events A, B, and C where

$$A = \{Eff_{SysTest} \le 240\}$$
$$B = \{Eff_{SysTest} \le 240 \mid X \le 12\}$$
$$C = \{\{Eff_{SysTest} \le 240\} \cap \{X \le 12\} \cap \{Y \le 20\}\}$$

b. From part (a) compute $P(A)$, $P(B)$, and $P(C)$.

5.2 In Example 5.3, $Eff_{SW} = X/Y$ and X and Y have a joint PDF

$$f(x,y) = \begin{cases} \dfrac{1}{5(10^6)}, & 50{,}000 \le x \le 100{,}000,\ 100 \le y \le 200 \\ 0, & \text{otherwise} \end{cases}$$

Find
a. $P(Eff_{SW} \le 313)$
b. $P(Eff_{SW} \le 410 \mid X \le 70{,}000)$
c. $P(\{Eff_{SW} \le 410\} \cap \{X \le 70{,}000\} \cap \{Y \ge 150\})$

5.3 Suppose $f(x, y) = \begin{cases} \dfrac{1}{240} & 5 \le x \le 15,\ 12 \le y \le 36 \\ 0 & \text{otherwise} \end{cases}$

Compute

a. $f_X(x)$ using Equation 5.10

b. $f_Y(y)$ using Equation 5.11

c. $P(X \le 10 \mid Y = 24)$

d. $P(Y > 24 \mid X = 10)$

e. Are X and Y dependent or independent random variables? Justify your answer.

5.4 a. If X and Y are random variables with means μ_X and μ_Y, show that

$$Cov(X, Y) = E(XY) - \mu_X\mu_Y$$

b. If X and Y are random variables, show that $Cov(X, Y) = Cov(Y, X)$. For any real numbers a, b, c, and d show that

$$Cov(aX + b, cY + d) = acCov(X, Y)$$

c. Show that $Cov(X, Y) = 0$ if X and Y are *independent* random variables.

d. Show that $Cov(X, X) = Var(X)$. Given this, show that $Corr(X, X) = 1$.

e. Show that $Var(X + Y) = Var(X) + Var(Y) + 2Cov(X, Y)$.

5.5 Suppose Y, X_1, and X_2 are independent random variables. If $Z = X_1 + X_2$ show that Y and Z are uncorrelated.

5.6 If $Y = X^{100}$ and $X \sim Unif(0, 1)$ show that $\rho_{X,Y} = 0.24$.

5.7 Let the total cost of a system's prime mission equipment (PME) be denoted by $Cost_{PME}$. Let

$$Cost_{PME} = X_1 + X_2$$

where X_1 is the total cost of the system's hardware and X_2 is the total cost of the system's software. Assume X_1 and X_2 are independent random variables. Suppose the cost to integrate and assemble the system's hardware and software is denoted by $Cost_{I\&A}$. If

$$Cost_{I\&A} = \frac{1}{10}X_1 + \frac{1}{5}X_2$$

a. Determine a general formula for $Corr\,(Cost_{PME}, Cost_{I\&A})$.

b. Compute $Corr\,(Cost_{PME}, Cost_{I\&A})$ when $\sigma_{X_1} = \sigma_{X_2}$.

5.8 Let $Cost_{PMP}$ denote the total cost of a system's prime mission product (PMP). Let

$$Cost_{PMP} = Cost_{PME} + Cost_{I\&A}$$

$$Cost_{PME} = X_1 + X_2$$

$$Cost_{I\&A} = \frac{1}{10}X_1 + \frac{1}{5}X_2$$

Let X_1 and X_2 denote the total costs ($M) of the system's hardware and software. Suppose X_1 and X_2 are independent random variables with $X_1 \sim Unif(5, 10)$ and $X_2 \sim Unif(30, 45)$. Compute

a. $E(Cost_{PMP})$

b. $Var(Cost_{PMP})$

c. $F_{Cost_{PME}}(x_1 + x_2)$

d. From part c) determine d such that $P(Cost_{PME} \leq d) = 0.75$.

5.9 Suppose X_1, X_2, X_3 are the cost element costs of an electronic system. Let the system's total cost be given by

$$Cost_{Sys} = X_1 + X_2 + X_3$$

where X_1, X_2, X_3 are given in the table below. Let X_1 and W be independent random variables.

Cost Element Name	Cost Element Cost X_i ($M)
Prime mission product (PMP)	$X_1 \sim N(12.5, 6.6)$
System eng. and prgm mgt (SEPM)	$X_2 = \frac{1}{2}X_1$
System test and evaluation (STE)	$X_3 = \frac{1}{4}X_1 + \frac{1}{8}X_2 + W$, where $W \sim Unif(0.60, 1)$

a. Write a general formula for $E(Cost_{Sys})$ and compute its value.

b. Show that

$$Var(Cost_{Sys}) = \frac{841}{256}Var(X_1) + Var(W)$$

from the expression

$$Var(Cost_{Sys}) = Var(X_1) + Var(X_2) + Var(X_3)$$
$$+ 2[Cov(X_1, X_2) + Cov(X_1, X_3) + Cov(X_2, X_3)]$$

c. Compute $Var(Cost_{Sys})$.

5.10 In Case Discussion 5.1, the K-S test revealed the normal distribution as a plausible model of the underlying distribution function for $Cost_{SSA}$. Use the K-S test on the data in Table 5.4 to show the lognormal distribution is also a plausible model.

5.11 In Example 5.7, X denoted the number of engineering staff required to test a new rocket propulsion system. The number of months Y required to design, conduct, and analyze the test was given by $Y = 2X + 3$. If X is uniformly distributed in the interval $5 \leq x \leq 15$, determine
a. $F_Y(y)$
b. $f_Y(y)$

5.12 In Example 5.10, verify that $F_W(1750) = 0.75 = F_{Hours}(87.67)$.

5.13 Suppose the direct engineering hours to design a new communication satellite is given by $Hours = 4 + 2\sqrt{W}$, where W is the satellite's weight, in lbs. Suppose the uncertainty in the satellite's weight is captured by a triangular distribution; that is, $W \sim Trng(1000, 1500, 2000)$. Suppose the satellite design team assessed 1500 lbs to be the point estimate for weight; that is, $w_{PE} = 1500$.

a. Determine the CDF of *Hours*.

b. Compute $P(Hours \leq h_{PE})$, where $h_{PE} = 4 + 2\sqrt{w_{PE}}$.

c. Determine the PDF of *Hours*.

5.14 Suppose the development effort Eff_{SW} for a software project is defined by $Eff_{SW} = c_1 I^{c_2}$. If $I \sim Trng(a, m, b)$ derive $F_{Eff_{SW}}(s)$, $f_{Eff_{SW}}(s)$, $E(Eff_{SW})$, $Var(Eff_{SW})$.

5.15 Suppose the development schedule for a software project is defined by $T_{SW} = k_1(Eff_{SW})^{k_2}$, where $Eff_{SW} = c_1 I^{c_2}$. Answer the following:

a. If $I \sim Unif(a, b)$ derive $F_{T_{SW}}(t)$, $f_{T_{SW}}(t)$, $E(T_{SW})$, $Var(T_{SW})$.

b. If $I \sim Trng(a, m, b)$ derive $F_{T_{SW}}(t)$, $f_{T_{SW}}(t)$, $E(T_{SW})$, $Var(T_{SW})$.

5.16 In Example 5.13, the effort (staff months) to develop software for a new system was given by $Eff_{SW} = 2.8 I^{1.2}$. The development schedule (months) was given by $T_{SW} = 2.5(Eff_{SW})^{0.32}$. If $I \sim Unif(30, 80)$, use Theorem 5.11 to show the following:

a. $F_{Eff_{SW}}(518) = F_{T_{SW}}(18.5) = 0.95$

b. $F_{T_{SW}}(t) = \dfrac{1}{50}\left[\left(\dfrac{t}{3.48}\right)^{\frac{1}{0.384}} - 30\right]$, $\quad 12.8 \leq t \leq 18.7$

5.17 The uncertainties in the amount of code to develop for the radar system in Example 5.16, was represented by the independent random variables $I_1, I_2, I_3, \ldots, I_{14}$. Let $I_{Total} = I_1 + I_2 + I_3 + \cdots + I_{14}$, where each I is in KDSI. From the information in Table 5.5, use the central limit theorem to determine the 0.25-fractile and the 0.75-fractile of $Eff_{SW} = 2.8(I_{Total})^{1.2}$.

5.18 Refer to Example 5.2 and use Theorem 5.12 to find the general formula for the PDF of $Eff_{SysTest}$.

5.19 a. Let X and Y be independent random variables with $Z = (X + Y)^2$. Show that $E(Z) = M_X(3) + M_Y(3) + 2M_X(2)M_Y(2)$.

     b. Suppose $X \sim Unif(a, b)$. Use Theorem 5.13 and the definition of the Mellin transform to show that

$$E\left(\frac{1}{X}\right) = \frac{1}{b-a}\ln\left(\frac{b}{a}\right) \text{ and } Var\left(\frac{1}{X}\right) = \frac{1}{ba} - \left(\frac{1}{b-a}\ln\left(\frac{b}{a}\right)\right)^2$$

5.20 In Example 5.19, a new software application was being developed that consisted of a mixture of new code I_{New} and reused code I_{Reused}. Suppose I_{New} and I_{Reused} are independent random variables with PDFs given in Example 5.19. If the effort Eff_{SW} associated with developing the application is a function of the equivalent size I_{Equiv}, where

$$I_{Equiv} = I_{New} + I_{Reused}^{0.857}$$

and

$$Eff_{SW} = 2.8\left(\frac{1}{1000}I_{Equiv}\right)^{1.2}$$

use the Mellin transform technique to approximate $E\left(Eff_{SW}\right)$. *Hint: Use the first three terms of the binomial series expansion of* $(I_{Equiv})^{1.2}$*, given by*

$$(I_{Equiv})^{1.2} \approx (I_{New})^{1.2} + 1.2(I_{New})^{0.2}I_{Reused}^{0.857} + 0.12(I_{New})^{-0.8}I_{Reused}^{1.714}$$

References

Boehm, B. W. 1981. *Software Engineering Economics*. Englewood Cliffs, NJ: Prentice-Hall, Inc.

Conte, S. D., H. E. Dunsmore, and V. Y. Shen. 1986. *Software Engineering Metrics and Models*. Menlo Park, CA: The Benjamin/Cummings Publishing Company, Inc.

Cramer, H. 1966. *Mathematical Methods of Statistics*. Princeton, NJ: Princeton University Press.

Epstein, B. 1948. Some Applications of the Mellin Transform in Statistics. *Annals of Mathematical Statistics*, 19, 370–379.

Garvey, P. R. and F. D. Powell. 1989. Three methods for quantifying software development effort uncertainty, in B. W. Boehm (ed.). *Software Risk Management*, pp. 292–306. Washington, DC: IEEE Computer Society Press.

Giffin, W. C. 1975. *Transform Techniques for Probability Modeling*. New York: Academic Press, Inc.

Law, A. M. and W. D. Kelton. 1991. *Simulation Modeling and Analysis*, 2nd edn. New York: McGraw-Hill, Inc.

Lurie, P. M. and M. S. Goldberg. 1993. *A Handbook of Cost Risk Analysis Methods*, P-2734. Alexandria, VA: The Institute for Defense Analyses.

Mood, A. M., F. A. Graybill, and D. C. Boes. 1974. *Introduction to the Theory of Statistics*, 3rd edn. New York: McGraw-Hill, Inc.

Additional Reading

Garvey, P. R. 1990. A general analytic approach to system cost uncertainty analysis, in W. R. Greer, Jr., and D. A. Nussbaum (eds.). *Cost Estimating and Analysis: Tools and Techniques*, pp. 161–181. New York: Springer-Verlag.

Rice, J. A. 1995. *Mathematical Statistics and Data Analysis*, 2nd edn. Belmont, CA: Duxbury Press.

Quirin, W. L. 1978. *Probability and Statistics*. New York: Harper and Row, Publishers, Inc.

6

System Cost Uncertainty Analysis

This chapter illustrates how key concepts developed thus far combine to produce the probability distribution of a system's total cost. Chapter 7 will extend this discussion to the joint and conditional distributions of a system's total cost and schedule. Chapter 6 begins with an introduction to the work breakdown structure, a primary method for organizing a system's total cost.

6.1 Work Breakdown Structures

The work breakdown structure (WBS) is a framework for identifying all elements of cost that relate to the tasks and activities of developing, producing, deploying, sustaining, and disposing a system. Work breakdown structures are unique to the system under consideration. They are developed according to the specific requirements and functions the system has to perform. WBS are defined for classes of systems. These classes include electronic systems, aircraft systems, surface vehicles, ship systems, spacecraft systems, and information technology systems (Blanchard and Fabrycky 1990, United States Department of Defense 2011).

Work breakdown structures are tiered by a hierarchy of cost elements. A typical electronic system WBS is illustrated in Figure 6.1. Shown are four hierarchies, or indenture levels. The first level represents the entire system (e.g., the air traffic control radar system). The second level reflects the major cost elements of the system. In Figure 6.1, these elements include prime mission product (PMP), system engineering, program management, and system test and evaluation. Each level 2 cost element is defined as follows:

- *Prime Mission Product (PMP)*: This element refers to the hardware and software used to accomplish the primary mission of the system. It includes the engineering effort and management activities associated with the system's individual hardware components and software functions, as well as the effort to integrate, assemble, test, and checkout the system's hardware and software.

- *Systems Engineering*: This element encompasses the overall engineering effort to define and deploy the system. It includes integrating the technical efforts of design engineering, specialty engineering (e.g.,

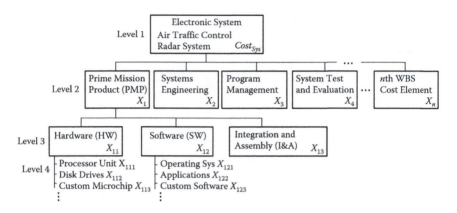

FIGURE 6.1
An illustrative electronic system WBS.

reliability engineering, security engineering), production engineering, and integrated test planning to produce an operational system.

- *Program Management*: This element includes all the efforts associated with the business and administrative management of the system. This includes cost, schedule, and performance measurement, as well as contract administration, data management, and customer/user liaison activities.

- *System Test and Evaluation*: This element includes all test engineering, test planning, and related technical efforts (test mockups, prototypes) to insure the deployed system has been tested against its requirements.

In Figure 6.1, the PMP cost element is divided into its level 3 cost elements. At this level, the radar's hardware, software, and integration cost elements are defined. A further division of PMP into its level 4 cost elements is also shown in Figure 6.1. Here, the individual cost elements of the system's hardware and software are defined. In practice, the number of levels specified in a system's WBS reflects the extent the system itself is defined. In most instances, cost elements are seldom specified below level 6 in a system's WBS.

Certain cost elements in a WBS qualify as *configuration items*. A configuration item is an aggregation of hardware or software that satisfies a particular end-use function of the system. A custom-made microchip or developed software applications are typically designated as configuration items. This designation means the item is subject to *configuration management*. Configuration management is the process of documenting, monitoring, and controlling change to the configuration item's technical baseline. Cost elements placed under configuration management typically begin to appear at level 4 of a WBS.

The WBS is the definitive cost element structure of a system. It is the basis upon which the system's cost is determined (or modeled). From a WBS perspective, a system's total cost (which we will denote by $Cost_{Sys}$) is a summation of cost element costs, summed across the levels of the WBS. In Figure 6.1,

$$Cost_{Sys} = X_1 + X_2 + X_3 + X_4 + \cdots + X_n \qquad (6.1)$$

where the first term in Equation 6.1, X_1, is

$$X_1 = X_{11} + X_{12} + X_{13} + \cdots + X_{1k} \qquad (6.2)$$

and k is the number of level 3 cost elements associated with X_1. Similarly,

$$X_{11} = X_{111} + X_{112} + X_{113} + \cdots + X_{11j} \qquad (6.3)$$

where j is the number of level 4 cost elements associated with X_{11}. The other terms in Equation 6.1, $X_2, X_3, X_4, \ldots, X_n$, are defined in a similar manner. This layered sum of cost element costs is often referred to as the "roll-up" cost. Cost elements of a WBS are specific to the system class. Cost elements* of a satellite system are illustrated in Figure 6.2.

1 Satellite System
   1.1 Launch Vehicle Segment
   1.2 Space Segment
      1.2.1 Satellite Integration, Assembly, and Test
      1.2.2 Spacecraft Bus
         1.2.2.1 Spacecraft Bus Integration, Assembly, and Test
         1.2.2.2 Structures and Mechanical Assembly Subsystem
         1.2.2.3 Attitude Determination and Control Subsystem
         1.2.2.4 Thermal Control Subsystem
         1.2.2.5 Electrical Power Subsystem
         1.2.2.6 Telemetry and Communication Subsystem
         1.2.2.7 Propulsion Subsystem
      1.2.3 Payload
         1.2.3.1 Payload Hardware
         1.2.3.2 Payload Software
   1.3 Command, Control, and Communications Segment
   1.4 Systems Engineering and Program Management
   1.5 System Test and Evaluation
   1.6 Peculiar Support Equipment
   1.7 Common Support Equipment
   1.8 Operations and Support
   1.9 Flight Support Operations
   1.10 Program Office

FIGURE 6.2
Illustrative spacecraft WBS.

---

* Cost element indenture levels are identified by numbering conventions that may or may not incorporate decimals. The convention used is a matter of presentation style.

Note the difference between these cost elements and those of the electronic system WBS, shown in Figure 6.1. In the satellite system, its cost elements are grouped into segments. Within segments, these elements are divided into levels. Levels can reflect subsystems, such as the spacecraft bus (platform) in Figure 6.2. For context, the spacecraft bus elements are defined here.

- *Spacecraft Bus Integration, Assembly, and Test*: This element refers to all efforts associated with the cost of integrating, assembling, and testing the individual subsystems that constitute the spacecraft bus.

- *Structures and Mechanical Assembly Subsystem*: This element (subsystem) refers to the central frame of the spacecraft that provides support and mounting surfaces for all equipment. It includes deployment mechanisms, the solar array boom, experimental booms, antenna supports, and mechanical design equipment.

- *Attitude Determination and Control Subsystem*: This element (subsystem) measures and maintains the orientation of the space vehicle relative to an inertial or external reference. Attitude determination components include inertial measurement devices (e.g., gyroscopes, accelerometers), earth sensors, sun sensors, horizon sensors, and magnetometers. Attitude control adjusts and maintains the space vehicle's attitude and stabilization. Attitude control components include fuel lines, fuel tanks, thrusters, inertia wheels, and any associated electronics.

- *Thermal Control Subsystem*: This element (subsystem) maintains the temperature of the spacecraft and mission payload through heat transfer between space vehicle elements. Thermal control techniques may be passive or active. Passive techniques include special paint, mirrors, and insulation. Active techniques include heat pipes, louvers, and heaters.

- *Electrical Power Subsystem* (EPS): This element (subsystem) generates, converts, regulates, stores, and distributes electrical power between major space vehicle subsystems. Two common types of EPS's are solar and electrochemical. Typical components of the EPS include solar array for power generation, batteries for power storage, as well as wiring harnesses, regulators, switching electronics, converters, and components for power conditioning and distribution.

- *Telemetry and Communication Subsystem*: This element (subsystem) measures the space vehicle's conditions (health and status), processes health and status data and mission data, stores and transmits data to ground receivers, as well as receives, processes, and initiates commands from ground controllers. This subsystem also maintains the track of the space vehicle; typical components include data processors, transmitters, receivers, antennas, decoders, amplifiers, and tape recorders.

- *Propulsion Subsystem*: This element (subsystem), also referred to as apogee kick motor (AKM), provides reaction force for the final maneuver into orbit and for orbit changes. Typical components include solid rocket motor and explosive squibs, nozzle control mechanisms, thrust sensing and shut-down controls, as well as any required cabling, wiring, and plumbing.

As mentioned earlier, a system's WBS is tailored from general work breakdown structures specific to the system's class. The satellite system WBS in Figure 6.2 was tailored from the general WBS for the unmanned space vehicle cost model (USCM) (United States Air Force 1994). The USCM WBS is presented in Figure 6.3.

```
1 Space Vehicle
    1.1 Integration, Assembly, & System Test (IA&T)
    1.2 Spacecraft
        1.2.1 Structure, Interstage/Adapter
        1.2.2 Thermal Control
        1.2.3 Attitude Determination Control System (ADCS)
            1.2.3.1    Attitude Determination
            1.2.3.2    Reaction Control System
        1.2.4 Electrical Power Supply (EPS)
            1.2.4.1    Power Generation
            1.2.4.2    Power Storage
            1.2.4.3    Power Conditioning and Distribution (PCD)
        1.2.5 Telemetry, Tracking, and Command
            1.2.5.1    Transmitter
            1.2.5.2    Receiver/Exciter
            1.2.5.3    Transponder
            1.2.5.4    Digital Electronics (Signal/Data Processor)
            1.2.5.5    Analog Electronics
            1.2.5.6    Antennas
            1.2.5.7    RF Distribution
    1.3 Communications Payload
        1.3.1 Transmitter
        1.3.2 Receiver/Exciter
        1.3.3 Transponder
        1.3.4 Digital Electronics (Signal/Data Processor)
        1.3.5 Analog Electronics
        1.3.6 Antennas
        1.3.7 RF Distribution
    1.4 Program-Level
        1.4.1 Program Management
        1.4.2 Systems Engineering
        1.4.3 Data
2 Aerospace Ground Equipment
3 Launch and Orbital Operations and Support
```

FIGURE 6.3
Unmanned spacecraft WBS. (From United States Air Force, *Unmanned Spacecraft Cost Model (USCM)*, 7th edn., Los Angeles, CA, 1994.)

TABLE 6.1

Illustrative CERs for Spacecraft Cost Elements (Nonrecurring Development Costs)

Cost Element	Input Parameters	CER
Attitude Control-Attitude Determination	$Z_1 = $ Dry Weight (kg)	$c_1 Z_1^{0.46}$
Telemetry, Tracking, and Command	$Z_1 = $ Weight (kg)	$c_2 + c_3 Z_1$
Structure/Thermal	$Z_1 = $ Weight (kg)	$c_4 + c_5 Z_1^{0.66}$
Electrical Power Supply (EPS)	$Z_1 = $ EPS Weight (kg) $Z_2 = $ BOLP (watts)	$c_6 + c_7 (Z_1 Z_2)^{0.97}$
Payload Communication Electronics	$Z_1 = $ Weight (kg)	$c_8 Z_1^{0.70}$

Source: Larson, W.J. and Wertz, J.R. (eds.), *Space Mission Analysis and Design*, 2nd edn., Kluwer Academic Press, Norwell, MA, 1995.

Work breakdown structures can be quite complex. They may involve many segments and levels, as well as numerous cost elements. Because the WBS is the basis for deriving a system's cost, it also contain a variety of mathematical relationships. These relationships are traditionally known as cost estimating relationships (CERs).* Their primary purpose is to generate point estimate costs of various WBS cost elements. Table 6.1 illustrates some spacecraft-related CERs.

In summary, a WBS provides the framework for developing a system's cost. It further serves as the framework for an analysis of the system's cost uncertainty. The complexity of these analyses is dictated by the complexity of the WBS and its associated CERs.

The following illustrates how probability methods are applied to the problem of quantifying a system's cost uncertainty within the framework of the WBS. Case discussions are presented that link theory to practice.

6.2 Analytical Framework

This section focuses on the application of probability methods for quantifying the uncertainty in a system's cost. The WBS will provide the analytical framework for quantifying this uncertainty, which is expressed as a probability distribution. Analytical methods from probability theory are stressed.

---

* Most CERs are statistically derived from data on cost and technical characteristics. This book uses the term CER to include those that are logically based, as well as those developed by statistical methods.

Analytical methods provide insight into problem structure and subtleties not always apparent from empirically based methods, such as Monte Carlo simulations.*

6.2.1 Computing the System Cost Mean and Variance

From Equation 6.1, we see that system cost, denoted by $Cost_{Sys}$, is a summation of work breakdown structure cost element costs, where each X_i ($i = 1, \ldots, n$) is in dollars. Illustrated in Figure 6.4, we define $Cost_{Sys}$ as

$$Cost_{Sys} = X_1 + X_2 + X_3 + \cdots + X_n \tag{6.4}$$

Inputs: Probability distributions for each cost element cost in a system's work breakdown structure

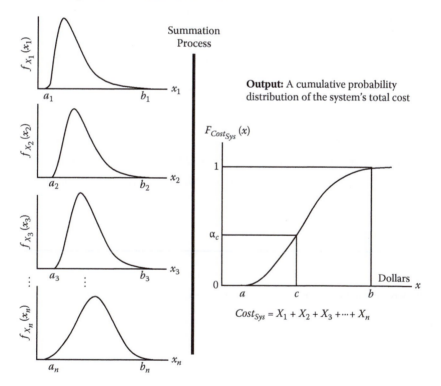

FIGURE 6.4
Cumulative probability distribution of $Cost_{Sys}$.

---

* Monte Carlo simulation is an empirical method often used for quantifying cost uncertainty. The concept underlying this method is discussed in Section 6.3.

If $X_1, X_2, X_3, \ldots, X_n$ are independent, then from Theorems 5.7 and 5.8

$$E(Cost_{Sys}) = E(X_1) + E(X_2) + E(X_3) + \cdots + E(X_n) \tag{6.5}$$

$$Var(Cost_{Sys}) = Var(X_1) + Var(X_2) + Var(X_3) + \cdots + Var(X_n) \tag{6.6}$$

If $X_1, X_2, X_3, \ldots, X_n$ are *not independent*, then

$$Var(Cost_{Sys}) = \sum_{i=1}^{n} Var(X_i) + 2 \sum_{i=1}^{n-1} \sum_{j=i+1}^{n} \rho_{X_i, X_j} \sigma_{X_i} \sigma_{X_j} \tag{6.7}$$

Equations 6.5 through 6.7 are the formal expressions for the mean and variance of $Cost_{Sys}$. The following case discussions illustrate how these expressions are used.

Case Discussion 6.1: [5] Suppose the cost element costs $X_1, X_2, X_3, \ldots, X_{10}$ of an electronic system are given by the WBS in Table 6.2. Let

$$Cost_{Sys} = X_1 + X_2 + X_3 + \cdots + X_{10}$$

Suppose the random variables $X_1, W, X_5, X_7, X_8, X_9$ (defined in Table 6.2) are independent.

  a. Compute $E(Cost_{Sys})$ and $Var(Cost_{Sys})$.
  b. What distribution function approximates the distribution of $Cost_{Sys}$?
  c. Find the value of $Cost_{Sys}$ that has a 5% chance of being exceeded.

TABLE 6.2

WBS for Case Discussion 6.1

Cost Element Name	Cost Element Cost X_i ($M)	Distribution of X_i or the Applicable Functional Relationship
Prime Mission Product (PMP)	X_1	$N(12.5, 6.6)$
System Engineering and Program Management (SEPM)	X_2	$X_2 = \frac{1}{2}X_1$
System Test and Evaluation (STE)	X_3	$X_3 = \frac{1}{4}X_1 + \frac{1}{8}X_2 + W$, where $W \sim Unif(0.6, 1.0)$
Data and Technical Orders	X_4	$X_4 = \frac{1}{10}X_1$
Site Survey and Activation	X_5	$Trng(5.1, 6.6, 12.1)$
Initial Spares	X_6	$X_6 = \frac{1}{10}X_1$
System Warranty	X_7	$Unif(0.9, 1.3)$
Early Prototype Phase	X_8	$Trng(1.0, 1.5, 2.4)$
Operations Support	X_9	$Trng(0.9, 1.2, 1.6)$
System Training	X_{10}	$X_{10} = \frac{1}{4}X_1$

a. It is given that

$$Cost_{Sys} = X_1 + X_2 + X_3 + \cdots + X_{10} \tag{6.8}$$

Using the relationships in Table 6.2, Equation 6.8 becomes

$$Cost_{Sys} = X_1 + \frac{1}{2}X_1 + \left(\frac{1}{4}X_1 + \frac{1}{8}X_2 + W\right) + \frac{1}{10}X_1 + X_5 + \frac{1}{10}X_1$$
$$+ X_7 + X_8 + X_9 + \frac{1}{4}X_1$$

Combining these terms yields

$$Cost_{Sys} = \frac{181}{80}X_1 + W + X_5 + X_7 + X_8 + X_9 \tag{6.9}$$

From Theorem 5.7 (and Equation 6.5)

$$E(Cost_{Sys}) = \frac{181}{80}E(X_1) + E(W) + E(X_5) + E(X_7) + E(X_8) + E(X_9) \tag{6.10}$$

From Theorem 5.8 (and Equation 6.6)

$$Var(Cost_{Sys}) = \left(\frac{181}{80}\right)^2 Var(X_1) + Var(W) + Var(X_5)$$
$$+ Var(X_7) + Var(X_8) + Var(X_9) \tag{6.11}$$

since X_1, W, X_5, X_7, X_8, and X_9 are independent random variables. To compute the mean and variance of $Cost_{Sys}$ we need the means and variances of X_1, W, X_5, X_7, X_8, and X_9. Table 6.3 presents these statistics.

TABLE 6.3

Cost Statistics for X_1, W, X_5, X_7, X_8, and X_9

Cost Element Cost X_i ($M)	$E(X_i)$ ($M)	$Var(X_i)$ ($M)^2
X_1	12.500	6.6
W	0.800	0.16/12
X_5	7.933	40.75/18
X_7	1.100	0.16/12
X_8	1.633	1.51/18
X_9	1.233	0.37/18

The statistics in Table 6.3 were determined by distribution-specific formulas given in Chapter 4. For instance, since $X_1 \sim N(12.5, 6.6)$, we know from Theorem 4.6 that $E(X_1) = 12.5$ and $Var(X_1) = 6.6$. Since W is a uniform distribution, from Theorem 4.2

$$E(W) = \frac{0.6 + 1}{2} = 0.8 \text{ and } Var(W) = \frac{(1 - 0.6)^2}{12} = \frac{0.16}{12} = 0.01\overline{333}$$

Since X_5 is a triangular distribution, from Theorem 4.3

$$E(X_5) = \frac{1}{3}(5.1 + 6.6 + 12.1) = 7.933$$

$$Var(X_5) = \frac{1}{18}\left[(6.6 - 5.1)(6.6 - 12.1) + (12.1 - 5.1)^2\right] = 40.75/18$$

Substituting the data in Table 6.3 into Equations 6.10 and 6.11 we obtain

$$E(Cost_{Sys}) = 40.98(\$M) \tag{6.12}$$

$$Var(Cost_{Sys}) = 36.18(\$M)^2 \tag{6.13}$$

b. To approximate the distribution function of $Cost_{Sys}$, observe the following. First, the random variables X_1, W, X_5, X_7, X_8, and X_9 are independent. Hence, the central limit theorem will *affect* the shape of the distribution of $Cost_{Sys}$. Second, the random variables X_2, X_3, X_4, X_6, and X_{10} are highly correlated to X_1, which is given in Table 6.2 to be $N(12.5, 6.6)$. It can be shown that

$$\rho_{X_v, X_1} = 1 (v = 2, 4, 6, 10) \text{ and } \rho_{X_3, X_1} = 0.9898$$

Thus, it is reasonable to conclude (for this case) the distribution function for $Cost_{Sys}$ is approximately normal—with mean and variance given by Equations 6.12 and 6.13, respectively. The cumulative distribution function (CDF) for $Cost_{Sys}$, assumed to be approximately normal, is shown in Figure 6.5.

c. In Figure 6.5, note that $P(Cost_{Sys} \leq 50.87) = 0.95$. This means a value of 50.87 (\$M) for $Cost_{Sys}$ has only a 5% chance of being exceeded. To arrive at this value, it is necessary to find x such that $P(Cost_{Sys} \leq x) = 0.95$. From Equation 4.24

$$P\left(\frac{Cost_{Sys} - E(Cost_{Sys})}{\sigma} \leq \frac{x - E(Cost_{Sys})}{\sigma}\right) = 0.95$$

$$= P\left(Z \leq \frac{x - 40.98}{6.015}\right) = 0.95 \tag{6.14}$$

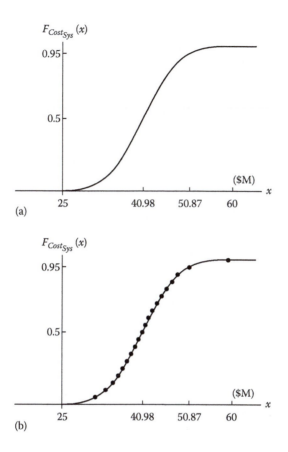

FIGURE 6.5
(a) Assumed normal CDF for $Cost_{Sys}$ (defined by the solid line) vs. (b) the simulated CDF (defined by the points).

Since $Cost_{Sys} \sim N(40.98, 36.18)$, from Table A.1

$$\frac{x - 40.98}{6.015} = 1.645$$

and $x = 50.87$. Thus, a value of 50.87 ($M) for $Cost_{Sys}$ has only a 5% chance of *being exceeded*. Equivalently, 50.87 ($M) is the 0.95-fractile (i.e., $x_{0.95} = 50.87$) of $Cost_{Sys}$. Furthermore, we can say the cost reserve (refer to Chapter 1) needed for a 95% chance of *not exceeding* 50.87 ($M) is 9.9 ($M) above the expected cost of the system.

Further Considerations: In Case Discussion 6.1, it was assumed the distribution function for $Cost_{Sys}$ could be approximated by a normal distribution.

How reasonable is this assumption? A series of 20 "points" is shown in Figure 6.5. These points reflect random statistical samples (values) of $Cost_{Sys}$, sampled by Monte Carlo simulation (explained in Section 6.3). The curve implied by these "points" represents the simulated distribution function* for $Cost_{Sys}$. The curve given by the solid line in Figure 6.5 is the assumed normal distribution for $Cost_{Sys}$. With this in mind, observe in Figure 6.5 how closely the simulated distribution for $Cost_{Sys}$ matches the assumed normal distribution.

The closeness with which these "points" fall along the curve given by the solid line, in Figure 6.5, *visually suggests* the reasonableness of the assumption that the distribution function for $Cost_{Sys}$ can be *approximated* by a normal. Although this is a practical conclusion, it remains an informal one. A more formal conclusion could be derived from the Kolmogorov–Smirnov (K-S) test, illustrated in Case Discussion 5.1. This would reveal whether the normal distribution is a *statistically plausible model* for the underlying distribution function of $Cost_{Sys}$, in this case.

In Case Discussion 6.1, a significant amount of correlation exists between certain pairs of cost element costs. In Table 6.2, the five cost element costs $X_2, X_3, X_4, X_6,$ and X_{10} were functionally related to X_1, the system's PMP cost. In particular, $X_2, X_4, X_6,$ and X_{10} are linearly related to X_1 by the expression

$$X_v = a_v X_1 \tag{6.15}$$

where $v = 2, 4, 6, 10, a_2 = 1/2, a_4 = a_6 = 1/10,$ and $a_{10} = 1/4$. In Table 6.2, cost element cost X_3 was a linear combination of $X_1, X_2,$ and W; specifically,

$$X_3 = \frac{1}{4}X_1 + \frac{1}{8}X_2 + W \tag{6.16}$$

where X_1 and W were given to be independent random variables and $X_2 = \frac{1}{2}X_1$. The functional relationships given by Equations 6.15 and 6.16 imply the following correlations.

$\rho_{X_v, X_1} = 1$ for $v = 2, 4, 6, 10$

$\rho_{X_1, W} = 0$ from Theorem 5.3

$\rho_{X_2, W} = \rho_{\frac{1}{2}X_1, W} = \rho_{X_1, W} = 0$ from Theorems 5.6 and 5.3

$\rho_{X_3, X_1} = \rho_{\frac{5}{16}X_1 + W, X_1} = 0.9898$ from Theorem 6.1

$\rho_{X_3, X_2} = \rho_{X_3, \frac{1}{2}X_1} = \rho_{X_3, X_1} = 0.9898$ from Theorem 5.6

$\rho_{X_3, W} = \rho_{\frac{5}{16}X_1 + W, W} = 0.1424$ from Theorem 6.1

---

* The simulated distribution is an empirically developed distribution. In establishing this distribution, no assumption is made that the distribution function for $Cost_{Sys}$ is normal.

Theorem 6.1 *If* $Y = aX + Z$ *where a is a real number and X and Z are independent random variables, then*

$$\rho_{Y,X} = a \frac{\sigma_X}{\sigma_Y} \text{ and } \rho_{Y,Z} = \frac{\sigma_Z}{\sigma_Y}$$

A proof of this theorem can be developed from Equation 5.29.

The existence of these correlations is hard to notice when $Cost_{Sys}$ is expressed in the form

$$Cost_{Sys} = \frac{181}{80} X_1 + W + X_5 + X_7 + X_8 + X_9$$

In this equation, $Cost_{Sys}$ is now written as the sum of six *independent* random variables instead of the sum of ten random variables (Equation 6.8). Capturing the combined effect of these correlations on the distribution function of $Cost_{Sys}$ is accounted for by the coefficient 181/80. This case discussion illustrates how correlation can exist in a WBS, by virtue of the functional relationships defined among the cost element costs. Functional relationships such as those in this WBS (Table 6.2) are *very* common in cost analysis. Although these relationships are primarily defined for developing the point estimate of $Cost_{Sys}$, such relationships come along with implied correlations. Cost analysts *must* be aware of this implication so as not to inadvertently *induce* correlation (or consider it absent) when it is already present. This concludes Case Discussion 6.1.

Many cost elements in Case Discussion 6.1 were a function of a single random variable. Thus, computing $E(Cost_{Sys})$ and $Var(Cost_{Sys})$ was "relatively" straightforward. More complex relationships are given in Case Discussion 6.2, which illustrates the computation of $E(Cost_{Sys})$ and $Var(Cost_{Sys})$ when cost elements are functions of two or more random variables. In addition, it will be seen how a program's schedule can be incorporated into cost estimating relationships. Case Discussion 6.2 also lays the groundwork for studying cost-schedule probability trade-offs and will be revisited in Chapter 7.

Case Discussion 6.2: Suppose the government is acquiring a new digital information system. The system consists of 3 large screen displays for "situation rooms," 47 display workstations, 2 support processors, and a suite of electronic communications equipment. Suppose the system requires new software to be developed for the large screen display and the display workstation. The WBS for this system is given in Figure 6.6. Cost element data for this WBS are provided in Table 6.4. Additional information about these data follows.

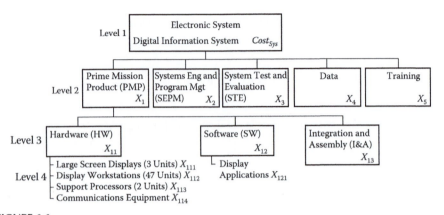

FIGURE 6.6

Case Discussion 6.2 work breakdown structure.

TABLE 6.4

Cost Element Data for Case Discussion 6.2

WBS Cost Element Cost ($K)	Functional Relationship (If Applicable)	Distribution (If Applicable)	Distributions of Random Variables (If Applicable)
X_{111}		$Unif(700, 750)$	
X_{112}		$Unif(3200, 4000)$	
X_{113}		$Unif(200, 250)$	
X_{114}		$Unif(350, 380)$	
X_{121}	$\ell_{r_{SW}}\left(2.8I^{1.2}\right)$		$\ell_{r_{SW}} \sim Unif(10, 15)$ $I \sim Trng(80, 100, 150)$
X_{13}	$0.05\left(X_{121} + \sum_{s=1}^{4} X_{11s}\right)$		
X_2	$\left(\ell_{r_{SEPM}}\right) \cdot$ $\left(SL_{SEPM}\right)\left(PrgmSched\right)$		$\ell_{r_{SEPM}} \sim Unif(20, 25)$ $SL_{SEPM} \sim Trng(12, 15, 25)$ $PrgmSched \sim N(33.36, 1.94)$
X_3	$\left(\ell_{r_{STE}}\right) \cdot$ $\left(SL_{STE}\right)\left(PrgmSched\right)$		$\ell_{r_{STE}} \sim Unif(15, 20)$ $SL_{STE} \sim Unif(4, 7)$ $PrgmSched \sim N(33.36, 1.94)$
X_4	$0.05\left(X_{13} + X_{121} + \sum_{s=1}^{4} X_{11s}\right)$		
X_5	$0.02\left(X_{13} + X_{121} + \sum_{s=1}^{4} X_{11s}\right)$		

In Figure 6.6, the total cost of the digital information system is

$$Cost_{Sys} = X_1 + X_2 + X_3 + X_4 + X_5 \tag{6.17}$$

Furthermore, assume in Table 6.4 that X_{111}, X_{112}, X_{113}, X_{114}, $\ell_{r_{SW}}$, I, $\ell_{r_{SEPM}}$, SL_{SEPM}, $PrgmSched$, $\ell_{r_{STE}}$, and SL_{STE} are *independent random variables*. In Table 6.4, we have the following random variable definitions.

- ℓ_{rSW}, ℓ_{rSEPM}, and ℓ_{rSTE} are labor rates for software (SW) development, systems engineering and program management (SEPM), and system test and evaluation (STE), respectively; the units are in ($K) per staff month.
- I denotes the number of delivered source instructions (DSI) to be developed. The units are in thousands (K); that is, I is expressed in terms of KDSI (as discussed in Chapter 5).
- SL_{SEPM} and SL_{STE} represents staff-levels (i.e., the number of persons) for the SEPM and STE activities, respectively.
- *PrgmSched* denotes the total number of months to complete the development of the digital information system.

From the information given in this case discussion, find the following:

a. Determine $E(Cost_{Sys})$ and $Var(Cost_{Sys})$.
b. Discuss correlations implied by the relationships in Table 6.4.
c. Identify distribution function(s) that approximate $F_{Cost_{Sys}}(x)$.

Preliminaries: Before beginning part (a), a simplified expression for $Cost_{Sys}$ will be developed. Recall from Equation 6.17, the system's total cost is given by

$$Cost_{Sys} = X_1 + X_2 + X_3 + X_4 + X_5$$

This can be written as

$$Cost_{Sys} = Cost_{PMP} + X_2 + X_3 + X_4 + X_5 \tag{6.18}$$

where

$$Cost_{PMP} = X_1 = X_{11} + X_{12} + X_{13} \tag{6.19}$$

From Figure 6.6 and Equation 6.19

$$Cost_{PMP} = X_1 = X_{11} + X_{12} + X_{13}$$
$$= (X_{111} + X_{112} + X_{113} + X_{114}) + (X_{121}) + X_{13}$$

From Table 6.4

$$X_{13} = 0.05 \left(X_{121} + \sum_{s=1}^{4} X_{11s} \right)$$

Combining these relationships

$$Cost_{PMP} = 1.05 \, (X_{111} + X_{112} + X_{113} + X_{114} + X_{121}) \qquad (6.20)$$

$$Cost_{PMP} = 1.05 \, (X_{11} + X_{12}) \qquad (6.21)$$

In electronic systems, the sum $(X_{11} + X_{12})$ is known as the *prime mission equipment* (PME) cost, that is, the total cost of just the system's hardware and software. Thus, Equation 6.21 is equivalent to

$$Cost_{PMP} = 1.05 \, Cost_{PME} \qquad (6.22)$$

Equation 6.22 will be used later in this case discussion. Returning to Equation 6.18, we have

$$Cost_{Sys} = Cost_{PMP} + (X_2 + X_3) + (X_4 + X_5) \qquad (6.23)$$

From Table 6.4, X_4 and X_5 can be written as

$$X_4 = 0.05 \left(X_{13} + X_{121} + \sum_{s=1}^{4} X_{11s} \right) = 0.05 X_1 = 0.05 \, Cost_{PMP}$$

$$X_5 = 0.02 \left(X_{13} + X_{121} + \sum_{s=1}^{4} X_{11s} \right) = 0.02 X_1 = 0.02 \, Cost_{PMP}$$

This simplifies $Cost_{Sys}$ (Equation 6.23) to

$$Cost_{Sys} = 1.07 \, Cost_{PMP} + (X_2 + X_3)$$
$$Cost_{Sys} = 1.07 \, Cost_{PMP} + Q \qquad (6.24)$$

where $Q = (X_2 + X_3)$. We will now work with Equation 6.24 to determine the mean and variance of $Cost_{Sys}$.

Part A. Mean and Variance: From Theorem 5.7, $E(Cost_{Sys})$ is

$$E(Cost_{Sys}) = 1.07 \, E(Cost_{PMP}) + E(Q) \qquad (6.25)$$

It can be shown, in this case, that $Cov(Cost_{PMP}, Q) = 0$. From Theorem 5.8, $Var(Cost_{Sys})$ is

$$Var(Cost_{Sys}) = (1.07)^2 \, Var(Cost_{PMP}) + Var(Q) \qquad (6.26)$$

To compute $E(Cost_{Sys})$ and $Var(Cost_{Sys})$, it is necessary to determine the means and variances of $Cost_{PMP}$ and Q. Since these computations are lengthy, part (a) is separated into three sections. They are defined as follows: (1) computing the mean and variance of $Cost_{PMP}$, (2) computing the mean and variance of Q, and (3) Combining (1) and (2) to determine the mean and variance of $Cost_{Sys}$.

Mean and Variance of $Cost_{PMP}$: To compute $E(Cost_{PMP})$ and $Var(Cost_{PMP})$, recall from Equation 6.22

$$Cost_{PMP} = 1.05\, Cost_{PME} \tag{6.27}$$

where

$$Cost_{PME} = X_{11} + X_{12} = (X_{111} + X_{112} + X_{113} + X_{114}) + X_{121} \tag{6.28}$$

From Equations 6.27 and 6.28

$$E(Cost_{PMP}) = 1.05\, E(Cost_{PME})$$
$$= 1.05\, E((X_{111} + X_{112} + X_{113} + X_{114}) + X_{121})$$
$$= 1.05\, [E(X_{111}) + E(X_{112}) + E(X_{113}) + E(X_{114}) + E(X_{121})] \tag{6.29}$$

Since it is assumed (refer to Figure 6.6), in this case discussion, X_{111}, X_{112}, X_{113}, X_{114}, and X_{121} are independent random variables, we can write

$$Var\,(Cost_{PMP}) = 1.05^2\, Var(Cost_{PME})$$
$$= 1.05^2\, Var((X_{111} + X_{112} + X_{113} + X_{114}) + X_{121})$$
$$= 1.05^2\, [Var(X_{111}) + Var(X_{112}) + Var(X_{113})$$
$$+ Var(X_{114}) + Var(X_{121})] \tag{6.30}$$

From Table 6.4, $X_{111} \sim Unif(700, 750)$; therefore, from Theorem 4.2

$$E(X_{111}) = \frac{700 + 750}{2} = 725 \text{ and } Var(X_{111}) = \frac{(750 - 700)^2}{12} = 208.333$$

Similarly, for X_{112}, X_{113}, and X_{114}

$$E(X_{112}) = \frac{3200 + 4000}{2} = 3600 \text{ and } Var(X_{112}) = \frac{(4000 - 3200)^2}{12} = 53,333.333$$

$$E(X_{113}) = \frac{200 + 250}{2} = 225 \text{ and } Var(X_{113}) = \frac{(250 - 200)^2}{12} = 208.333$$

$$E(X_{114}) = \frac{350 + 380}{2} = 365 \text{ and } Var(X_{114}) = \frac{(380 - 350)^2}{12} = 75$$

To complete the calculation of $E(Cost_{PMP})$ and $Var(Cost_{PMP})$, it is necessary to compute mean and variance of X_{121} (the cost to develop the display software). Two methods from Chapter 5 will show ways this can be done.

Method 1—Transformation of Variables Approach: In this method, transformation formulas developed in Chapter 5 are used. From Table 6.4, the software cost, denoted by X_{121}, is

$$X_{121} = \ell_{rSW} \left(2.8 I^{1.2} \right) \tag{6.31}$$

It was given the random variables ℓ_{rSW} and I are independent. From Theorem 5.5

$$E(X_{121}) = E(\ell_{rSW}) E \left(2.8 I^{1.2} \right)$$

Since $\ell_{rSW} \sim Unif(10, 15)$, from Theorem 4.2 $E(\ell_{rSW}) = 12.5$. Therefore,

$$E(X_{121}) = 12.5 \left[E \left(2.8 I^{1.2} \right) \right] \tag{6.32}$$

Recall if $Eff_{SW} = c_1 I^{c_2}$, and $I \sim Trng(a, m, b)$, then from Equation 5.76

$$E(Eff_{SW}) = c_1 \frac{2}{b - a} \cdot \frac{1}{m - a} \left[\frac{m^{c_2+2} - a^{c_2+2}}{c_2 + 2} + \frac{a^{c_2+2} - am^{c_2+1}}{c_2 + 1} \right]$$

$$+ c_1 \frac{2}{b - a} \cdot \frac{1}{m - b} \left[\frac{b^{c_2+2} - m^{c_2+2}}{c_2 + 2} + \frac{bm^{c_2+1} - b^{c_2+2}}{c_2 + 1} \right] \tag{6.33}$$

Relating Equation 6.33 to this case, $c_1 = 2.8$, $c_2 = 1.2$, $a = 80$, $m = 100$, and $b = 150$. Substituting these values into Equation 6.33 yields $E(2.8 I^{1.2}) = 790.23$. Therefore,

$$\boxed{E(X_{121}) = 12.5[790.23] = 9877.875} \tag{6.34}$$

We next compute $Var(X_{121})$. From Theorem 3.10 and from Equation 6.34

$$Var(X_{121}) = E\left(X_{121}^2\right) - [E(X_{121})]^2 = E\left(X_{121}^2\right) - [9877.875]^2 \quad (6.35)$$

To determine $Var(X_{121})$, it remains to determine $E\left(X_{121}^2\right)$ in Equation 6.35. Now,

$$E\left(X_{121}^2\right) = E\left(\ell_{rsw}^2 \cdot \left(2.8I^{1.2}\right)^2\right) = E\left(\ell_{rsw}^2 \left(7.84I^{2.4}\right)\right) \quad (6.36)$$

Since the random variables ℓ_{rsw} and I are independent

$$E\left(X_{121}^2\right) = E\left(\ell_{rsw}^2\right) E\left(7.84I^{2.4}\right) \quad (6.37)$$

We will take the following approach to compute $E\left(\ell_{rsw}^2\right)$. Since

$$Var\left(\ell_{rsw}\right) = E\left(\ell_{rsw}^2\right) - [E\left(\ell_{rsw}\right)]^2$$

it follows that

$$E\left(\ell_{rsw}^2\right) = Var\left(\ell_{rsw}\right) + [E\left(\ell_{rsw}\right)]^2 \quad (6.38)$$

Since $\ell_{rsw} \sim Unif(10, 15)$, from Theorem 4.2

$$E(\ell_{rsw}) = 12.5 \text{ and } Var\left(\ell_{rsw}\right) = \frac{(15 - 10)^2}{12} = \frac{25}{12}$$

Substituting these values into Equation 6.38 yields $E\left(\ell_{rsw}^2\right) = 158\frac{1}{3}$. Therefore, Equation 6.37 becomes

$$E\left(X_{121}^2\right) = 158\frac{1}{3} E\left(7.84I^{2.4}\right) \quad (6.39)$$

To compute $E\left(7.84I^{2.4}\right)$, Equation 6.33 will be used again with $c_1 = 7.84$, $c_2 = 2.4$, and $a = 80$, $m = 100$, and $b = 150$. Substituting these values into Equation 6.33 yields $E\left(7.84I^{2.4}\right) = 640,626.866$. Therefore,

$$E\left(X_{121}^2\right) = 158\frac{1}{3}(640,626.866) = 101,432,587.1 \quad (6.40)$$

Hence, Equation 6.35 becomes

$$\boxed{Var(X_{121}) = 101,432,587.1 - [9877.875]^2 = 3,860,172.585}$$ (6.41)

and

$$\sigma_{X_{121}} = \sqrt{Var(X_{121})} = 1964.732$$

Method 2—Mellin Transform Approach: In this method, the Mellin transform (refer to Section 5.5) is used to illustrate an alternative approach to computing $E(X_{121})$ and $Var(X_{121})$. Recall that

$$X_{121} = \ell_{r_{SW}}\left(2.8I^{1.2}\right)$$ (6.42)

From Theorem 5.13, the Mellin transform of X_{121} is

$$M_{X_{121}}(s) = M_{\ell_{r_{SW}}}(s)(2.8)^{s-1}M_I(1.2s - 1.2 + 1)$$ (6.43)

From Equation 5.98

$$E(X_{121}) = M_{X_{121}}(2) = M_{\ell_{r_{SW}}}(2)(2.8)^{2-1}M_I(1.2(2) - 1.2 + 1)$$
$$E(X_{121}) = 2.8M_{\ell_{r_{SW}}}(2)M_I(2.2)$$ (6.44)

Since $\ell_{r_{SW}} \sim Unif(10, 15)$, from Equation 5.108

$$M_{\ell_{r_{SW}}}(2) = \frac{1}{2} \cdot \frac{1}{(15-10)}\left(15^2 - 10^2\right) = 12.5$$

Since $I \sim Trng(80, 100, 150)$, from Equation 5.109, with $s = 2.2$, $a = 80$, $m = 100$, and $b = 150$ we have

$$M_I(2.2) = 282.225$$

therefore

$$\boxed{E(X_{121}) = 2.8(12.5)(282.225) = 9877.875}$$ (6.45)

To compute $Var(X_{121})$, we have

$$Var(X_{121}) = E\left(X_{121}^2\right) - [E(X_{121})]^2 = E\left(X_{121}^2\right) - [9877.875]^2$$ (6.46)

From Equation 5.99

$$E\left(X_{121}^2\right) = M_{X_{121}}(3)$$

$$= M_{\ell_{r_{SW}}}(3)\,(2.8)^{3-1}M_I(1.2(3) - 1.2 + 1)$$

$$= (2.8)^2 M_{\ell_{r_{SW}}}(3)\,M_I(3.4) \tag{6.47}$$

where

$$M_{\ell_{r_{SW}}}(3) = \frac{1}{3} \cdot \frac{1}{(15-10)}\left(15^3 - 10^3\right) = 158\frac{1}{3}$$

and

$$M_I(3.4) = 81,712.61045$$

Substituting these values into Equation 6.47 yields $E\left(X_{121}^2\right) = 101,432,587.1$. Therefore

$$\boxed{Var(X_{121}) = 101,432,587.1 - [9877.875]^2 = 3,860,172.585} \tag{6.48}$$

and

$$\sigma_{X_{121}} = \sqrt{Var\,(X_{121})} = 1964.732\blacklozenge$$

All the information needed to complete the computation of $E\,(Cost_{PMP})$ and $Var\,(Cost_{PMP})$ is now available. From Equation 6.29, recall that

$$E(Cost_{PMP}) = 1.05\,E\,(Cost_{PME})$$

$$= 1.05\,[E(X_{111}) + E(X_{112}) + E(X_{113}) + E(X_{114}) + E(X_{121})] \tag{6.49}$$

Substituting the expected value computations developed in the above discussions into Equation 6.49 yields

$$E\,(Cost_{PMP}) = 1.05\,[725 + 3600 + 225 + 365 + 9877.875]$$

$$= 15,532.52\ (\$K) \tag{6.50}$$

From Equation 6.30

$$Var\,(Cost_{PMP}) = 1.05^2\,Var\,(Cost_{PME})$$

$$= 1.05^2\left[\begin{matrix} Var(X_{111}) + Var(X_{112}) \\ +Var(X_{113}) + Var(X_{114}) + Var(X_{121}) \end{matrix}\right] \tag{6.51}$$

Substituting the variance computations developed in these discussions into Equation 6.51 yields

$$Var\,(Cost_{PMP}) = 1.05^2 \begin{bmatrix} 208.333 + 53,333.333 \\ +208.333 + 75 + 3,860,172.585 \end{bmatrix}$$

$$= 4,315,182.336 \quad (\$K)^2 \tag{6.52}$$

and

$$\sigma_{Cost_{PMP}} = \sqrt{Var\,(Cost_{PMP})} = 2077.3\ (\$K)$$

Mean and Variance of Q: The previous discussion presented the mean and variance of the system's prime mission product cost. To complete the computation of $E(Cost_{Sys})$ and $Var(Cost_{Sys})$, defined by Equations 6.25 and 6.26, the values of $E(Q)$ and $Var(Q)$, where $Q = X_2 + X_3$, must be determined.

From Table 6.4, observe that X_2 and X_3 are *not independent* random variables. They are both a function of the random variable *PrgmSched*. From Theorem 5.7, $E(Q)$ is the sum of the means of X_2 and X_3 regardless of whether or not the two random variables are independent. Hence,

$$E(Q) = E(X_2 + X_3) = E(X_2) + E(X_3) \tag{6.53}$$

However, because X_2 and X_3 are not independent, $Var(Q)$ is *not* just the sum of their respective variances. Applying Theorem 5.8 to this particular case,

$$Var(Q) = Var(X_2) + Var(X_3) + 2\rho_{X_2,X_3}\,\sigma_{X_2}\,\sigma_{X_3} \tag{6.54}$$

The following presents the computations for the means and variances of X_2 and X_3, as well as ρ_{X_2,X_3}, their correlation coefficient.

Mean and Variance of X_2: From the WBS in Figure 6.6, recall that the cost of SEPM is denoted by X_2. From Table 6.4, X_2 is a function of three random variables, specifically,

$$X_2 = \ell_{r_{SEPM}}\,(SL_{SEPM})\,(PrgmSched) \tag{6.55}$$

Given $\ell_{r_{SEPM}}$, SL_{SEPM}, and *PrgmSched* are independent random variables

$$E(X_2) = E\left(\ell_{r_{SEPM}}\right) E\,(SL_{SEPM})\,E\,(PrgmSched) \tag{6.56}$$

From the distribution functions for $\ell_{r_{SEPM}}$, SL_{SEPM}, and *PrgmSched* in Table 6.4, it can be shown that

$$\boxed{E(X_2) = (22.5)(17\tfrac{1}{3})(33.36) = 13,010.4} \tag{6.57}$$

The variance of X_2 is

$$Var(X_2) = E(X_2^2) - [E(X_2)]^2$$

which is equivalent to

$$Var(X_2) = E\left(\ell_{rSEPM}^2 \, SL_{SEPM}^2 \, PrgmSched^2\right) - [13,010.4]^2 \qquad (6.58)$$

To compute $Var(X_2)$, it remains to determine $E(X_2^2)$. Again, since ℓ_{rSEPM}, SL_{SEPM}, and $PrgmSched$ are independent

$$E\left(X_2^2\right) = E\left(\ell_{rSEPM}^2 \, SL_{SEPM}^2 \, PrgmSched^2\right)$$

$$= E\left(\ell_{rSEPM}^2\right) E\left(SL_{SEPM}^2\right) E\left(PrgmSched^2\right) \qquad (6.59)$$

Similar to the previous calculations involving ℓ_{rSW}, it is left to the reader to show that

$$E\left(\ell_{rSEPM}^2\right) = 508\frac{1}{3} \qquad (6.60)$$

since $\ell_{rSEPM} \sim Unif(20,25)$.

From Table 6.4, the distribution function for SEPM staff-level is triangular, specifically $SL_{SEPM} \sim Trng(12,15,25)$. To determine $E\left(SL_{SEPM}^2\right)$, the following relationship is used:

$$E\left(SL_{SEPM}^2\right) = Var(SL_{SEPM}) + [E(SL_{SEPM})]^2 \qquad (6.61)$$

From Theorem 4.3, it can be shown that

$$Var(SL_{SEPM}) = 7.7222 \text{ and } E(SL_{SEPM}) = 17\frac{1}{3}$$

therefore,

$$E\left(SL_{SEPM}^2\right) = 7.7222 + \left[17\frac{1}{3}\right]^2 = 308.166 \qquad (6.62)$$

The last term in Equation 6.59 is $E\left(PrgmSched^2\right)$. To compute this expected value, note that

$$E\left(PrgmSched^2\right) = Var\left(PrgmSched\right) + \left[E\left(PrgmSched\right)\right]^2 \qquad (6.63)$$

From Table 6.4, $PrgmSched \sim N(33.36, 1.94)$. Therefore

$$E\left(PrgmSched^2\right) = 1.94 + [33.36]^2 = 1114.829 \tag{6.64}$$

The expected value of each term in Equation 6.59 has now been determined. Thus,

$$E\left(\ell_{r_{SEPM}}^2 SL_{SEPM}^2 \, PrgmSched^2\right) = \left(508\frac{1}{3}\right)(308.166)(1114.829)$$

$$= 174,639,133.4 \tag{6.65}$$

Combining these results, it follows that

$$\boxed{Var(X_2) = 174,639,133.4 - (13,010.4)^2 = 5,368,625.24} \tag{6.66}$$

and

$$\sigma_{X_2} = \sqrt{Var(X_2)} = 2317.03$$

Mean and Variance of X_3: From the WBS in Figure 6.6, recall the cost of STE is denoted by X_3. From Table 6.4, X_3 is a function of three independent random variables, specifically,

$$X_3 = \ell_{r_{STE}} (SL_{STE}) (PrgmSched) \tag{6.67}$$

The same approach to determine the mean and variance of the cost of SEPM can be used to determine the mean and variance of the cost of STE. For this reason, it is left to the reader to verify the following:

$$\boxed{E(X_3) = 3210.9} \tag{6.68}$$

Since $\ell_{r_{STE}} \sim Unif(15, 20)$ and $SL_{STE} \approx Unif(4, 7)$, it follows that

$$E\left(X_3^2\right) = E\left(\ell_{r_{STE}}^2\right) E\left(SL_{STE}^2\right) E\left(PrgmSched^2\right)$$

$$= \left(308\frac{1}{3}\right)(31)(1114.829) = 10,655,907.19 \tag{6.69}$$

With

$$Var(X_3) = E\left(X_3^2\right) - [E(X_3)]^2 \tag{6.70}$$

Substituting the results from Equations 6.68 and 6.69 into Equation 6.70 yields

$$\boxed{Var(X_3) = 346,028.38} \tag{6.71}$$

and

$$\sigma_{X_3} = \sqrt{Var(X_3)} = 588.242$$

Correlation Between X_2 and X_3: By definition (Equations 5.29 and 5.30), the correlation between X_2 and X_3 is

$$\rho_{X_2,X_3} = \frac{Cov(X_2, X_3)}{\sigma_{X_2}\sigma_{X_3}} = \frac{E(X_2X_3) - E(X_2)E(X_3)}{\sigma_{X_2}\sigma_{X_3}} \tag{6.72}$$

From Table 6.4, it was given that

$$X_2 = \ell_{r_{SEPM}} SL_{SEPM} PrgmSched \tag{6.73}$$
$$X_3 = \ell_{r_{STE}} SL_{STE} PrgmSched \tag{6.74}$$

All the terms in Equation 6.72, except for $E(X_2X_3)$, have been determined from the above computations. The term $E(X_2X_3)$ is

$$E(X_2X_3) = E\left(\ell_{r_{SEPM}} SL_{SEPM} PrgmSched \cdot \ell_{r_{STE}} SL_{STE} PrgmSched\right)$$
$$= E\left(\ell_{r_{SEPM}} \ell_{r_{STE}} SL_{SEPM} SL_{STE} PrgmSched^2\right) \tag{6.75}$$

Since $\ell_{r_{SEPM}}$, $\ell_{r_{STE}}$, SL_{SEPM}, SL_{STE}, and $PrgmSched$ were given to be independent random variables, Equation 6.75 can be written as

$$E(X_2X_3) = E(\ell_{r_{SEPM}})E(\ell_{r_{STE}})E(SL_{SEPM})E(SL_{STE})E(PrgmSched^2) \tag{6.76}$$

It can be determined that

$$E(X_2X_3) = (22.5)(17.5)\left(17\frac{1}{3}\right)(5.5)(1114.829) = 41,847,893.59 \tag{6.77}$$

In Equations 6.76 and 6.77, the term $E(PrgmSched^2) = 1114.829$ comes from Equation 6.64. Substituting the result from Equation 6.77 into Equation 6.72 yields

$$\rho_{X_2,X_3} = \frac{41,847,893.59 - (13,010.4)(3210.9)}{(2317.03)(588.242)} = 0.0534 \qquad (6.78)$$

All the terms necessary to complete the computation of $E(Cost_{Sys})$ and $Var(Cost_{Sys})$ have now been determined.

Mean and Variance of Cost$_{Sys}$: From Equation 6.25

$$
\begin{aligned}
E(Cost_{Sys}) &= 1.07\,E(Cost_{PMP}) + E(Q) \\
&= 1.07\,E(Cost_{PMP}) + E(X_2 + X_3) \\
&= 1.07\,E(Cost_{PMP}) + E(X_2) + E(X_3) \\
&= 1.07\,(15,532.52) + 13,010.4 + 3210.9 \\
&= 32,841.1\ (\$K) \qquad (6.79)
\end{aligned}
$$

From Equation 6.26

$$
\begin{aligned}
Var(Cost_{Sys}) &= (1.07)^2\,Var(Cost_{PMP}) + Var(Q) \\
&= (1.07)^2\,Var(Cost_{PMP}) + Var(X_2 + X_3) \\
&= (1.07)^2\,Var(Cost_{PMP}) + Var(X_2) + Var(X_3) \\
&\quad + 2\rho_{X_2,X_3}\sigma_{X_2}\sigma_{X_3} \\
&= (1.07)^2\,(4,315,182.336) + 5,368,625.24 + 346,028.38 \\
&\quad + 2(0.0534)(2317.03)(588.242) \\
&= 10,800,671.5(\$K)^2 \qquad (6.80)
\end{aligned}
$$

which implies

$$\sigma_{Cost_{Sys}} = \sqrt{Var(Cost_{Sys})} = 3286.44 \qquad (6.81)$$

In summary, the mean cost of the digital information system is 32.8 ($M) and the standard deviation is 3.3 ($M). This concludes part (a) of this case discussion.

Part B. Some Implied Correlations: This section discusses the correlations implied by some of the cost relationships in this WBS. The correlation between cost element costs X_i, for $i = 1, \ldots, 5$, is best explored from the relationships given in Table 6.4. From Equation 6.22, we have

$$Cost_{PMP} = 1.05\, Cost_{PME}$$

Since $Cost_{PMP}$ is a linear function of $Cost_{PME}$ (with positive slope), the correlation between $Cost_{PMP}$ and $Cost_{PME}$ is unity. In Table 6.4, we are also given that

$$X_{13} = 0.05 \left(X_{121} + \sum_{s=1}^{4} X_{11s} \right) = 0.05\, Cost_{PME}$$

Thus, the correlation between X_{13} (the integration and assembly cost) and $Cost_{PME}$ is unity. A perfect correlation also exists between the $Cost_{PMP}$ and other cost element costs in this WBS. From Table 6.4 and this case discussion, we can write

$$X_4 = 0.05\, Cost_{PMP} \text{ and } X_5 = 0.02\, Cost_{PMP}$$

Thus, there are implied correlations between X_4 and $Cost_{PMP}$ and X_5 and $Cost_{PMP}$ because of these functional (mathematical) relationships. Here, the correlation between the cost of Data, denoted by X_4, and $Cost_{PMP}$ is unity. Similarly, the correlation between the cost of Training, denoted by X_5, and $Cost_{PMP}$ is unity. These relationships illustrate "logical" or "factor-based" cost relationships, which are common in electronic systems cost analyses.

Last, there is another important correlation in this case discussion. Notice the costs of SEPM and STE, denoted by X_2 and X_3, are functions of *PrgmSched*—the system's development schedule. As a result, a positive correlation exists between $Cost_{Sys}$ and *PrgmSched*. The following presents a derivation of this correlation.

From Equation 6.24, recall that

$$Cost_{Sys} = 1.07\, Cost_{PMP} + (X_2 + X_3) = 1.07\, Cost_{PMP} + Q \qquad (6.82)$$

To simplify the notation, let $C \equiv Cost_{Sys}$ and $P \equiv PrgmSched$. The correlation between the system's total cost C and its development schedule P will be determined. By definition, this correlation is

$$\rho_{C,P} = \frac{E(CP) - E(C)E(P)}{\sigma_C \sigma_P} \qquad (6.83)$$

where

$$E(C) = 32,841.1 \text{ (from Equation 6.79)}$$

$$E(P) = 33.36 \text{ (seen in Table 6.4)}$$

$$\sigma_C = 3286.44 \text{ (from Equation 6.81)}$$

$$\sigma_P = \sqrt{1.94} = 1.39283 \text{ (seen from Table 6.4)}$$

To determine $\rho_{C,P}$ we need $E(CP)$. Multiplying Equation 6.82 by P, we can write

$$E(CP) = E[(1.07\,Cost_{PMP} + Q)P]$$
$$= 1.07\,E(Cost_{PMP}P) + E(QP)$$

It can be shown, in this case, that $Cov(Cost_{PMP}, P) = 0$. Therefore, from Theorem 5.1

$$E(Cost_{PMP}P) - E(Cost_{PMP})E(P) = 0 \Rightarrow E(Cost_{PMP}P) = E(Cost_{PMP})E(P)$$

Thus,

$$E(CP) = 1.07E(Cost_{PMP})E(P) + E(QP)$$
$$E(CP) = 1.07(15,532.52)(33.36) + E(QP)$$

$$(6.84)$$

To complete the computation of $E(CP)$, it remains to determine $E(QP)$. Given the specifics of this case discussion, the random variables Q and P are *not* independent so $E(QP) \neq E(Q)E(P)$. The computation of $E(QP)$ proceeds as follows:

$$E(QP) = E[(X_2 + X_3)P] = E\left[\ell_{r_{SEPM}}SL_{SEPM}P^2 + \ell_{r_{STE}}SL_{STE}P^2\right]$$
$$= E\left[\ell_{r_{SEPM}}SL_{SEPM}P^2\right] + E\left[\ell_{r_{STE}}SL_{STE}P^2\right]$$

$$(6.85)$$

Since the random variables $\ell_{r_{SEPM}}$, $\ell_{r_{STE}}$, SL_{SEPM}, SL_{STE}, and P were given to be independent, Equation 6.85 can be written as

$$E(QP) = \left(E\left[\ell_{r_{SEPM}}\right]E[SL_{SEPM}] + E\left[\ell_{r_{STE}}\right]E[SL_{STE}]\right)E\left(P^2\right)$$
$$= \left(22.5\left(17\frac{1}{3}\right) + 17.5(5.5)\right)1114.829$$
$$= 542,085.6013$$

$$(6.86)$$

Therefore

$$E(CP) = 1.07(15,532.52)(33.36) + 542,085.6013 = 1,096,522.009$$

and

$$\rho_{C,P} = \frac{E(CP) - E(C)E(P)}{\sigma_C \sigma_P} = \frac{1,096,522.009 - (32,841.1)(33.36)}{(3286.44)(1.39283)} = 0.206$$

(6.87)

Part C. Distribution Function Approximation to $F_{Cost_{Sys}}(x)$: Figure 6.7 presents distributions that approximate the CDF of the system's total cost. The curves defined by the two solid lines reflect two assumed theoretical distributions. They are a normal distribution (Figure 6.7a) and a lognormal

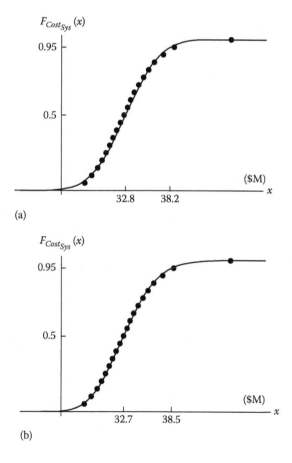

(a)

(b)

FIGURE 6.7
Assumed theoretical CDFs vs simulated CDFs for $Cost_{Sys}$. (a) Normal CDF vs the simulated CDF, (b) Lognormal CDF vs the simulated CDF.

distribution (Figure 6.7b), each with mean 32.8 ($M) and standard deviation 3.3 ($M).

A third distribution is shown in Figure 6.7 by a series of 20 "points." These points reflect random statistical samples (values) of $Cost_{Sys}$, sampled by Monte Carlo simulation (explained in Section 6.3). In Figure 6.7, the curve implied by these "points" is the simulated distribution function for $Cost_{Sys}$. In Figure 6.7, observe how closely the simulated distribution matches the assumed normal and lognormal distribution for $Cost_{Sys}$. The closeness with which these "points" fall along the two curves (each defined by the solid lines in Figure 6.7) *visually suggests* the reasonableness of the assumption that the distribution function for $Cost_{Sys}$ can be approximated by a normal or by a lognormal. Although this is a practical conclusion, it remains an informal one. A formal statistical conclusion could be derived from the K-S test, illustrated in Case Discussion 5.1. This would reveal whether the normal and the lognormal distributions are *statistically plausible models* for the underlying distribution function of $Cost_{Sys}$, in this case.

6.2.2 Approximating the Distribution Function of System Cost

This section provides *guidance* for approximating the distribution function of a system's total cost. Some of this guidance reflects mathematical theory; some of it reflects observations from numerous project applications.

In the examples and case discussions presented in this book, the normal distribution often approximates the distribution function of a system's total cost. There are many reasons for this. Primary among them is $Cost_{Sys}$ (a system's total cost) is a summation of WBS cost element costs. Within the WBS, it is typical to have a mixture of independent and correlated cost element costs. The greater the number of independent cost element costs, the more likely it is that the distribution function of $Cost_{Sys}$ is approximately normal. Why is this?

It is essentially the phenomenon described by the central limit theorem (Theorem 5.10). An seen in this book, the central limit theorem is very powerful. It does not take many independent cost element costs for the distribution of $Cost_{Sys}$ to move toward normality. Such a move is evidenced when (1) a sufficient number of independent cost element costs are summed, and (2) no cost element's cost distribution has a much larger standard deviation than the standard deviations of the other cost element cost distributions. When conditions in the WBS result in $Cost_{Sys}$ being positively skewed (i.e., a nonnormal distribution function), then the lognormal often (Abramson and Young 1997, Garvey 1996)* analytically approximates the distribution function of $Cost_{Sys}$.

What drives the distribution of $Cost_{Sys}$ to be normal or to be skewed? To address this, cost relationships that frequently occur in a system's WBS are examined. The electronic system is used to provide a context for

---

* Many practitioners (Black and Wilder 1982, McNichols 1984, Neimeier 1994, Sobel 1965) have empirically shown the beta distribution also approximates the distribution of $Cost_{Sys}$ well.

the discussion. Work breakdown structures associated with other system classes (e.g., spacecraft systems) can also exhibit properties similar to those discussed here.

From the electronic system WBS in Figure 6.8, $Cost_{Sys}$ is defined by

$$Cost_{Sys} = X_1 + X_2 + X_3 + X_4 + \cdots + X_n \tag{6.88}$$

where $X_1, X_2, X_3, X_4, \ldots, X_n$ denote the n costs of the system's level 2 cost elements. These elements include (but are not limited to) the system's PMP, as well as the system's systems engineering, program management, and system test. Referring to Figure 6.8, Equation 6.88 can also be written as

$$Cost_{Sys} = Cost_{PMP} + \sum_{i=2}^{n} X_i \tag{6.89}$$

where $Cost_{PMP} = X_1$ and $X_1 = X_{11} + X_{12} + X_{13}$.

In the cost analysis of electronic systems, the distribution function of $Cost_{Sys}$ is often *observed* to be approximately normal. Situations specific to cost analysis contribute to this observation. The following cases describe the most common of these situations. In each case, the distribution functions for $Cost_{PMP}, X_2, X_3, X_4, \ldots, X_n$ are assumed to be "well-behaved" (e.g., unimodal, continuous).

Case A *If (in Equation 6.89), the distribution function of $Cost_{PMP}$ is normal and $X_2, X_3, X_4, \ldots, X_n$ are linear functions of $Cost_{PMP}$, such as $X_i = a_i Cost_{PMP}$ where $a_i \geq 0$ $(i = 2, \ldots, n)$, then the distribution function of $Cost_{Sys}$ is normal with mean*

$$E(Cost_{Sys}) = (1 + a_2 + a_3 + \cdots + a_n)E\,(Cost_{PMP})$$

and variance

$$Var(Cost_{Sys}) = (1 + a_2 + a_3 + \cdots + a_n)^2 Var\,(Cost_{PMP})$$

Case A is a direct consequence of the following proposition.

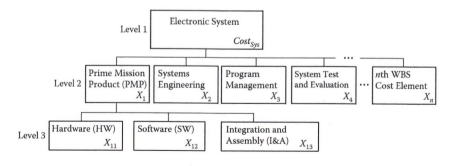

FIGURE 6.8
An Electronic system WBS.

Proposition 6.1   *If X is a normal random variable and $Y = aX$, where a is a constant, then the distribution function for Y is normal with mean $a\,E(X)$ and variance $a^2\,Var(X)$.*

Case B *If (in Equation 6.89) $Cost_{PMP}$ and $X_2, X_3, X_4, \ldots, X_n$ are independent random variables and each are normally distributed, then the distribution function of $Cost_{Sys}$ is normal with mean*

$$E(Cost_{Sys}) = E\left(Cost_{PMP}\right) + \sum_{i=2}^{n} E\left(X_i\right)$$

and variance

$$Var(Cost_{Sys}) = Var\left(Cost_{PMP}\right) + \sum_{i=2}^{n} Var\left(X_i\right)$$

Case B is a direct consequence of the following proposition.

Proposition 6.2   *If $X_1, X_2, X_3, \ldots, X_k$ are independent normally distributed random variables and $Y = X_1 + X_2 + X_3 + \cdots + X_k$, then, regardless of the size of k, the distribution function of Y is normal with mean $\sum_{i=1}^{k} E(X_i)$ and variance $\sum_{i=1}^{k} Var(X_i)$.*

Case C *Suppose (in Equation 6.89) $Cost_{PMP}, X_2, X_3, X_4, \ldots, X_n$ are independent random variables. Furthermore, suppose $Cost_{PMP}, X_2, X_3, X_4, \ldots, X_n$ are not necessarily each normally distributed. If the number of cost element costs in the sequence $Cost_{PMP}, X_2, X_3, X_4, \ldots, X_n$ is sufficiently large with none dominating in standard deviation, then (by the central limit theorem) the distribution function of $Cost_{Sys}$ is approximately normal with mean*

$$E(Cost_{Sys}) = E\left(Cost_{PMP}\right) + \sum_{i=2}^{n} E\left(X_i\right)$$

and variance

$$Var(Cost_{Sys}) = Var\left(Cost_{PMP}\right) + \sum_{i=2}^{n} Var\left(X_i\right)$$

These three cases stem from mathematical theory. The next two cases originate from observations. They are not intended to be rigorous findings; rather, they reflect results often seen in practice.

Case D *Suppose (in Equation 6.89) $Cost_{PMP}$ is normal and $Cost_{PMP}, X_2, X_3,$ $X_4, \ldots, X_n$ are independent random variables. Furthermore, suppose $X_2, X_3,$ $X_4, \ldots, X_n$ are not necessarily each normally distributed. If the number of cost element costs in the sequence $X_2, X_3, X_4, \ldots, X_n$ is sufficiently large with no X_i $(i = 2, \ldots, n)$ dominating in standard deviation, then the distribution function of $Cost_{Sys}$ is approximately normal with mean*

$$E(Cost_{Sys}) = E\,(Cost_{PMP}) + \sum_{i=2}^{n} E\,(X_i)$$

and variance

$$Var(Cost_{Sys}) = Var\,(Cost_{PMP}) + \sum_{i=2}^{n} Var\,(X_i)$$

Case D stems from the influences of the central limit theorem and Proposition 6.2. To see this, recall from Equation 6.89 $Cost_{Sys}$ is given by

$$Cost_{Sys} = Cost_{PMP} + \sum_{i=2}^{n} X_i$$

If the distribution function for $Cost_{PMP}$ is normal and the distribution function of the sum $\sum_{i=2}^{n} X_i$ is *approximately* normal (by the central limit theorem), then $Cost_{Sys}$ is *approximately* the sum of two normally distributed random variables. In Case D, $Cost_{PMP}$ and $\sum_{i=2}^{n} X_i$ are independent. Thus, from Proposition 6.2, the distribution function of $Cost_{Sys}$ is *approximately* normal.

Case E *Suppose (in Equation 6.89) $Cost_{PMP}$ is normal. Suppose the sequence $X_2, X_3, X_4, \ldots, X_n$ contains some cost element costs correlated to $Cost_{PMP}$ (with correlation coefficient ρ_{Cost_{PMP}, X_i}) and some that are uncorrelated to $Cost_{PMP}$. Suppose $X_2, X_3, X_4, \ldots, X_n$ are mutually independent random variables. If the number of $X_i's$ $(i \geq 2)$ uncorrelated to $Cost_{PMP}$ is sufficiently large, with none of the $X_i's$ (correlated or uncorrelated to $Cost_{PMP}$) dominating in standard deviation, then the distribution function of $Cost_{Sys}$ is approximately normal with mean*

$$E(Cost_{Sys}) = E\,(Cost_{PMP}) + \sum_{i=2}^{n} E\,(X_i)$$

and variance

$$Var(Cost_{Sys}) = Var\,(Cost_{PMP}) + \sum_{i=2}^{n} Var\,(X_i) + 2\sum_{i=2}^{n} \rho_{Cost_{PMP}, X_i}\, \sigma_{Cost_{PMP}}\, \sigma_{X_i}$$

In all but Case C, the distribution function for $Cost_{PMP}$ was given to be normal. This is common in electronic systems. The normality of $Cost_{PMP}$

is primarily driven by the central limit theorem, where $Cost_{PMP}$ typically reflects the sum of many *independent* hardware and software costs.

Normality of $Cost_{PMP}$: In electronic systems, $Cost_{PMP}$ is typically defined as the sum of three cost element costs; specifically,

$$Cost_{PMP} = X_1 = X_{11} + X_{12} + X_{13} \tag{6.90}$$

Equation 6.90 can also be written as

$$Cost_{PMP} = Cost_{PME} + X_{13} \tag{6.91}$$

where $Cost_{PME}$ is the system's prime mission equipment cost. It represents the total cost of the system's hardware and software; that is,

$$Cost_{PME} = X_{11} + X_{12} \tag{6.92}$$

The normality of $Cost_{PMP}$ will be discussed by examining the distribution functions that frequently characterize X_{11}, X_{12}, and X_{13}.

Distribution Function of Hardware Cost: Typically, a system's total hardware cost X_{11} is the sum of the individual hardware item costs. Referring to Figure 6.8, suppose

$$X_{11} = X_{111} + X_{112} + X_{113} + \cdots + X_{11j} \tag{6.93}$$

where X_{11i} $(i = 1, 2, \ldots, j)$ are independent random variables representing the costs of the individual hardware items. Under appropriate conditions, the distribution function of X_{11} can be approximately normal by the central limit theorem (Theorem 5.10); that is, $X_{11} \sim N(E(X_{11}), Var(X_{11}))$ with

$$E(X_{11}) = E(X_{111}) + E(X_{112}) + E(X_{113}) + \cdots + E(X_{11j})$$
$$Var(X_{11}) = Var(X_{111}) + Var(X_{112}) + Var(X_{113}) + \cdots + Var(X_{11j})$$

If the distribution functions for X_{11i} are well behaved, then the approximation (in most cases) is good for small j (e.g., not less than or equal to $j = 5$ hardware items). The more asymmetric (skewed) the distribution functions are for X_{11i}, the larger j must be for X_{11} to become approximately normal. In practice, it is *very* common to see the normal distribution approximate X_{11}, particularly in systems designed around the use of *commercial hardware items*. The uncertainty in the cost of such items tends to vary independently and cost analysts often describe these uncertainties by distribution functions that are well behaved.

The cost distribution functions of hardware items that require *custom development* may be *asymmetric*. In practice, this asymmetry typically reflects a positive skew. The presence of asymmetry in the distribution functions for X_{11i} $(i = 1, 2, \ldots, j)$ *will* affect how well (or how quickly) the normal distribution approximates X_{11}. If j (in Equation 6.93) is sufficiently large and the asymmetry is isolated to just a few hardware items whose cost standard deviations contribute only a small amount to the standard deviation of X_{11}, then the distribution of X_{11} may still be approximately normal. If X_{11} is the sum of just a few asymmetric distributions (i.e., j is small), then the distribution of X_{11} may indeed be nonnormal. In such circumstances, the lognormal (or beta distribution) might well approximate the distribution function of X_{11}. It is a good exercise for the reader to study this further. After reading Section 6.3, use the Monte Carlo simulation technique to study the reasonableness of certain distribution function approximations of X_{11}. Do this using various symmetric and asymmetric distributions for the costs of the hardware items X_{11i} $(i = 1, 2, \ldots, j)$.

Distribution Function of Software Cost: Can the distribution function of software cost also be approximated by the normal distribution? The answer depends on *how* software cost is determined. Cost analysts sometimes determine software cost according to the equation

$$X_{12} = \ell_{r_{SW}} \left[c_1 (I_{X_{121}})^{c_2} + c_1 (I_{X_{122}})^{c_2} + \cdots + c_1 (I_{X_{12k}})^{c_2} \right] \tag{6.94}$$

where $I_{X_{12i}} (i = 1, 2, \ldots, k)$ is the number of KDSI to be developed for the ith software function in the system, and c_1, c_2, and $\ell_{r_{SW}}$ are constants (discussed in Section 5.4.2). Equation 6.94 is traditionally applied in cases where the individual software functions are independently developed. Such functions would have minimal to no interdependencies. They would integrate and execute in the system in a highly modular fashion. Under this formulation, if $\ell_{r_{SW}}$ is a constant, k is sufficiently large, and $I_{X_{121}}, I_{X_{122}}, \ldots, I_{X_{12k}}$ are independent random variables, then, by the central limit theorem, the distribution function of X_{12} will be approximately normal. This result is dependent on the way X_{12} is *mathematically defined*. Other definitions for X_{12} *may* yield distribution functions for X_{12} that are skewed. Two such definitions are given by Equations 6.95 and 6.96.

$$X_{12} = \ell_{r_{SW}} \frac{I}{P_r} \tag{6.95}$$

$$X_{12} = \ell_{r_{sw}} (Y_1(I) + Y_2(I)) \tag{6.96}$$

where

$$Y_1(I) = c_1 \left(I_{X_{121}}\right)^{c_2} + c_1 \left(I_{x_{122}}\right)^{c_2} + \cdots + c_1 \left(I_{x_{12m}}\right)^{c_2}$$

$$Y_2(I) = c_1 \left(I_{x_{12(m+1)}} + I_{X_{12(m+2)}} + \cdots + I_{X_{12(m+k)}}\right)^{c_2}$$

In Equation 6.96, $Y_1(I)$ could represent software functions that have *independent* development efforts while $Y_2(I)$ could represent software functions that have *dependent* development efforts. Equation 6.96 is traditionally applied when a combination of independently developed software functions, and a set of software functions that share functionality, characterizes the system.

Equation 6.95 (refer to Chapter 5) *might* be used when software cost is based on the total size I (in DSI) of the software to be developed and its development productivity rate P_r (i.e., DSI per staff month). Here, I and P_r may or may not be independent random variables.

In the definitions for X_{12} (given by Equations 6.94 through 6.96), it would be reasonable to consider ℓ_{rSW} a random variable instead of a constant. This consideration also affects whether the distribution function of X_{12} can be approximated by a normal distribution. The reader is encouraged to explore these questions further, using the Monte Carlo simulation technique discussed in Section 6.3.

Distribution Function of Integration and Assembly (I&A): Similar to the previous discussion, the distribution function for X_{13}—the cost to integrate, assemble, and checkout the system's hardware and software (known in the cost analysis community as I&A)—is also driven by how X_{13} is mathematically defined. The following approaches are commonly used to define X_{13}.

Approach 1—Cost Factor: Cost analysts often define X_{13} as a scalar multiple of *Cost_PME*, that is,

$$X_{13} = a Cost_{PME} \tag{6.97}$$

where $a > 0$. For electronic systems, a typical value for a is 0.05. If *Cost_PME* is normally distributed, then from Proposition 6.1

$$X_{13} \sim N\left(aE(Cost_{PME}), a^2 Var(Cost_{PME})\right) \tag{6.98}$$

Under this approach, the correlation between X_{13} and *Cost_PME* is unity.

Approach 2—Level of Effort: Another way cost analysts define X_{13} is by a level-of-effort formulation, that is,

$$X_{13} = \ell_{rX_{13}} SL_{X_{13}} T_{X_{13}} \tag{6.99}$$

where $\ell_{rX_{13}}$ is a labor rate (e.g., dollars per staff month), $SL_{X_{13}}$ is the staff-level (i.e., the number of persons) needed for I&A, and $T_{X_{13}}$ is the number of months needed for I&A activities. As discussed in Chapter 5, if n is sufficiently large, then the distribution function of a product of n-independent random variables is approximately lognormal. If $\ell_{rX_{13}}$, $SL_{X_{13}}$, and $T_{X_{13}}$ are independent, then X_{13} is the product of three independent random variables. Are three independent random variables enough for the distribution function of X_{13} to be well approximated by the lognormal? After reading Section 6.3, the Monte Carlo simulation technique can be used to explore this question.

To summarize, conditions can occur in the WBS that drive the distribution functions for X_{11}, X_{12}, X_{13} to be normal (or approximately normal). Recall $Cost_{PMP}$ is defined by

$$Cost_{PMP} = X_{11} + X_{12} + X_{13} = Cost_{PME} + X_{13} \tag{6.100}$$

where

$$Cost_{PME} = X_{11} + X_{12} \tag{6.101}$$

If X_{11} and X_{12} are independent normal random variables, then the distribution function for $Cost_{PME}$ is normal with mean

$$E(Cost_{PME}) = E(X_{11}) + E(X_{12})$$

and variance

$$Var(Cost_{PME}) = Var(X_{11}) + Var(X_{12})$$

Furthermore, if $Cost_{PME}$ is normally distributed and $X_{13} = aCost_{PME}$, with $a > 0$, then $Cost_{PMP}$ is normally distributed (by Proposition 6.1) with mean

$$E(Cost_{PMP}) = (1+a)\left[E(X_{11}) + E(X_{12})\right]$$

and variance

$$Var(Cost_{PMP}) = (1+a)^2 \left[Var(X_{11}) + Var(X_{12})\right]$$

Even if X_{13} is not normal, the distribution function of $Cost_{PMP}$ may still be approximately normal. However, this depends on the extent the distribution of $Cost_{PME}$ influences the overall distribution of $Cost_{PMP}$. If $Cost_{PME}$ is normal with a standard deviation *significantly* larger than the standard deviation of X_{13} and X_{13} is independent of $Cost_{PME}$, it is possible that the normal distribution approximates the distribution of $Cost_{PMP}$. Again, it is a worthwhile exercise for the reader to empirically explore cases when this is (and is not) true.

From these discussions, it is seen how frequently the distribution function for $Cost_{Sys}$ can become approximately normal. This is *not* to argue that $Cost_{Sys}$ is always normally distributed. Rather, it is to encourage cost analysts to *study the mathematical relationships* they define in a work breakdown structure to see whether analytical approximations to the distribution function of $Cost_{Sys}$ can be argued. Where possible, analytical forms of the distribution function of $Cost_{Sys}$ are desirable. They can reveal much information about the "cost-behavior" in a system's work breakdown structure. They offer analysts and decision-makers insight about this behavior, so potential areas for cost-reductions and trade-offs might be easily seen.

6.3 Monte Carlo Simulation

Throughout the many examples and case discussions presented in this book, analytical techniques have been used to develop (or approximate) the probability distribution of a system's cost. As previously stressed, analytical solutions to these types of problems are recommended. However, at times there are limitations when using analytical techniques. A system's work breakdown structure cost model can contain cost estimating relationships too complex for strict analytical study. In such circumstances, a technique known as the Monte Carlo method is frequently used. This section provides an introduction to this method.

The Monte Carlo method falls into a class of techniques known as simulation. Simulation has varying definitions among practitioners. For instance, Winston (1994) defines *simulation* as a technique that imitates the operation of a real-world system as it evolves over time. Rubinstein (1981) offers a definition close to the context of this book:

> Simulation is a numerical technique for conducting experiments on a digital computer, which involves certain types of mathematical and logical models that describe the behavior of a business or economic system (or some component thereof) over extended periods of real time.

With easy access to powerful computing and applications software, simulation is a widely used problem-solving technique in management science and operations research.

The Monte Carlo method involves the generation of random variables from known, or assumed, probability distributions. The process of generating random variables from such distributions is known as *random variate generation* or *Monte Carlo sampling*. Simulations driven by Monte Carlo sampling are known as *Monte Carlo simulations*. As mentioned in Chapter 1, one of the earliest applications of Monte Carlo simulation to cost analysis problems was at the RAND Corporation (Dienemann 1966). Since then, Monte Carlo simulation became (and remains) a popular approach for studying cost uncertainty, as well as for evaluating the cost-effectiveness of a system's design alternatives.

For cost uncertainty analysis, Monte Carlo simulation can be used to develop the empirical distribution of a system's cost. In concert with Rubinstein's definition, the WBS serves as the mathematical/logical cost model of the system within which to conduct the simulation. In this context, the steps in a Monte Carlo simulation are as follows:

- For each random variable defined in the system's WBS, randomly select (sample) a value from its distribution function, which is known (or assumed).

- Once a set of feasible values for each random variable has been established, combine these values according to the mathematical relationships specified across the WBS (such as the relationships given in Case Discussions 6.1 and 6.2). This process produces a single value for the system's total cost.

- Repeat these two steps *n*-times (e.g., ten thousand times). This produces *n*-values each representing a possible (i.e., feasible) value for the system's total cost.

- Develop a frequency distribution from these *n*-values. This is the simulated (i.e., empirical) distribution of the total system cost.

To illustrate the concept of Monte Carlo sampling, consider the problem of determining the mean effort (staff months) to develop a software application. For discussion purposes, assume effort Eff_{SW} (see Chapter 5) is given by

$$Eff_{SW} = \frac{I}{P_r} \tag{6.102}$$

where the distribution functions for I and P_r are as given in Figure 6.9.

In the Monte Carlo method, samples for I and P_r are randomly drawn from their distribution functions. These samples are Monte Carlo samples. For each sample (value) of I and P_r, a value for Eff_{SW} is computed according to Equation 6.102. This process of sampling I and P_r and computing the associated Eff_{SW} is repeated thousands of times. From the many sampled values of Eff_{SW}, a simulated (empirical) probability distribution of Eff_{SW} is determined. In addition, various statistical measures such as the mean of Eff_{SW}

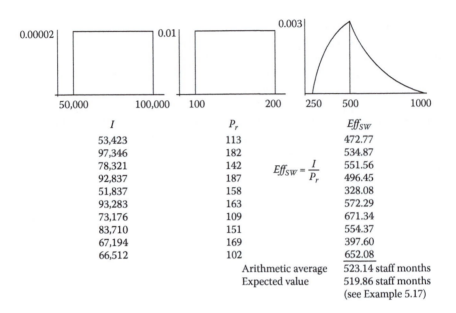

I	P_r	Eff_{SW}
53,423	113	472.77
97,346	182	534.87
78,321	142	551.56
92,837	187	496.45
51,837	158	328.08
93,283	163	572.29
73,176	109	671.34
83,710	151	554.37
67,194	169	397.60
66,512	102	652.08

$$Eff_{SW} = \frac{I}{P_r}$$

Arithmetic average   523.14 staff months
Expected value   519.86 staff months
                                   (see Example 5.17)

FIGURE 6.9
Monte Carlo sampling—ten random samples drawn from the distribution functions for I and P_r.

can be computed from these sampled values. In Figure 6.9, ten random samples of I and P_r are shown along with the associated values of Eff_{SW}. From these samples, an average value of Eff_{SW} is computed. After only ten Monte Carlo samples, this average is close to the computed expected value of Eff_{SW} (refer to Example 5.17).

A way to randomly sample values from a given distribution function is essential to the Monte Carlo method. There are a number of well-established techniques for randomly sampling values. One method is the inverse transform method, which is presented in the following section. For a full discussion of random variate generation techniques, as well as the general topic of modeling and simulation, the reader is directed to Rubinstein (1981) and Law and Kelton (1991).

In cost uncertainty analysis, Monte Carlo simulations are generally static simulations. *Static simulations* are those used to study the behavior of a system (or model) at a specific point in time. In contrast, *dynamic simulations* are those used to study such behavior as it changes over time.

6.3.1 Inverse Transform Method

The inverse transform method (ITM) is a popular technique for generating random variates from continuous distributions. It is a relatively straightforward method for distribution functions that exist in closed form, such as the

uniform or triangular distributions (see Chapter 4). Alternative random variate generation techniques, such as those described in Law and Kelton (1991),
are recommended for working with distribution functions that are not in
closed form. The following illustrates the ITM.

> Suppose a set of random variates for the size of a software application
> must be generated, where the distribution function for size (expressed as
> delivered source instructions I) is given by
>
> $$I \sim Unif(50,000, 100,000)$$
>
> From Equation 4.5, the CDF for I is
>
> $$F_I(t) = \frac{t - 50,000}{50,000} \quad 50,000 \le t \le 100,000 \tag{6.103}$$
>
> To apply the ITM, a random number η, where $0 \le \eta \le 1$, is generated.
> Next, a value for t that satisfies $\eta = F_I(t)$ is found. Repeating this process
> for various η produces Monte Carlo samples that stem from the given
> distribution function. In this case, Monte Carlo samples of I whose under-
> lying distribution function is Equation 6.103 are generated. For example,
> if a random number generator (discussed next) produces $\eta = 0.06846$,
> then the value of t such that
>
> $$0.06846 = \frac{t - 50,000}{50,000}$$
>
> is 53,423, which is the first value of I shown in Figure 6.9. Generalizing
> further, this expression can be solved for any η; this yields
>
> $$t = 50,000(\eta + 1) \tag{6.104}$$
>
> Equation 6.104 is known as the *random variate generator* for I. In particular,
> note that if $\eta = 0$, $\eta = \frac{1}{2}$, and $\eta = 1$, then Equation 6.104 generates
> $t = 50,000$, $t = 75,000$ (which is the median of I), and $t = 100,000$,
> respectively. Thus, for any random number η the random variate gen-
> erator given by Equation 6.104 will produce Monte Carlo samples whose
> underlying distribution function is precisely that given by Equation 6.103.

Essential to random variate generators is the generation of random numbers
identified in the above discussion by η. In general, *random numbers* are independent random variables uniformly distributed over the unit interval. In
Monte Carlo sampling, independent random samples are drawn from the
standard uniform distribution, defined by Equation 6.105.

$$f_X(x) = \begin{cases} 1, & 0 \le x \le 1 \\ 0, & \text{otherwise} \end{cases} \tag{6.105}$$

The statistical literature offers a number of algorithms for generating random numbers. One such generator, commonly available in many present-day software applications, is given by the recursive relationship

$$x_{i+1} = (ax_i + c)(\bmod m), \quad (i = 0, 1, 2, \ldots) \tag{6.106}$$

where a (the *multiplier*), c (the *increment*), and m (the *modulus*) are nonnegative integers. Generators that produce random numbers by Equation 6.106 are known as *linear congruential generators* (Law and Kelton 1991, Rubinstein 1981). They produce a sequence of integers between 0 and $m - 1$. Equation 6.106 is equivalent to

$$x_{i+1} = ax_i + c - m\kappa_i \tag{6.107}$$

where $\kappa_i = [(ax_i + c)/m]$ is the largest *integer* less than or equal to $(ax_i + c)/m$. For each x_i ($i \geq 1$), the associated random number between 0 and 1 is generated by $\eta_{i+1} = (x_{i+1})/m$. For example, suppose $a = 75$, $c = 50$, $m = 5000$, and $x_0 = 20$. The term x_0 is known as the initial value or seed. It is assigned arbitrarily to the random number generator. Using Equation 6.107, the first two random numbers, η_1 and η_2, associated with the sequence of integers $x_1, x_2, \ldots, x_{4999}$ are

$$x_1 = 75(20) + 50 - 5000\kappa_0 = 1550 - 5000(0) = 1550$$
$$x_2 = 75(1550) + 50 - 5000\kappa_1 = 116,300 - 5000(23) = 1300$$

where

$$\kappa_0 = [(75(20) + 50)/5000] = 0$$
$$\kappa_1 = [(75(1550) + 50)/5000] = [23.26] = 23$$

Thus,

$$\eta_1 = \frac{1550}{5000} = 0.310 \text{ and } \eta_2 = \frac{1300}{5000} = 0.260$$

In a strict sense, random numbers generated by recursive relationships are not "purely random." Since they are produced by a deterministic procedure, with results that can be replicated, such numbers are considered "pseudorandom." In practice, the values of a, c, m, and x_0 are selected in a way to create a sequence of x_i's such that their corresponding η_i's appear to be statistically independent uniformly distributed random variates in the unit interval.

6.3.2 Sample Size for Monte Carlo Simulations

In Monte Carlo simulations, a question frequently asked is "How many trials (the sample size) are necessary to have confidence in the outputs of the simulation?" Morgan and Henrion (1990) provide a guideline for determining sample size as a function of the precision desired in the outputs of a Monte Carlo simulation. Specifically, formulas are presented to address the question: "What sample size is needed so that, with probability α, a true fractile of the underlying distribution falls between a pair of fractiles estimated from the Monte Carlo sample?"

Morgan–Henrion Guideline (Morgan and Henrion 1990): Define m as the sample size and let x_p be the p-fractile of X (the underlying distribution); that is, $P(X \leq x_p) = p$. Let c satisfy the probability $P(-c \leq Z \leq c) = \alpha$, where $Z \sim N(0, 1)$. Then, the pair of fractiles $(\hat{x}_i, \hat{x}_k)$ estimated from a Monte Carlo sample with

$$i = \frac{mp - c\sqrt{mp(1 - p)}}{m} = p - c\sqrt{\frac{p(1 - p)}{m}} \tag{6.108}$$

$$k = \frac{mp + c\sqrt{mp(1 - p)}}{m} = p + c\sqrt{\frac{p(1 - p)}{m}} \tag{6.109}$$

contains x_p with probability α. For different sample sizes m, Figure 6.10 illustrates, with probability 0.95 ($c \approx 2$), the values of i and k such that the true median of the distribution falls between $(\hat{x}_i, \hat{x}_k)$. The lower and upper curves in Figure 6.10 are generated from Equations 6.108 to 6.109. As the sample size increases, the difference between the lower and upper curves decreases

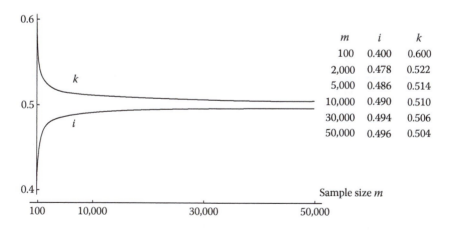

m	i	k
100	0.400	0.600
2,000	0.478	0.522
5,000	0.486	0.514
10,000	0.490	0.510
30,000	0.494	0.506
50,000	0.496	0.504

FIGURE 6.10
Sample size for Monte Carlo simulations.

dramatically. With 100 samples you can be 95% confident the true median $\hat{x}_{0.50}$ falls between the estimated fractiles $\hat{x}_{0.40}$ and $\hat{x}_{0.60}$. Increasing that sample size by a factor of 100 ($m = 10,000$) brings the same degree of confidence to within $\hat{x}_{0.49}$ to $\hat{x}_{0.51}$. As a guideline, 10,000 trials (Monte Carlo samples) should be sufficient to meet the precision requirements for most Monte Carlo simulations, particularly those conducted for cost uncertainty analyses.

Exercises

Exercises 6.1 through 6.4 refer to Case Discussion 6.1.

6.1 Review Case Discussion 6.1 and verify the computations that led to $E(Cost_{Sys})$ and $Var(Cost_{Sys})$.

6.2 Prove Theorem 6.1.

6.3 Referring to Case Discussion 6.1, use Theorem 6.1 to show that
a. $\rho_{X_3,X_1} = 0.9898$
b. $\rho_{X_3,W} = 0.1424$

6.4 The coordinates listed here are the 20 points shown in Figure 6.5. They are values for $(x, F_{Cost_{Sys}}(x))$ determined by the Monte Carlo simulation. The simulation was run with a sample size of $n = 5000$.

$(31.01, 0.05), (33.225, 0.10), (34.76, 0.15), (35.885, 0.20), (36.849, 0.25),$

$(37.785, 0.30), (38.67, 0.35), (39.563, 0.40), (40.272, 0.45), (41.069, 0.50),$

$(41.728, 0.55), (42.326, 0.60), (43.191, 0.65), (44.183, 0.70), (45.151, 0.75),$

$(46.208, 0.80), (47.368, 0.85), (48.548, 0.90), (51.028, 0.95), (59.235, 1)$

Using these values for $(x, F_{Cost_{Sys}}(x))$, apply the K-S test (Chapter 5) to show $Cost_{Sys} \sim N(40.98, 36.18)$ is a statistically plausible model for the distribution function of $Cost_{Sys}$.

Exercises 6.5 through 6.9 refer to Case Discussion 6.2.

6.5 Review Case Discussion 6.2 and verify the computations that led to $E(Cost_{Sys})$ and $Var(Cost_{Sys})$.

6.6 Referring to Table 6.4 and Equation 6.20, show that
a. $Cov(Cost_{PMP}, Q) = 0$, where $Q = X_2 + X_3$
b. $Cov(Cost_{PMP}, P) = 0$, where $P = PrgmSched$

6.7 Use the Mellin transform technique to verify, in Case Discussion 6.2, the mean and variance of the cost of STE, which was denoted by X_3.

6.8 Review Case Discussion 6.2 and verify the computations that led to the correlation between $Cost_{Sys}$ and $PrgmSched$.

6.9 The coordinates listed below are the 20 points shown in Figure 6.7. They are values for $(x, F_{Cost_{Sys}}(x))$ determined by the Monte Carlo simulation. The simulation was run with a sample size of $n = 5000$.

$(27.88, 0.05), (28.72, 0.10), (29.44, 0.15), (29.97, 0.20), (30.45, 0.25),$

$(30.9, 0.30), (31.3, 0.35), (31.74, 0.40), (32.2, 0.45), (32.64, 0.50),$

$(33.07, 0.55), (33.43, 0.60), (33.87, 0.65), (34.41, 0.70), (34.99, 0.75),$

$(35.6, 0.80), (36.29, 0.85), (37.39, 0.90), (38.72, 0.95), (45.71, 1)$

Using these values for $(x, F_{Cost_{Sys}}(x))$, apply the K-S test to show that a normal distribution and a lognormal distribution, each with mean 32.8 ($M) and standard deviation 3.3 ($M), are statistically plausible models for the distribution functions of $Cost_{Sys}$.

6.10 Use the ITM (Section 6.3) to develop a random number generator that produces triangularly distributed random variables.

References

Abramson, R. L. and P. H. Young. 1997 (Spring). FRISKEM–Formal Risk Evaluation Methodology. *The Journal of Cost Analysis*, 14(1), 29–38.

Black R. L. and J. J. Wilder. 1982. Probabilistic cost approximations when inputs are dependent. In *Proceedings of the 1982 International Society of Parametric Analysts (ISPA) Conference*.

Blanchard, B. S. and W. J. Fabrycky. 1990. *Systems Engineering and Analysis*, 2nd edn. Englewood Cliffs, NJ: Prentice-Hall, Inc.

Dienemann, P. F. 1966. *Estimating Uncertainty Using Monte Carlo Techniques*, RM-4854-PR. Santa Monica, CA: The RAND Corporation.

Garvey, P. R. 1990. A general analytic approach to system cost uncertainty analysis, in W. R. Greer, Jr., and D. A. Nussbaum (eds.). *Cost Estimating and Analysis: Tools and Techniques*, pp. 161–181. New York: Springer-Verlag.

Garvey, P. R. 1996 (Spring). Modeling cost and schedule uncertainties—A work breakdown structure perspective. *Military Operations Research*, 2(1), 37–43.

Larson, W. J. and J. R. Wertz (eds.). 1995. *Space Mission Analysis and Design*, 2nd edn. Norwell, MA: Kluwer Academic Press.

Law, A. M. and W. D. Kelton. 1991. *Simulation Modeling and Analysis*, 2nd edn. New York: McGraw-Hill, Inc.

McNichols, G. R. 1984 (Spring). The state of the art of cost uncertainty analysis. *Journal of Cost Analysis*, 1(1), 149–174.

Morgan, M. G. and M. Henrion. 1990. *Uncertainty: A Guide to Dealing with Uncertainty in Quantitative Risk and Policy Analysis*. New York: Cambridge University Press.

Neimeier, H. 1994. Analytic uncertainty modeling. In *International System Dynamics Conference Proceedings*, Stirling, Scotland.

Rubinstein, R. Y. 1981. *Simulation and the Monte Carlo Method*. New York: John Wiley & Sons, Inc.

Sobel, S. 1965. *A Computerized Technique to Express Uncertainty in Advanced System Cost Estimates*, ESD-TR-65-79. Bedford, MA: The MITRE Corporation.

United States Air Force. 1994. *Unmanned Spacecraft Cost Model* (USCM), 7th edn. Los Angeles, CA.

United States Department of Defense. 2011. Work breakdown structures for defense material items, MIL-STD-881, Arlington, VA.

Winston, W. L. 1994. *Operations Research—Applications and Algorithms*. Belmont, CA: Duxbury Press.

7

Modeling Cost and Schedule Uncertainties: An Application of Joint Probability Theory

7.1 Introduction

When cost uncertainty analyses are presented to decision-makers, questions often asked are "What is the chance the system can be delivered within cost and schedule?" "How likely might the point estimate cost be exceeded for a given schedule?" "How are cost reserve recommendations affected by schedule risk?" During the past 30 years, techniques from univariate probability theory have been widely applied to provide insight into $P(Cost \leq x_1)$ and $P(Schedule \leq x_2)$. Although it has long been recognized that a system's cost and schedule are correlated, little has been applied from multivariate probability theory to study joint cost-schedule distributions. A multivariate probability model would provide analysts and decision-makers visibility into joint and conditional cost-schedule probabilities, such as

$$P(Cost \leq x_1 \text{ and } Schedule \leq x_2)$$

and

$$P(Cost \leq x_1 \mid Schedule = x_2)$$

This chapter introduces modeling cost and schedule uncertainties by joint probability distributions. A family of joint distributions (Garvey 1996) has been developed for this purpose. This family consists of the classical bivariate normal and two lesser known joint distributions, the bivariate normal–lognormal and the bivariate lognormal. Experiences with Monte Carlo simulations suggest these distributions are plausible models for computing joint and conditional cost-schedule probabilities. Appendixes B and C summarize key statistical formulas associated with the bivariate normal–lognormal and bivariate lognormal distributions. Formulas for the bivariate normal distribution are well known and are summarized in this chapter.

7.2 Joint Probability Models for Cost-Schedule

As mentioned, decision-makers often require understanding how uncertainties between a system's cost and schedule interact. A decision-maker might bet on a "high-risk" schedule in the hopes of keeping the system's cost within requirements. On the other hand, the decision-maker may be willing to assume "more cost" for a schedule with a small chance of being exceeded. This is a common trade-off faced by decision-makers on systems engineering projects. This is illustrated in Figure 7.1.

Suppose the cumulative distribution functions (CDFs) for a system's cost and schedule are shown in Figure 7.1. The CDF for schedule (the left-side of Figure 7.1) indicates a 20% chance of delivering the system within 43 months. However, there is slightly better than an 80% chance of doing so in 53 months. Given this information, a decision-maker might ask, *What is the cost tradeoff given these two possible schedule outcomes?* To answer this question, we need the distribution function of the system's cost *conditioned* on schedule. Three CDFs for the system's cost are shown on the right-side of Figure 7.1. The left CDF is the cost distribution *conditioned* on a schedule of 43 months. The right CDF is the cost distribution *conditioned* on a schedule of 53 months. The middle CDF is the overall cost distribution conditioned across the *entire* schedule distribution (i.e., not conditioned on a *specific* schedule outcome). The difference between the conditional median cost (107.8 ($M)) given a schedule of 53 months and the conditional median cost (87.4 ($M)) given a

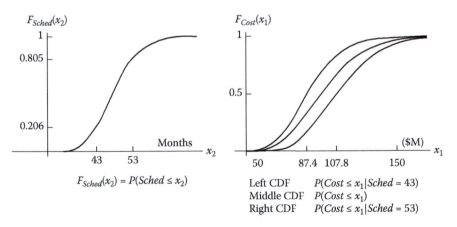

FIGURE 7.1
Illustrative distributions for a system's cost and schedule.

"high-risk" schedule of 43 months is 20.4 ($M).* In the context of Figure 7.1, this difference in cost is certainly significant for any cost-schedule trade-offs under consideration. This discussion highlights how joint probability models can be used to analyze cost-schedule interactions and reveal important trade-offs between them.

The following presents a family of bivariate probability distributions for modeling cost-schedule uncertainty. This family of distributions are candidate theoretical models that may be assumed by an analyst, when joint or conditional cost-schedule probabilities are needed. These distributions have key features desirable for cost analysis. First, they can directly incorporate correlation between cost and schedule on a given system. Second, we will see that their marginal distributions are either both normal, normal and lognormal, or both lognormal. Shown throughout this book, marginal distributions such as these are frequently observed in Monte Carlo simulations (Abramson and Young 1997, Garvey and Taub 1997) of system cost and schedule.

7.2.1 Bivariate Normal

This section presents the classical bivariate normal distribution and summarizes its major characteristics. An important feature of this distribution is its marginal distributions, which are both univariate normal.

In cost analysis, normal distributions can arise when a system's cost is the sum of many independent work breakdown structure cost element costs. Normal distributions can also occur in schedule analyses. For instance, a system's schedule is approximately normal if it is the sum of many independent activities in a schedule network. If normal distributions characterize a system's cost and schedule, then the bivariate normal could serve as an *assumed*[†] model of their joint distribution.

Mathematical Definition: Suppose X_1 and X_2 are two random variables defined on $-\infty < x_1 < \infty$ and $-\infty < x_2 < \infty$. Let

$$E(X_1) = \mu_{X_1} = \mu_1 \tag{7.1}$$

$$E(X_2) = \mu_{X_2} = \mu_2 \tag{7.2}$$

$$Var(X_1) = \sigma_{X_1}^2 = \sigma_1^2 \tag{7.3}$$

$$Var(X_2) = \sigma_{X_2}^2 = \sigma_2^2 \tag{7.4}$$

---

* Example 7.4 will discuss Figure 7.1 further and show how these conditional median costs are determined.
† In general, the true joint distribution of (X_1, X_2) cannot be *uniquely* determined from the marginal distributions of X_1 and X_2. Only when random variables are *independent* can their joint distribution be obtained from their marginal distributions. From Section 5.1.2 recall that two random variables X_1 and X_2 are independent *if and only if* $f_{X_1,X_2}(x_1,x_2) = f_{X_1}(x_1)f_{X_2}(x_2)$.

The pair of random variables

$$(X_1, X_2) \sim Bivariate\ N\left((\mu_1, \mu_2), \left(\sigma_1^2, \sigma_2^2, \rho_{1,2}\right)\right) \tag{7.5}$$

has a bivariate normal distribution if

$$f_{X_1,X_2}(x_1, x_2) = \frac{1}{(2\pi)\sigma_1\sigma_2\sqrt{1 - \rho_{1,2}^2}} e^{-\frac{1}{2}w} \tag{7.6}$$

where

$$w = \frac{1}{1 - \rho_{1,2}^2} \left\{ \left(\frac{x_1 - \mu_1}{\sigma_1}\right)^2 - 2\rho_{1,2}\left(\frac{x_1 - \mu_1}{\sigma_1}\right)\left(\frac{x_2 - \mu_2}{\sigma_2}\right) + \left(\frac{x_2 - \mu_2}{\sigma_2}\right)^2 \right\}$$

for $-\infty < x_1 < \infty$ and $-\infty < x_2 < \infty$. The terms μ_i and σ_i^2 ($i = 1, 2$) in this expression are given by Equations 7.1 through 7.4. The correlation term $\rho_{1,2}$ in Equation 7.6 is

$$\rho_{1,2} = \rho_{X_1, X_2} \tag{7.7}$$

The admissible values for $\rho_{1,2}$ are given by the interval

$$-1 < \rho_{1,2} < 1$$

If two continuous random variables X_1 and X_2 have a bivariate normal distribution, then

$$P(a_1 \le X_1 \le b_1 \text{ and } a_2 \le X_2 \le b_2) = \int_{a_2}^{b_2} \int_{a_1}^{b_1} f_{X_1,X_2}(x_1, x_2)\ dx_1\, dx_2 \tag{7.8}$$

where $f_{X_1,X_2}(x_1, x_2)$ is given by Equation 7.6.

Marginal and Conditional Distributions: A characteristic of the bivariate normal distribution is the distribution of X_1 and the distribution of X_2 are each univariate normal. These are the marginal distributions. They are given by

$$f_1(x_1) = \frac{1}{\sqrt{2\pi}\,\sigma_1} e^{-\frac{1}{2}[(x_1 - \mu_1)^2/\sigma_1^2]} \tag{7.9}$$

$$f_2(x_2) = \frac{1}{\sqrt{2\pi}\,\sigma_2} e^{-\frac{1}{2}[(x_2 - \mu_2)^2/\sigma_2^2]} \tag{7.10}$$

Important trade-offs in cost analysis often involve assessing the impact a given set of schedules has on the probability that system cost will not exceed a required threshold. To make these assessments, the conditional probability distribution is needed. Conditional distributions provide probabilities of the type $P(X_1 \leq a \mid X_2 = b)$. If two continuous random variables X_1 and X_2 have a bivariate normal distribution, then the conditional probability density function of X_1 given $X_2 = x_2$, denoted by $f_{X_1 \mid x_2}(x_1)$, is normally distributed. That is,

$$X_1 \mid x_2 \sim N\left(\mu_1 + \frac{\sigma_1}{\sigma_2}\rho_{1,2}(x_2 - \mu_2),\ \sigma_1^2\left(1 - \rho_{1,2}^2\right)\right) \tag{7.11}$$

Similarly

$$X_2 \mid x_1 \sim N\left(\mu_2 + \frac{\sigma_2}{\sigma_1}\rho_{1,2}(x_1 - \mu_1),\ \sigma_2^2\left(1 - \rho_{1,2}^2\right)\right) \tag{7.12}$$

From Equations 7.11 and 7.12, the conditional means and variances of the bivariate normal distribution are

$$E(X_1 \mid x_2) = \mu_1 + \frac{\sigma_1}{\sigma_2}\rho_{1,2}(x_2 - \mu_2) \tag{7.13}$$

$$E(X_2 \mid x_1) = \mu_2 + \frac{\sigma_2}{\sigma_1}\rho_{1,2}(x_1 - \mu_1) \tag{7.14}$$

$$Var(X_1 \mid x_2) = \sigma_1^2(1 - \rho_{1,2}^2) \tag{7.15}$$

$$Var(X_2 \mid x_1) = \sigma_2^2(1 - \rho_{1,2}^2) \tag{7.16}$$

Views of the Bivariate Normal: Figures 7.2 and 7.3 provide views of a bivariate normal density function. These figures are plots of

$$(X_1, X_2) \sim Bivariate\ N((100, 48), (625, 36, 0.5))$$

Figure 7.2 is a surface view of this function, which has a "hill-like" appearance. The marginal distributions of X_1 and X_2, viewed from the sides of the surface, are both univariate normal. The peak of the bivariate normal density function occurs at $x_1 = \mu_1$ and $x_2 = \mu_2$. In particular,

$$f_{X_1, X_2}(\mu_1, \mu_2) = \frac{1}{(2\pi)\sigma_1\sigma_2\sqrt{1 - \rho_{1,2}^2}}$$

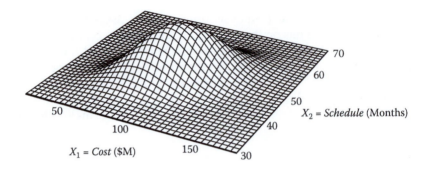

FIGURE 7.2
A bivariate normal density $(X_1, X_2) \sim$ *Bivariate* $N((100, 48), (625, 36, 0.5))$.

Another way to view the bivariate normal is to look at its topography, also known as its *contours*. Contours of constant probability density h are produced by finding x_1 and x_2 such that $h = f_{X_1, X_2}(x_1, x_2)$. In general, contours of the *bivariate normal* are ellipses concentric at (μ_1, μ_2). Figure 7.3 illustrates a set of contours for the bivariate normal density specified in Figure 7.2. The innermost ellipse corresponds to $h = 0.001$, the middle ellipse corresponds to $h = 0.0005$, and the outer ellipse corresponds to $h = 0.0001$. The contour associated with the peak of the bivariate normal is given by the single point (μ_1, μ_2).

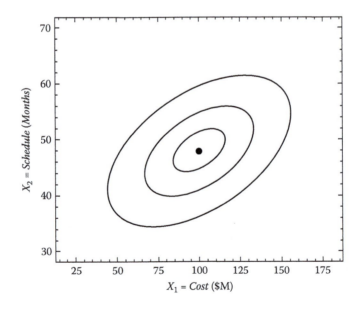

FIGURE 7.3
Contours of a bivariate normal density $(X_1, X_2) \sim$ *Bivariate* $N((100, 48), (625, 36, 0.5))$.

Example 7.1 *Prove that the function given by Equation 7.6 is indeed a joint probability density function.*

Solution To prove this, it is necessary to show

$$\int_{-\infty}^{\infty} \int_{-\infty}^{\infty} f_{X_1,X_2}(x_1,x_2)\, dx_2\, dx_1 = 1 \qquad (7.17)$$

With some algebra, the density function $f_{X_1,X_2}(x_1,x_2)$ (Equation 7.6) can be factored as

$$f_{X_1,X_2}(x_1,x_2) = \left\{ \frac{1}{\sqrt{2\pi}\sigma_1} e^{-(x_1-\mu_1)^2/2\sigma_1^2} \right\} Q(x_1,x_2)$$

where

$$Q(x_1,x_2) = \left\{ \frac{1}{\sqrt{2\pi}(\sigma_2\sqrt{1-\rho_{1,2}^2})} e^{-(x_2-b)^2/2\sigma_2^2(1-\rho_{1,2}^2)} \right\}$$

and $b = \mu_2 + \frac{\sigma_2}{\sigma_1}\rho_{1,2}(x_1 - \mu_1)$. Substituting this factorization into Equation 7.17 yields

$$\int_{-\infty}^{\infty} \frac{1}{\sqrt{2\pi}\sigma_1} e^{-(x_1-\mu_1)^2/2\sigma_1^2}$$

$$\times \left\{ \int_{-\infty}^{\infty} \frac{1}{\sqrt{2\pi}(\sigma_2\sqrt{1-\rho_{1,2}^2})} e^{-(x_2-b)^2/2\sigma_2^2(1-\rho_{1,2}^2)}\, dx_2 \right\} dx_1$$

The right-most integrand in this expression is the probability density function of a $N(b, \sigma_2^2(1 - \rho_{1,2}^2))$ random variable, which by definition has integral equal to unity. Similarly, the left-most integrand in the expression is the probability density of a $N(\mu_1, \sigma_1^2)$ random variable. Therefore,

$$\int_{-\infty}^{\infty} \int_{-\infty}^{\infty} f_{X_1,X_2}(x_1,x_2)\, dx_2\, dx_1 = \int_{-\infty}^{\infty} \frac{1}{\sqrt{2\pi}\sigma_1} e^{-(x_1-\mu_1)^2/2\sigma_1^2} dx_1 = 1$$

Example 7.2 *Suppose the joint probability density function of a system's cost and schedule is a bivariate normal given by*

$$(X_1, X_2) \sim Bivariate\ N((100, 48), (625, 36, 0.5))$$

where X_1 is the random variable that denotes the system's cost (M) and X_2 is the random variable that denotes the system's schedule (months). Determine the median cost of the system conditioned on a schedule of 53 months.

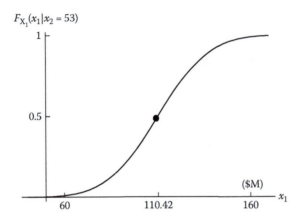

FIGURE 7.4
Cumulative conditional cost distribution $(X_1, X_2) \sim$ *Bivariate* $N((100, 48), (625, 36, 0.5))$.

Solution Following the notation specific to expression (7.5)

$$(X_1, X_2) \sim \text{Bivariate } N((100, 48), (625, 36, 0.5))$$

implies $\mu_1 = 100$, $\mu_2 = 48$, $\sigma_1^2 = 625$, $\sigma_2^2 = 36$, and $\rho_{1,2} = 0.5$. The median system cost conditioned on a schedule of 53 months is found by computing $Med(X_1 \mid x_2 = 53)$. From expression (7.11), the conditional distribution of $X_1 \mid x_2$ is

$$X_1 \mid x_2 \sim N\left(\mu_1 + \frac{\sigma_1}{\sigma_2}\rho_{1,2}(x_2 - \mu_2),\ \sigma_1^2\left(1 - \rho_{1,2}^2\right)\right)$$

Given the parameters $\mu_1 = 100$, $\mu_2 = 48$, $\sigma_1^2 = 625$, $\sigma_2^2 = 36$, and $\rho_{1,2} = 0.5$

$$X_1 \mid x_2 \sim N\left(100 + 2.0833(x_2 - 48),\ 625\left(1 - (0.5)^2\right)\right)$$

$$X_1 \mid 53 \sim N(110.42, 468.75)$$

Since the conditional distribution of system cost $X_1 \mid x_2$ is normal,

$$Med(X_1 \mid 53) = E(X_1 \mid 53) = 110.42\ (\$M)$$

Figure 7.4 depicts the cumulative conditional cost distribution of $X_1 \mid 53$. The "point" shown along the distribution is aligned to $Med(X_1 \mid 53)$.

7.2.2 Bivariate Normal–Lognormal

This section presents the bivariate normal–lognormal distribution and summarizes its major characteristics. An important feature of this distribution is its marginal distributions. One is normal and the other is lognormal.

In cost analysis, it is common for the distribution functions of a system's cost and schedule to be normal and lognormal, respectively. In particular, a system's schedule is often observed (from Monte Carlo simulations) to be lognormal if it is the sum of many positively correlated schedule activities in an overall schedule network. Thus, if normal and lognormal distributions characterize a system's cost and schedule (or vice versa), then the bivariate normal–lognormal could serve as an *assumed* model of their joint distribution.

Mathematical Definition: Suppose $Y_1 = X_1$ and $Y_2 = \ln X_2$ are two random variables where X_1 and X_2 are defined on $-\infty < x_1 < \infty$ and $0 < x_2 < \infty$. If Y_1 and Y_2 each have a normal distribution, then the mean and variance of Y_i $(i = 1, 2)$ are

$$E(Y_1) = \mu_{Y_1} = \mu_{X_1} = \mu_1 \tag{7.18}$$

$$Var(Y_1) = \sigma_{Y_1}^2 = \sigma_{X_1}^2 = \sigma_1^2 \tag{7.19}$$

$$E(Y_2) = \mu_{Y_2} = \mu_2 = \frac{1}{2} \ln \left[\frac{(\mu_{X_2})^4}{(\mu_{X_2})^2 + \sigma_{X_2}^2} \right] \tag{7.20}$$

$$Var(Y_2) = \sigma_{Y_2}^2 = \sigma_2^2 = \ln \left[\frac{(\mu_{X_2})^2 + \sigma_{X_2}^2}{(\mu_{X_2})^2} \right] \tag{7.21}$$

The pair of random variables

$$(X_1, X_2) \sim Bivariate\ NLogN\ ((\mu_1, \mu_2), (\sigma_1^2, \sigma_2^2, \rho_{1,2})) \tag{7.22}$$

has a bivariate normal–lognormal distribution if

$$f_{X_1,X_2}(x_1, x_2) = \frac{1}{(2\pi)\sigma_1\sigma_2\sqrt{1 - \rho_{1,2}^2}\, x_2} e^{-\frac{1}{2}w} \tag{7.23}$$

where

$$w = \frac{1}{1 - \rho_{1,2}^2} \left\{ \left(\frac{x_1 - \mu_1}{\sigma_1} \right)^2 - 2\rho_{1,2} \left(\frac{x_1 - \mu_1}{\sigma_1} \right) \left(\frac{\ln x_2 - \mu_2}{\sigma_2} \right) + \left(\frac{\ln x_2 - \mu_2}{\sigma_2} \right)^2 \right\}$$

for $-\infty < x_1 < \infty$ and $0 < x_2 < \infty$. The terms μ_i and σ_i^2 $(i = 1, 2)$ in this expression are specifically given by Equations 7.18 through 7.21. The correlation term $\rho_{1,2}$ in Equation 7.23 (derived in Appendix B) is

$$\rho_{1,2} = \rho_{Y_1,Y_2} = \rho_{X_1, \ln X_2} = \rho_{X_1, X_2} \frac{(e^{\sigma_2^2} - 1)^{1/2}}{\sigma_2} \tag{7.24}$$

The admissible values for $\rho_{1,2}$ are given by the interval $-1 < \rho_{1,2} < 1$, therefore, admissible values for ρ_{X_1,X_2} (in Equation 7.24) must be *restricted* to the interval

$$\frac{-\sigma_2}{\sqrt{e^{\sigma_2^2} - 1}} < \rho_{X_1,X_2} < \frac{\sigma_2}{\sqrt{e^{\sigma_2^2} - 1}} \tag{7.25}$$

If two continuous random variables X_1 and X_2 have a bivariate normal–lognormal distribution, then

$$P(a_1 \le X_1 \le b_1 \text{ and } a_2 \le X_2 \le b_2) = \int_{a_2}^{b_2} \int_{a_1}^{b_1} f_{X_1,X_2}(x_1, x_2) \, dx_1 \, dx_2 \tag{7.26}$$

where $f_{X_1,X_2}(x_1, x_2)$ is given by Equation 7.23.

Marginal and Conditional Distributions: For the bivariate normal–lognormal distribution given by Equation 7.23, the distribution of X_1 is normal and the distribution of X_2 is lognormal. These are the marginal distributions. They are given by

$$f_1(x_1) = \frac{1}{\sqrt{2\pi}\,\sigma_1} e^{-\frac{1}{2}[(x_1 - \mu_1)^2/\sigma_1^2]} \tag{7.27}$$

$$f_2(x_2) = \frac{1}{\sqrt{2\pi}\,\sigma_2 x_2} e^{-\frac{1}{2}[(\ln x_2 - \mu_2)^2/\sigma_2^2]} \tag{7.28}$$

The conditional distributions of the bivariate normal–lognormal distribution are normal and lognormal. In particular,

$$X_1 \,|\, x_2 \sim N\left(\mu_1 + \frac{\sigma_1}{\sigma_2}\rho_{1,2}(\ln x_2 - \mu_2), \ \sigma_1^2\left(1 - \rho_{1,2}^2\right)\right) \tag{7.29}$$

and

$$X_2 \,|\, x_1 \sim LogN\left(\mu_2 + \frac{\sigma_2}{\sigma_1}\rho_{1,2}(x_1 - \mu_1), \ \sigma_2^2\left(1 - \rho_{1,2}^2\right)\right) \tag{7.30}$$

From these conditional distributions, it can be readily shown (left for the reader) that

$$E(X_1 \,|\, x_2) = \mu_1 + \frac{\sigma_1}{\sigma_2}\rho_{1,2}(\ln x_2 - \mu_2) \tag{7.31}$$

$$E(X_2 \,|\, x_1) = e^{\mu_2 + \frac{\sigma_2}{\sigma_1}\rho_{1,2}(x_1 - \mu_1) + \frac{1}{2}\sigma_2^2(1 - \rho_{1,2}^2)} \tag{7.32}$$

and

$$Var(X_1 \mid x_2) = \sigma_1^2(1 - \rho_{1,2}^2) \tag{7.33}$$

$$Var(X_2 \mid x_1) = e^{2(\mu_2 + \frac{\sigma_2}{\sigma_1}\rho_{1,2}(x_1 - \mu_1))}e^z(e^z - 1) \tag{7.34}$$

where $z = \sigma_2^2\left(1 - \rho_{1,2}^2\right)$.

Views of the Bivariate Normal–Lognormal: Figures 7.5 and 7.6 provide views of a bivariate normal–lognormal density function. These figures are plots of

$$(X_1, X_2) \sim Bivariate\ NLogN((100, 3.86345), (625, 0.0155, 0.502))$$

Figure 7.5 is a surface view of the function, which has a "hill-like" appearance. The marginal distributions of X_1 and X_2, when viewed from the sides of the surface, are univariate normal and univariate lognormal, respectively. A topographic view of a bivariate normal–lognormal density function in Figure 7.5 is shown in Figure 7.6. In Figure 7.6, the innermost contour corresponds to $h = 0.001$, the middle contour corresponds to $h = 0.0005$, and the outer contour corresponds to $h = 0.0001$. The point $(\mu_{X_1}, \mu_{X_2}) = (100, 48)$, shown in Figure 7.6, stems from

$$(X_1, X_2) \sim Bivariate\ NLogN((100, 3.86345), (625, 0.0155, 0.502))$$

This is seen in the following example.

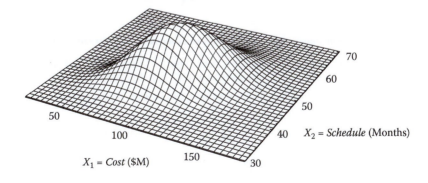

FIGURE 7.5
A bivariate normal–lognormal density $(X_1, X_2) \sim Bivariate\ NLogN((100, 3.86345), (625, 0.0155, 0.502))$.

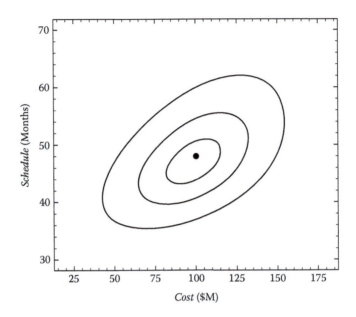

FIGURE 7.6
Contours of a bivariate normal–lognormal density $(X_1, X_2) \sim$ *Bivariate NLogN*$((100, 3.86345),$
$(625, 0.0155, 0.502))$.

Example 7.3 *Assume the joint probability density function of a system's cost X_1 and schedule X_2 is bivariate normal–lognormal with density function given by Equation 7.23. Suppose X_1 has mean 100 ($M) and variance 625 ($M)². Suppose X_2 has mean 48 (months) and variance 36 (months)². If the correlation between the system's cost and schedule is*

$$\rho_{X_1, X_2} = 0.5$$

determine the median system cost conditioned on a schedule of 53 months.

Solution First, determine the five parameters that specify the bivariate normal–lognormal defined by expression (7.22). Since $\mu_{X_1} = 100$, $\sigma^2_{X_1} = 625$, $\mu_{X_2} = 48$, and $\sigma^2_{X_2} = 36$, Equations 7.18 through 7.21 give

$$E(Y_1) = \mu_{Y_1} = \mu_{X_1} = \mu_1 = 100 \tag{7.35}$$

$$Var(Y_1) = \sigma^2_{Y_1} = \sigma^2_{X_1} = \sigma^2_1 = 625 \tag{7.36}$$

$$E(Y_2) = \mu_{Y_2} = \mu_2 = \frac{1}{2} \ln \left[\frac{(\mu_{X_2})^4}{(\mu_{X_2})^2 + \sigma^2_{X_2}} \right] = 3.86345 \tag{7.37}$$

$$Var(Y_2) = \sigma_{Y_2}^2 = \sigma_2^2 = \ln\left[\frac{(\mu_{X_2})^2 + \sigma_{X_2}^2}{(\mu_{X_2})^2}\right] = 0.0155 \qquad (7.38)$$

$$\rho_{1,2} = \rho_{Y_1,Y_2} = \rho_{X_1,\ln X_2} = \rho_{X_1,X_2}\frac{(e^{\sigma_2^2} - 1)^{1/2}}{\sigma_2} = 0.502 \qquad (7.39)$$

From expression 7.25, the interval for the correlation between X_1 and X_2, in this example, is restricted to

$$-0.996126 < \rho_{X_1,X_2} < 0.996126$$

Thus, the correlation given between the system's cost and schedule is admissible, since $-0.996126 < 0.5 < 0.996126$. From these computations, the parameters of the bivariate normal–lognormal distribution are

$$(X_1, X_2) \sim Bivariate\ NLogN((100, 3.86345), (625, 0.0155, 0.502)) \qquad (7.40)$$

The median system cost conditioned on a schedule of 53 months is found by computing $Med(X_1 \mid x_2 = 53)$. From expression (7.29), the conditional distribution of $X_1 \mid x_2$ is

$$X_1 \mid x_2 \sim N\left(\mu_1 + \frac{\sigma_1}{\sigma_2}\rho_{1,2}(\ln x_2 - \mu_2),\ \sigma_1^2\left(1 - \rho_{1,2}^2\right)\right)$$

From Equations 7.35 through 7.39

$$X_1 \mid x_2 \sim N\left(100 + 100.8(\ln x_2 - 3.86345),\ 625\left(1 - (0.502)^2\right)\right)$$

Therefore,

$$X_1 \mid 53 \sim N(110.8, 467.5)$$

Since the conditional distribution of the system cost $X_1 \mid x_2$ is normal

$$Med(X_1 \mid 53) = E(X_1 \mid 53) = 110.8\ (\$M)$$

Figure 7.7 depicts the cumulative conditional cost distribution of $X_1 \mid 53$. The "point" shown along the distribution is aligned to $Med(X_1 \mid 53)$.

7.2.3 Bivariate Lognormal

This section presents the bivariate lognormal and summarizes its major characteristics. From a practical perspective, if the distribution functions of a system's cost and schedule are lognormal, then the bivariate lognormal could serve as an *assumed* model of their joint distribution. However, it again must be emphasized that this is indeed an assumption. In general, the true joint distribution of a pair of random variables (X_1, X_2) cannot be *uniquely* determined from the marginal distributions of X_1 and X_2. Only when random variables are *independent* can their joint distribution be obtained from their marginal distributions.

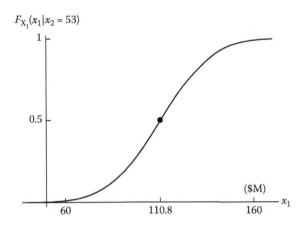

FIGURE 7.7
Cumulative conditional cost distribution $(X_1, X_2) \sim Bivariate\ NLogN((100, 3.86345), (625, 0.0155, 0.502))$.

Mathematical Definition: Suppose $Y_1 = \ln X_1$ and $Y_2 = \ln X_2$ are two random variables where X_1 and X_2 are defined on $0 < x_1 < \infty$ and $0 < x_2 < \infty$. If Y_1 and Y_2 each have a normal distribution, then the mean and variance of Y_i $(i = 1, 2)$ are

$$E(Y_i) = \mu_{Y_i} = \mu_i = \frac{1}{2} \ln \left[\frac{(\mu_{X_i})^4}{(\mu_{X_i})^2 + \sigma_{X_i}^2} \right] \qquad (7.41)$$

$$Var(Y_i) = \sigma_{Y_i}^2 = \sigma_i^2 = \ln \left[\frac{(\mu_{X_i})^2 + \sigma_{X_i}^2}{(\mu_{X_i})^2} \right] \qquad (7.42)$$

The pair of random variables

$$(X_1, X_2) \sim Bivariate\ LogN((\mu_1, \mu_2), (\sigma_1^2, \sigma_2^2, \rho_{1,2})) \qquad (7.43)$$

has a bivariate lognormal distribution if

$$f_{X_1, X_2}(x_1, x_2) = \frac{1}{(2\pi)\sigma_1\sigma_2\sqrt{1 - \rho_{1,2}^2}\, x_1 x_2} e^{-\frac{1}{2}w} \qquad (7.44)$$

where

$$w = \frac{1}{1 - \rho_{1,2}^2} \left\{ \left(\frac{\ln x_1 - \mu_1}{\sigma_1} \right)^2 - 2\rho_{1,2}\left(\frac{\ln x_1 - \mu_1}{\sigma_1} \right)\left(\frac{\ln x_2 - \mu_2}{\sigma_2} \right) + \left(\frac{\ln x_2 - \mu_2}{\sigma_2} \right)^2 \right\}$$

for $0 < x_1 < \infty$ and $0 < x_2 < \infty$. The terms μ_i and σ_i^2 ($i = 1, 2$) in this expression are given by Equations 7.41 and 7.42. The correlation term $\rho_{1,2}$ in Equation 7.44 (derived in Appendix C) is

$$\rho_{1,2} = \frac{1}{\sigma_1 \sigma_2} \ln\left[1 + \rho_{X_1,X_2}\sqrt{e^{\sigma_1^2} - 1}\sqrt{e^{\sigma_2^2} - 1}\right] \tag{7.45}$$

The admissible values for $\rho_{1,2}$ are given by the interval $-1 < \rho_{1,2} < 1$. From this, it can be shown that admissible values for ρ_{X_1,X_2} (in Equation 7.45) are *restricted* to the interval

$$\frac{e^{-\sigma_1\sigma_2} - 1}{\sqrt{e^{\sigma_1^2} - 1}\sqrt{e^{\sigma_2^2} - 1}} < \rho_{X_1,X_2} < \frac{e^{\sigma_1\sigma_2} - 1}{\sqrt{e^{\sigma_1^2} - 1}\sqrt{e^{\sigma_2^2} - 1}} \tag{7.46}$$

If two continuous random variables X_1 and X_2 have a bivariate lognormal distribution, then

$$P(a_1 \leq X_1 \leq b_1 \text{ and } a_2 \leq X_2 \leq b_2) = \int_{a_2}^{b_2} \int_{a_1}^{b_1} f_{X_1,X_2}(x_1, x_2)\, dx_1\, dx_2 \tag{7.47}$$

where $f_{X_1,X_2}(x_1, x_2)$ is given by Equation 7.44.

Marginal and Conditional Distributions: For the bivariate lognormal distribution (given by Equation 7.44), the distribution of X_1 is lognormal and the distribution of X_2 is lognormal. The marginal distributions are given by

$$f_1(x_1) = \frac{1}{\sqrt{2\pi}\, \sigma_1 x_1} e^{-\frac{1}{2}[(\ln x_1 - \mu_1)^2/\sigma_1^2]} \tag{7.48}$$

$$f_2(x_2) = \frac{1}{\sqrt{2\pi}\, \sigma_2 x_2} e^{-\frac{1}{2}[(\ln x_2 - \mu_2)^2/\sigma_2^2]} \tag{7.49}$$

The conditional distributions of the bivariate lognormal distribution are both lognormal. In particular,

$$X_1|x_2 \sim LogN\left(\mu_1 + \frac{\sigma_1}{\sigma_2}\rho_{1,2}(\ln x_2 - \mu_2)\right), \sigma_1^2(1 - \rho_{1,2}^2)) \tag{7.50}$$

and

$$X_2|x_1 \sim LogN\left(\mu_2 + \frac{\sigma_2}{\sigma_1}\rho_{1,2}(\ln x_1 - \mu_1)\right), \sigma_2^2(1 - \rho_{1,2}^2)) \tag{7.51}$$

From these conditional distributions it can be readily shown (left for the reader) that

$$E(X_1|x_2) = x_2^{\frac{\sigma_1}{\sigma_2}\rho_{1,2}} e^{\mu_1 - \frac{\sigma_1}{\sigma_2}\rho_{1,2}\mu_2 + \frac{1}{2}\sigma_1^2(1-\rho_{1,2}^2)} \tag{7.52}$$

$$E(X_2|x_1) = x_1^{\frac{\sigma_2}{\sigma_1}\rho_{1,2}} e^{\mu_2 - \frac{\sigma_2}{\sigma_1}\rho_{1,2}\mu_1 + \frac{1}{2}\sigma_2^2(1-\rho_{1,2}^2)} \tag{7.53}$$

and

$$Var(X_1|x_2) = x_2^{2\frac{\sigma_1}{\sigma_2}\rho_{1,2}} e^{2(\mu_1 - \frac{\sigma_1}{\sigma_2}\rho_{1,2}\mu_2)} e^{z^*}(e^{z^*} - 1) \tag{7.54}$$

$$Var(X_2|x_2) = x_1^{2\frac{\sigma_2}{\sigma_1}\rho_{1,2}} e^{2(\mu_2 - \frac{\sigma_2}{\sigma_1}\rho_{1,2}\mu_1)} e^{z}(e^{z} - 1) \tag{7.55}$$

where

$$z^* = \sigma_1^2(1 - \rho_{1,2}^2) \text{ and } z = \sigma_2^2\left(1 - \rho_{1,2}^2\right).$$

Views of the Bivariate Lognormal: Figures 7.8 and 7.9 provide views of a bivariate lognormal density function. These figures are plots of

$$(X_1, X_2) \sim Bivariate\ LogN((4.57486, 3.86345), (0.0606246, 0.0155, 0.505708))$$

Figure 7.8 is a surface view of the function, which has a "hill-like" appearance. The marginal distributions of X_1 and X_2, viewed from the sides of the surface, are both univariate lognormal.

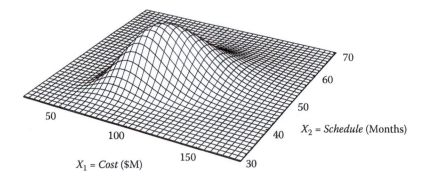

FIGURE 7.8
A bivariate lognormal density $(X_1, X_2) \sim Bivariate\ LogN((4.57, 3.86), (0.0606, 0.0155, 0.5057))$.

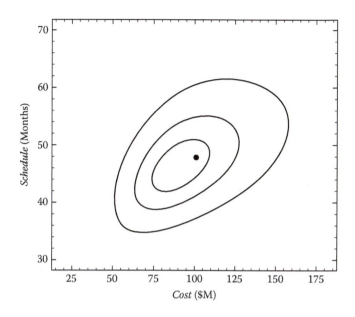

FIGURE 7.9
Contours of a bivariate lognormal density $(X_1, X_2) \sim$ *Bivariate LogN*((4.57486, 3.86345), (0.0606246, 0.0155, 0.505708)).

A topographic view of a bivariate lognormal density function in Figure 7.8 is shown in Figure 7.9. In Figure 7.9, the innermost contour corresponds to $h = 0.001$, the middle contour corresponds to $h = 0.0005$, and the outer contour corresponds to $h = 0.0001$. The point $(\mu_{X_1}, \mu_{X_2}) = (100, 48)$, shown in Figure 7.9, stems from

$$(X_1, X_2) \sim \textit{Bivariate LogN}((4.57486, 3.86345), (0.0606246, 0.0155, 0.505708))$$

This is seen in the following example.

Example 7.4 *Assume the joint probability density function of a system's cost X_1 and schedule X_2 is bivariate lognormal with density function given by Equation 7.44. Suppose X_1 has mean 100 (M) and variance 625 (M)^2. Suppose X_2 has mean 48 (months) and variance 36 (months)^2. Let cost and schedule have a correlation of 0.5. Show that the difference between the median system cost conditioned on a schedule with a 20% chance of being achieved and the median system cost conditioned on a schedule with an 80% chance of being achieved is 20.4 (M).*

Solution It is given that $\mu_{X_1} = 100$, $\sigma^2_{X_1} = 625$, $\mu_{X_2} = 48$, $\sigma^2_{X_2} = 36$, and $\rho_{X_1,X_2} = 0.5$. From Equations 7.41, 7.42, and 7.45, the parameters of the bivariate lognormal, given in expression (7.43), are

$$\mu_1 = \frac{1}{2}\ln\left[\frac{(\mu_{X_1})^4}{(\mu_{X_1})^2 + \sigma_{X_1}^2}\right] = 4.57486 \quad \sigma_1^2 = \ln\left[\frac{(\mu_{X_1})^2 + \sigma_{X_1}^2}{(\mu_{X_1})^2}\right] = 0.0606246$$

$$\mu_2 = \frac{1}{2}\ln\left[\frac{(\mu_{X_2})^4}{(\mu_{X_2})^2 + \sigma_{X_2}^2}\right] = 3.86345 \quad \sigma_2^2 = \ln\left[\frac{(\mu_{X_2})^2 + \sigma_{X_2}^2}{(\mu_{X_2})^2}\right] = 0.0155$$

$$\rho_{1,2} = \frac{1}{\sigma_1 \sigma_2}\ln\left[1 + \rho_{X_1,X_2}\sqrt{e^{\sigma_1^2}-1}\sqrt{e^{\sigma_2^2}-1}\right] = 0.50578$$

From expression (7.50), the cost distribution X_1 conditioned on a schedule of x_2 months is

$$X_1 \mid x_2 \sim LogN\left(\mu_1 + \frac{\sigma_1}{\sigma_2}\rho_{1,2}(\ln x_2 - \mu_2),\ \sigma_1^2\left(1 - \rho_{1,2}^2\right)\right)$$

so

$$X_1 \mid x_2 \sim LogN(4.57486 + (\ln x_2 - 3.86345), 0.0451204) \qquad (7.56)$$

Figure 7.1 illustrates the CDFs associated with this example. The schedule distribution is shown on the left-side of Figure 7.1. Since X_2 is lognormal with mean 48 (months) and variance 36 (months)^2, $X_2 \sim LogN(3.86345, 0.0155)$. It is left to the reader to show that the value of x_2 such that $P(X_2 \le x_2) = 0.20$ is 43 months (rounded). Similarly, the value of x_2 such that $P(X_2 \le x_2) = 0.80$ is 53 months (rounded). From expression (7.56), the conditional cost distribution given $x_2 = 43$ months is

$$X_1 \mid 43 \sim LogN(4.47, 0.045)$$

Likewise, the conditional cost distribution given $x_2 = 53$ months is

$$X_1 \mid 53 \sim LogN(4.68, 0.045)$$

Since $X_1 \mid x_2$ is lognormal, we know from Equation 4.41 that

$$Med(X_1 \mid 43) = e^{4.47} = 87.4 \ (\$M)$$

$$Med(X_1 \mid 53) = e^{4.68} = 107.8 \ (\$M)$$

Therefore, the difference between the median system cost conditioned on a schedule with a 20% chance of being achieved and the median system cost conditioned on a schedule with an 80% chance of being achieved is 20.4 (\$M).

Case Discussion 7.1: In Case Discussion 6.2, the random variable $Cost_{Sys}$ denoted the total cost (\$K) of a digital information system, and the random variable $PrgmSched$ represented the duration of its development (in months). Suppose the joint probability density function of $Cost_{Sys}$ and $PrgmSched$ is bivariate normal. Let b be the number of months such that $P(PrgmSched \leq b) = 0.95$, where $PrgmSched$ is normally distributed with $E(PrgmSched) = 33.36$ (months) and $Var(PrgmSched) = 1.94$ (months)^2. Determine a such that $P(Cost_{Sys} \leq a \mid PrgmSched = b) = 0.95$.

To determine a, we first find b such that $P(PrgmSched \leq b) = 0.95$. This probability can be written as $P(PrgmSched \leq b) = P(Z \leq v)$, where

$$v = \frac{b - E(PrgmSched)}{\sigma_{PrgmSched}} = \frac{b - 33.36}{1.39283}$$

From Table A.1, we have $P(Z \leq v) = 0.95$ if $v = \frac{b-33.36}{1.39283} = 1.645 \Rightarrow b = 35.65$. Now, it remains to determine a such that

$$P(Cost_{Sys} \leq a \mid PrgmSched = 35.65) = 0.95$$

Since the joint probability density function of $Cost_{Sys}$ and $PrgmSched$ is given to be bivariate normal, from expression (7.11), the distribution of $Cost_{Sys}$ conditioned on $PrgmSched$ is

$$Cost_{Sys} \mid PrgmSched = x_2 \sim N\left(\mu_1 + \frac{\sigma_1}{\sigma_2}\rho_{1,2}(x_2 - \mu_2),\ \sigma_1^2\left(1 - \rho_{1,2}^2\right)\right)$$

From Case Discussion 6.2

$$\mu_1 = E(Cost_{Sys}) = 32,841.1 \quad \sigma_1^2 = Var(Cost_{Sys}) = 10,800,698.3$$
$$\mu_2 = E(PrgmSched) = 33.36 \quad \sigma_2^2 = Var(PrgmSched) = 1.94$$

and

$$\rho_{1,2} = \rho_{Cost_{Sys}, PrgmSched} = 0.206$$

Therefore,

$$Cost_{Sys} \mid PrgmSched = x_2 \sim N(32,841.1 + 486.06(x_2 - \mu_2), 10,342,359.87)$$

At $x_2 = 35.65$ we have

$$Cost_{Sys} \mid PrgmSched = 35.65 \sim N(33,954.18, 10,342,359.87)$$

The density function of $Cost_{Sys}$ conditioned on a system schedule of 35.65 months is normal, with mean 33,954.18 ($K) and variance 10,342,359.87 ($K)^2. To find a such that $P(Cost_{Sys} \leq a \mid PrgmSched = 35.65) = 0.95$, let

$$P(Cost_{Sys} \leq a \mid PrgmSched = 35.65) = P(Z \leq \phi)$$

where $\phi = \frac{a-33,954.18}{\sqrt{10,342,359.87}}$. From Table A.1, $P(Z \leq \phi) = 0.95$ if

$$\phi = \frac{a - 33,954.18}{\sqrt{10,342,359.87}} = 1.645$$

This implies that $a = 39,244.4$. Thus, the cost of the digital information system that has only a 5% chance of being exceeded, when conditioned on a schedule having the same chance of being exceeded, is 39,244.4 ($K).

7.3 Summary

The family of distributions described in this chapter provides an analytical basis for computing joint and conditional cost-schedule probabilities. They are mathematical models that might be hypothesized for capturing the joint interactions between a system's cost and schedule.

Seen throughout this chapter, a parameter required by these models is the correlation between cost and schedule.* This can be a difficult value to determine. One approach is its direct computation as illustrated in Case Discussion 6.2. However, in some instances this might not be analytically possible or practical. Another approach is to obtain an estimate of the correlation, from sample values generated by Monte Carlo simulation. This is a reasonable method that can be used regardless of the complexity of the cost-schedule estimation relationships. Subjective assessments might be used. However, care must be taken to specify an *admissible correlation* for the particular pair of random variables. Furthermore, there may already exist an implied correlation by virtue of how the cost-schedule estimation relationships are mathematically defined (refer to Case Discussion 6.2). Subjectively specifying a correlation when one is already present (only its magnitude is unknown or yet to be determined) is double counting correlation. This must be avoided to ensure the mathematical integrity of the cost uncertainty analysis.

---

* Because these models treat cost and schedule as correlated random variables, it is important to recognize that *they do not capture causal impacts* that schedule compression or extension has on cost.

In summary, systems engineering typically takes place in environments of limited funds and challenging schedules. It is incumbent upon engineers and analysts to continually assess affordability relative to the chance of jointly meeting cost and schedule, or meeting cost for a given feasible schedule, against specific tradeoffs in system requirements, acquisition strategies, and post-development support. The distributions described in this chapter are one way such assessments may be made.

Exercises

7.1 Suppose the mean cost and mean schedule of a program is 100 ($M) and 48 months, respectively. Furthermore, suppose the program's cost and schedule variances are 625 (months)^2 and 36 (months)^2, respectively. If the correlation between the program's cost and schedule is 0.5, find x_1 such that

    a. $P(Cost \leq x_1 \mid x_2 = 53\,\text{months}) = 0.95$ if program cost and schedule have a bivariate normal distribution

    b. $P(Cost \leq x_1 \mid x_2 = 53\,\text{months}) = 0.95$ if program cost and schedule have a bivariate normal–lognormal distribution

    c. $P(Cost \leq x_1 \mid x_2 = 53\,\text{months}) = 0.95$ if program cost and schedule have a bivariate lognormal distribution

7.2 Suppose $(X_1, X_2) \sim$ *Bivariate NLogN* $((\mu_1, \mu_2), (\sigma_1^2, \sigma_2^2, \rho_{1,2}))$ where μ_1, μ_2, σ_1^2, σ_2^2, and $\rho_{1,2}$ are defined in Section 7.2.2. If $\mu_{X_2} = \sqrt{e}$ and $\sigma_{X_2}^2 = e(e-1)$, show that $-\frac{1}{\sqrt{e-1}} < \rho_{X_1, X_2} < \frac{1}{\sqrt{e-1}}$.

7.3 Suppose $(X_1, X_2) \sim$ *Bivariate LogN* $((\mu_1, \mu_2), (\sigma_1^2, \sigma_2^2, \rho_{1,2}))$ where μ_1, μ_2, σ_1^2, σ_2^2, and $\rho_{1,2}$ are defined in Section 7.2.3. If $\mu_{X_1} = \mu_{X_2} = \sqrt{e}$ and $\sigma_{X_1}^2 = \sigma_{X_2}^2 = e(e-1)$ show that $-\frac{1}{e} < \rho_{X_1, X_2} < 1$.

7.4 Assume the joint probability density function of program cost X_1 and schedule X_2 is bivariate normal–lognormal with density function given by Equation 7.23. Suppose X_1 has mean 100 ($M) and variance 625 ($M)^2. Suppose X_2 has mean 48 (months) and variance 36 (months)^2. Let the program cost and schedule have a correlation of 0.5. Compute the difference between the median program cost conditioned on a schedule with a 50% chance of being achieved, and the median program cost conditioned on a schedule with a 95% chance of being achieved.

7.5 Show that the functions given by Equations 7.23 and 7.44 are each joint probability density functions.

7.6 If $(X_1, X_2) \sim$ *Bivariate NLogN* $((\mu_1, \mu_2), (\sigma_1^2, \sigma_2^2, \rho_{1,2}))$, where μ_1, μ_2, σ_1^2, σ_2^2, and $\rho_{1,2}$ are defined in Section 7.2.2, show that

a. $E(X_1 \mid x_2) = \mu_1 + \frac{\sigma_1}{\sigma_2}\rho_{1,2}(\ln x_2 - \mu_2)$

b. $E(X_2 \mid x_1) = e^{\mu_2 + \frac{\sigma_2}{\sigma_1}\rho_{1,2}(x_1 - \mu_1) + \frac{1}{2}\sigma_2^2(1 - \rho_{1,2}^2)}$

c. $Var(X_1 \mid x_2) = \sigma_1^2(1 - \rho_{1,2}^2)$

d. $Var(X_2 \mid x_1) = e^{2(\mu_2 + \frac{\sigma_2}{\sigma_1}\rho_{1,2}(x_1 - \mu_1))}e^z(e^z - 1)$, where $z = \sigma_2^2(1 - \rho_{1,2}^2)$

7.7 If $(X_1, X_2) \sim$ *Bivariate LogN* $((\mu_1, \mu_2), (\sigma_1^2, \sigma_2^2, \rho_{1,2}))$, where μ_1, μ_2, σ_1^2, σ_2^2, and $\rho_{1,2}$ are defined in Section 7.2.3, show that

a. $E(X_1 \mid x_2) = x_2^{\frac{\sigma_1}{\sigma_2}\rho_{1,2}} e^{\mu_1 - \frac{\sigma_1}{\sigma_2}\rho_{1,2}\mu_2 + \frac{1}{2}\sigma_1^2(1 - \rho_{1,2}^2)}$

b. $E(X_2 \mid x_1) = x_1^{\frac{\sigma_2}{\sigma_1}\rho_{1,2}} e^{\mu_2 - \frac{\sigma_2}{\sigma_1}\rho_{1,2}\mu_1 + \frac{1}{2}\sigma_2^2(1 - \rho_{1,2}^2)}$

c. $Var(X_1 \mid x_2) = x_2^{2\frac{\sigma_1}{\sigma_2}\rho_{1,2}} e^{2(\mu_1 - \frac{\sigma_1}{\sigma_2}\rho_{1,2}\mu_2)} e^{z^*}(e^{z^*} - 1)$, $z^* = \sigma_1^2(1 - \rho_{1,2}^2)$

d. $Var(X_2 \mid x_1) = x_1^{2\frac{\sigma_2}{\sigma_1}\rho_{1,2}} e^{2(\mu_2 - \frac{\sigma_2}{\sigma_1}\rho_{1,2}\mu_1)} e^{z}(e^{z} - 1)$, $z = \sigma_2^2(1 - \rho_{1,2}^2)$

References

Abramson, R. L. and P. H. Young. 1997 (Spring). FRISKEM–Formal Risk Evaluation Methodology. *The Journal of Cost Analysis*, 14(1), 29–38.

Garvey, P. R. 1996 (Spring). Modeling cost and schedule uncertainties—A work breakdown structure perspective. *Military Operations Research*, 2(1), 37–43.

Garvey, P. R. and A. E. Taub. 1997 (Spring). A joint probability model for cost and schedule uncertainties. *The Journal of Cost Analysis*, 14(1), 3–27.

Section II

Practical Considerations and Applications

8

A Review of Cost Uncertainty Analysis

> It's not what you don't know that hurts you—it's what you *do* know that isn't true.
>
> **Dr. Stephen A. Book (1995)**

Section I of this book presented the theory and foundations of cost uncertainty analysis. As mentioned in the preface, the chapters in Section II focus on the applications of theory to problems encountered in practice. This chapter provides readers a condensed review and summary of key elements of cost uncertainty analysis. Materials in this chapter that were first discussed in Section I have been brought forward in Section II for renewed emphasis and for the convenience of the reader.

8.1 Introduction

This chapter begins with a review of core concepts and key terms. This includes a discussion on the scope of cost uncertainty analysis (what is captured, what is not captured), what it means to present and interpret cost as a probability distribution, and insights the analysis brings to decision-makers.

Cost uncertainty analysis methods fall into one (or a mix) of two worlds. They are Monte Carlo simulation methods and method of moments (analytical) approaches. This chapter reviews these methods and clarifies certain issues associated with them. These include the use of normal or lognormal distributions to derive confidence intervals around point estimates, as well as how best to deal with correlation between cost elements of a program. Common mistakes, pitfalls, and guidance on conducting a cost uncertainty analysis and ways to convey analysis findings are discussed.

In systems engineering, costs are estimated to reveal the economic significance of technical and programmatic choices that guide procuring a system that is affordable, cost-effective, and risk-managed. Identifying risks enables decision-makers to develop, execute, and monitor management actions based on the knowledge of potential cost consequences of inactions. Together, cost and cost uncertainty analysis are undertaken to address paramount considerations of affordability, cost-effectiveness, and risk. Affordability addresses the question: "Can the system be procured with the funds available?" Cost-effectiveness addresses the question: "Does the system represent the best use of funds?" Risk addresses the question: "What is the chance the planned or budgeted cost of the system will be exceeded?" Given this, the purpose of cost uncertainty analysis is to (1) enable the early and continuous identification of cost risk driving elements and (2) produce a defensible assessment of the level of cost to plan or budget, so there is reasonable confidence in assuring program affordability and cost-effectiveness.

Risk analysis is an inseparable part of cost analysis. Many different elements of a program's technical baseline and cost estimate are involved. This includes technology maturity, supply chain integrity, quantities, schedules, and acquisition considerations. The mathematics of risk analysis can involve concepts of correlation, probability distributions, and means and variances. Conveying risk analysis findings clearly and concisely to audiences with broad backgrounds is a challenging yet crucial aspect of the process. It is critically important to step back from the analysis to understand what the findings really mean, whether risks have been adequately captured, and what can be done to reduce their potential negative consequences.

Technology maturity can be a significant risk and cost growth driver. The U.S. Government Accountability Office (GAO) reported in the past that programs working to mature technologies after the start of development while concurrently attempting to mature a system's design and prepare for production are at higher risk of experiencing cost growth and schedule delays. GAO observed that those programs tend to have higher cost growth than programs that start system development with mature technologies. The GAO analysis indicates the average rate of development cost growth for those programs that started with immature technologies is 86%, while the average growth rate for development costs is about half that amount for programs that began with their critical technologies at least nearing maturity.*

A recent initiative to address cost growth in defense acquisitions is the Weapon Systems Acquisition Reform Act (WSARA), signed into law in 2009. One aim of WSARA is to improve cost realism by requiring acquisition programs to budget at a high degree of confidence, such that it can be completed without the need for significant cost adjustments at a later phase. The analysis methods described in this book are ways to support this aim.

Acquiring today's systems is more sophisticated and complex than ever before. Increasingly, systems are engineered by bringing together many individual systems, which, as a whole, provide a capability otherwise not possible. Systems are now richly connected. They involve and evolve webs of users, technologies, and systems-of-systems through environments that offer cross-boundary access to a wide array of resources and information

---

* Paragraph excerpted from GAO-13-294SP, p. 25, March 2013.

repositories. Today's systems create value by delivering capabilities over time that meet user needs for increased agility, robustness, and scalability. System architectures must be open to allow the insertion of innovation that advances the efficacies of capabilities and services to users.

Many systems no longer physically exist within well-defined boundaries. They are increasingly ubiquitous and operate as an enterprise of technologies and cooperating entities in a dynamic that can behave in unpredictable ways. Pervasive with these challenges are economic and budgetary realities that necessitate greater accuracy in the estimated life-cycle costs and cost risks of acquiring these systems.

Systems engineering is more than developing and employing inventive technologies. Designs must be adaptable to change, flexible to meet user needs, and resource-managed. They must be balanced with respect to performance and affordability goals while being continuously risk-managed throughout a system's life cycle. Systems engineers and managers must also understand the social, political, and economic environments within which a system operates. These factors can significantly influence risk, affordability, design options, and investment decisions.

Applied early and continuously, risk analysis can expose events that, if realized, might impede an acquisition program from achieving its required cost, schedule, and performance goals. Risk analysis is more than identifying and quantifying the consequences of unwanted events on an acquisition program and its cost. It provides a context for bringing realism to technical and program decisions that shape a program's acquisition strategy and the cost-effectiveness of its long-term performance.

Why are there risks? Pressures to meet cost, schedule, and technical performance are the practical realities in acquiring systems. Illustrated in Figure 8.1, risk becomes an increasing threat when stakeholder expectations push what is technically or economically feasible. Thus, managing risk is managing the inherent contention that exists within and across these dimensions.

For cost uncertainty analysis to shape and influence program decisions, it must provide insights into cost estimate confidence that are otherwise unseen. What does cost estimate confidence mean? In general, it is a statement of the sureness in an estimate along with a supporting rationale. The intent of cost uncertainty analysis is to enable statements on cost estimate confidence to be addressed, such as "there is an 80 percent chance the program's actual cost will not exceed $250M." How is cost estimate confidence measured?

Probability theory is an ideal formalism for deriving measures of cost estimate confidence. With it, a program's cost can be treated as an uncertain quantity—one sensitive to conditions and assumptions that change across its acquisition life cycle. Figure 8.2 illustrates the conceptual process for using probability theory to analyze cost uncertainty and produce measures of cost estimate confidence.

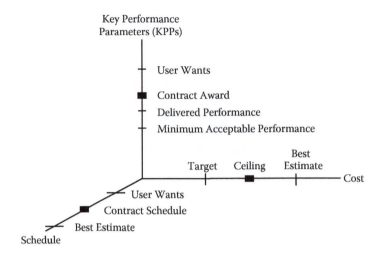

FIGURE 8.1
Pressures on a program manager's decision space.

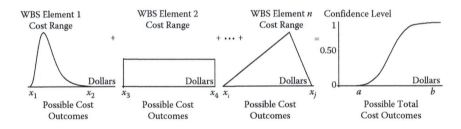

FIGURE 8.2
Cost estimate confidence: a statistical summation of cost uncertainty.

In Figure 8.2, the uncertainty in the cost of each work breakdown struc-ture* (WBS) element is expressed by a probability distribution to characterize its range of possible cost outcomes. These distributions are combined by probability methods to generate an overall distribution of the work break-down structure's total cost, hereafter denoted by the notation $Cost_{WBS}$. This distribution is the range of total cost outcomes possible for the WBS, as it represents a program or system. How does the output from this analytical process enable confidence levels to be determined? Consider Figure 8.3.

Figure 8.3 illustrates a cumulative probability distribution of a work break-down structure's total cost. It derives from an analysis like that in Figure 8.2. Cost estimate confidence is read from this distribution. For example, there is

---

* Refer to military standard *MIL-STD-881C Work Breakdown Structures for Defense Material Items*, October 3, 2011, for information about work breakdown structures, how they are designed, and their roles in the systems engineering life cycle.

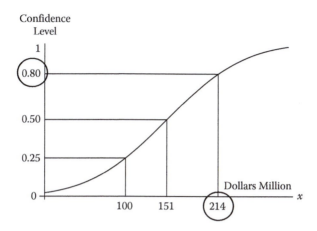

FIGURE 8.3
A distribution of cost estimate confidence.

a 25% chance the program will cost less than or equal to $100M, a 50% chance the program will cost less than or equal to $151M, and an 80% chance the program will cost less than or equal to $214M. These are examples of statistical measures of cost estimate confidence.

Figure 8.3 provides decision-makers an analytical basis for trade-offs between a program's point estimate cost* and its confidence. For example, if a program's point estimate cost is $100M, then Figure 8.3 reveals the amount of additional dollars needed to plan or budget the program at a desired or specified level of confidence. Clearly, the range of possible cost outcomes in Figure 8.3 is wide. Cost analysts can use such a finding to signal a review of the major cost risk drivers responsible for this variance, before settling too soon on a cost confidence level to plan or budget the program. Furthermore, the results presented in Figure 8.3 might spark a series of design options taken to reduce risk as measured by this variance. Figure 8.4 shows how a cost uncertainty analysis of a set of risk-reducing design alternatives can be portrayed.

Discovering these findings and bringing them to decision-makers is a key outcome of the analysis and how it best contributes to ensuring program affordability and cost-effectiveness. Expressing cost estimate confidence by a range of possible cost outcomes has high value to decision-makers. The extent of the cost range itself is a measure of cost uncertainty, which varies across the life cycle. One would expect uncertainty to be higher, and, therefore, the cost range to be wider, early in a program's life cycle. Identifying critical elements that drive a program's cost range offers opportunities for deploying risk mitigation actions in the early acquisition phases.

---

* The point estimate (PE) cost is the cost that does not include allowances for cost uncertainty. The PE cost is the sum of the WBS element costs summed across a program's work breakdown structure without adjustments for uncertainty.

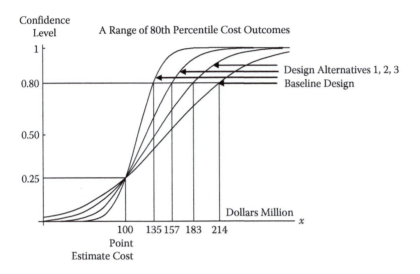

FIGURE 8.4
Reductions in cost risk from competing design options.

8.2 Cost as Probability Distribution

Cost estimates are highly sensitive to many conditions and assumptions that change frequently across a program's life cycle. Examining the change in cost subject to varying certain conditions, while holding others constant, is known as sensitivity analysis. Sensitivity analysis is an excellent way to isolate cost drivers, however, it is a deterministic procedure defined by a postulated set of what-if scenarios. Sensitivity analysis alone does not offer decision-makers insight into the question "What is the chance of exceeding a particular cost in the range of possible costs?" A probability distribution is an ideal way to address this question. In the context of cost uncertainty analysis, a probability distribution is a mathematical rule associating a probability to each cost in a range of possible cost outcomes.

There are two ways to present a probability distribution. It can be shown as a probability density function (PDF) or as a cumulative probability distribution,* as shown in Figure 8.5a and b, respectively.

In Figure 8.5, the range of possible cost outcomes for a program is given by the interval $a \le x \le b$. These distributions reveal the confidence that the true cost of a program will not exceed any cost in the range of possible outcomes. For example, suppose the probability that the true cost of a program will be

---

* The term cumulative probability distribution means the same as the term Cumulative Distribution Function (CDF).

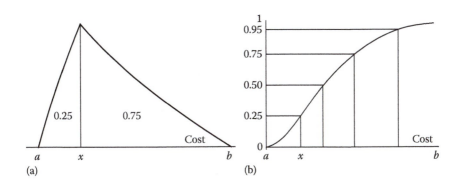

FIGURE 8.5
Ways to view a program's cost probability distribution. (a) Probability Density Function. (b) Cumulative Probability Distribution.

less than or equal to x is 25%. In Figure 8.5a, this probability is given by the area (0.25) under the curve. In Figure 8.5b, this probability is given by the value 0.25 along the vertical axis.

There is an important distinction between the terms *risk* and *uncertainty* and their use in cost analysis. In general, *risk* is the chance of loss or injury. In a situation that includes favorable and unfavorable events, risk is the probability an unfavorable event occurs. In systems engineering risk management, such events might be {failing to achieve performance objectives}, {overrunning the budgeted cost}, or {delivering the system too late to meet user needs}. *Uncertainty* is the indefiniteness about the outcome of a situation—it includes favorable and unfavorable events. We analyze uncertainty to measure risk. *Cost risk* is a measure of the chance that, due to unfavorable events, the planned or budgeted cost of a program will be exceeded. *Cost uncertainty analysis* is the process of measuring the cost impacts of uncertainties associated with a system's technical baseline and cost estimation methodologies.

A cost estimate is stochastic and, therefore, is merely one outcome in a probability distribution of cost outcomes. Mentioned earlier, a cost estimate developed without adjustments for uncertainty is called a point estimate (PE). Thus, a point estimate is just one outcome in a probability distribution of cost outcomes. Measuring the confidence in a point estimate is equivalent to addressing the question "What is the chance of exceeding the point estimate in the range of possible cost outcomes?"

Identifying and measuring confidence in a point estimate is a fundamental objective of cost uncertainty analysis, especially in the early life-cycle phases. Evidence from the community reveals that point estimates are highly uncertain. Recent studies continue to show that, in the early development milestones, the confidence that the final cost of a program falls below its point estimate is less than 50% (Garvey et al. 2012). Basing or planning a program's budget on the point estimate alone is a high-risk decision. GAO found that "budgeting programs to a risk-adjusted point estimate that reflects a program's risks is critical to its success in achieving objectives" (GAO 2009). GAO further observed that overly optimistic assumptions and unrealistic expectations are significant factors for cost growth above point estimates.

Developing a point estimate is traditionally done from a WBS. Shown in Figure 8.6, a WBS is a hierarchical framework that depicts all elements of cost associated with the tasks and activities needed to acquire a program or system.

Work breakdown structures can be complex. They may involve many segments and levels, as well as numerous cost elements. Work breakdown structures are unique to the system under consideration and the program's life-cycle phase. They are developed according to the specific requirements and functions the system has to perform.

The WBS is the cost element structure of the program, where the summation of its element costs across WBS levels forms an estimate of total program cost. Initially, this estimate is usually the point estimate cost. In a similar way, the WBS serves as a cost risk model of the program, where the summation of its element cost ranges across WBS levels forms a probability distribution of possible total cost outcomes, one of which is the point estimate.

Figure 8.7 illustrates using the WBS as a cost risk structure. Within this, analysts develop ranges around individual cost equation parameters or cost elements. These ranges are stochastically summed to produce a probability distribution of $Cost_{WBS}$, shown on the right of Figure 8.7. The vertical axis of the distribution provides the measures of cost estimate confidence. The confidence level of the point estimate can be found by locating where it falls in the range of other possible cost outcomes. In Figure 8.7, the point estimate

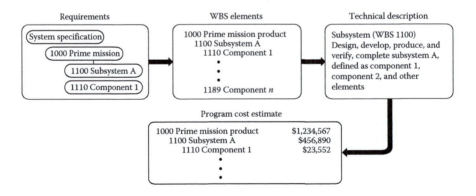

FIGURE 8.6
A WBS integrates requirements, technical, and cost perspectives. (From GAO, Cost estimating and assessment guide: Best practices for developing and managing capital program costs, GAO-09-3SP, March 2009.)

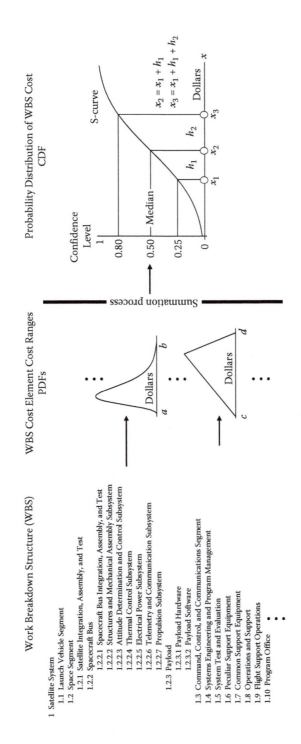

FIGURE 8.7
Measuring cost estimate confidence with probability distributions. Suppose x_1, x_2, and x_3 denote the point estimate, 50th percentile, and 80th percentile costs, respectively.

has a 25% confidence level. This means there is a 75% chance the true cost will exceed the point estimate due to risks identified and quantified in the manner described. In Figure 8.7, values to the right of x_1 are other possible cost outcomes along with their associated confidence levels.

The PDF is the most common form of a probability distribution used to characterize the cost uncertainties of elements that comprise a program's work breakdown structure. This is illustrated in Figure 8.7 by the elements shown on the left, which is the input side of a cost uncertainty analysis. The right side of Figure 8.7 shows the outputs of the analysis, where the CDF or S-curve is the most common form used to express percentile levels of confidence that the actual cost of a program is less than or equal to a value x.

The S-curve in Figure 8.7 provides decision-makers an analytical basis for trade-offs between a program's point estimate cost and its confidence. For example, if a program's point estimate cost is x_1 dollars, then the S-curve reveals the amount of additional dollars needed to plan or budget the program at a desired or specified level of confidence. Cost analysts can use the range of possible costs revealed by the S-curve to signal a review of the major cost risk drivers responsible for its variance, before settling too soon on a cost confidence level to plan or budget the program. Furthermore, the results presented by an S-curve can spark a series of design options taken to reduce risk evident in the variance of the distribution.

In Figure 8.7, x_1 denotes a program's point estimate cost. The difference between x_1 and cost outcomes greater than x_1 is the amount of risk dollars contained in that cost. For instance, if a program is budgeted to the 50th percentile cost x_2, then relative to x_1 there are h_1 risk dollars contained in x_2. If a program is budgeted to the 80th percentile cost x_3, then relative to x_1 there are $(h_1 + h_2)$ risk dollars contained in x_3. Although there are more risk dollars contained in x_3 than in x_2, there is less chance of a program cost overrun at the 80th percentile confidence level than at the 50th percentile—where an overrun has even odds of occurring. Analyzing the probability distribution of $Cost_{WBS}$ in this way provides an otherwise unseen trade-off between cost risk dollars and cost estimate confidence.

Mentioned earlier, risk analysis is an inseparable part of cost analysis. Many different elements of a system's technical baseline and cost estimate are involved, as well as individual experts. Recently, the cost analysis community developed steps to guide the conduct of a cost uncertainty analysis. They are summarized below and were excerpted from the referenced report (GAO 2009).

- Determine the program cost drivers and identify associated risks.
- Develop probability distributions to model the uncertainties affecting program cost.
- Account for correlation between WBS element costs to properly capture factors that influence cost and cost risk.

- Perform the cost uncertainty analysis using Monte Carlo simulation, various analytical approaches, or a combination of methods.
- Identify the probability level associated with the program's point estimate cost.
- Recommend a program budget sufficient to achieve targeted levels of cost estimate confidence.
- Allocate, phase, and convert a risk-adjusted cost estimate to then-year dollars and identify high-risk elements as candidates for targeted risk mitigation efforts.

Along with these steps, the community published guidelines (GAO 2009) to aid in the identification of potential sources of program cost estimate uncertainty. They are presented in Figure 8.8.

8.3 Monte Carlo Simulation and Method of Moments

This section presents the two primary worlds where cost uncertainty analysis methods fall. They are Monte Carlo simulation techniques and nonsimulation approaches categorized by a class of procedures known as method of moments. This discussion introduces these approaches, highlights differences, and provides guidelines on their use. Popular industry tools that execute these approaches are identified and nuances on their application are explained.

This section also identifies and clarifies issues with certain technical topics that arise in cost uncertainty analysis. These include capturing correlation between WBS element costs, use of normal or lognormal distributions to measure confidence intervals around point estimates, and sample sizes for Monte Carlo simulations.

8.3.1 Monte Carlo Method

The Monte Carlo method is a random sampling technique that empirically derives numerically feasible solutions to a mathematical problem. The technique is best applied to problems not amenable to closed form, analytical solutions derived by deterministic algebraic methods.

The Monte Carlo method falls into a class of techniques known as simulation. Simulation has varying definitions among practitioners. For instance, Winston (1994) defines simulation as a technique that imitates the operation of a real-world system as it evolves over time. Rubinstein (1981) offers the following: "simulation is a numerical technique for conducting experiments on a computer, which involves certain types of mathematical and logical models

Uncertainty	Definition	Example
Business or Economic	Variations from change in business or economic assumptions	Changes in labor rate assumptions—e.g., wages, overhead, general and administrative cost—supplier viability, inflation indexes, multiyear savings assumptions, market conditions, and competitive environment for future procurements
Cost Estimating	Variations in the cost estimate despite a fixed configuration baseline	Errors in historical data and cost estimating relationships, variation associated with input parameters, errors with analogies and data limitations, data extrapolation errors, optimistic learning rate assumptions, using the wrong estimating technique, omission or lack of data, misinterpretation of data, incorrect escalation factors, overoptimism in contractor capabilities, optimistic savings associated with new ways of doing business, inadequate time to develop a cost estimate
Program	Risks outside the program office control	Program decisions made at higher levels of authority, indirect events that adversely affect a program, directed funding cuts, multiple contractor teams, conflicting schedules and workload, lack of resources, organizational interface issues, lack of user input when developing requirements, personnel management issues, organization's ability to accept change, other program dependencies
Requirements	Variations in the cost estimate caused by change in the configuration baseline from unforeseen design shifts	Changes in system architecture (especially for system of systems programs), specifications, hardware and software requirements, deployment strategy, critical assumptions, program threat levels, procurement quantities, network security, data confidentiality
Schedule	Any event that changes the schedule: stretching it out may increase funding requirements, delay delivery, and reduce mission benefits	Amount of concurrent development, changes in configuration, delayed milestone approval, testing failures requiring rework, infeasible schedule with no margin, overly optimistic task durations, unnecessary activities, omission of critical reviews

FIGURE 8.8

Potential areas of program cost estimate uncertainty. (From GAO, Cost estimating and assessment guide: Best practices for developing and managing capital program costs, GAO-09-3SP, March 2009.)   (*Continued*)

Uncertainty	Definition	Example
Software	Cost growth from overly optimistic assumptions about software development	Underestimated software sizing, overly optimistic software productivity, optimistic savings associated with using commercial off-the-shelf software, underestimated integration effort, lack of commercial software documentation, underestimating the amount of glue code needed, configuration changes required to support commercial software upgrades, changes in licensing fees, lack of support for older software versions, lack of interface specification, lack of software metrics, low staff capability with development language and platform, underestimating software defects
Technology	Variations from problems associated with technology maturity or availability	Uncertainty associated with unproven technology, obsolete parts, optimistic hardware or software heritage assumptions, feasibility of producing large technology leaps, relying on lower reliability components, design error or omissions

FIGURE 8.8 (*Continued*)
Potential areas of program cost estimate uncertainty. (From GAO, Cost estimating and assessment guide: Best practices for developing and managing capital program costs, GAO-09-3SP, March 2009.)

that describe the behavior of a business or economic system over extended periods of real time."

The Monte Carlo method involves the generation of random samples from known or assumed probability distributions. The process of generating random samples from distributions is known as random variate generation or Monte Carlo sampling. Simulations driven by Monte Carlo sampling are Monte Carlo simulations. One of the earliest applications of Monte Carlo simulation to cost analysis was at the RAND Corporation (Dienemann 1996). Since then, Monte Carlo simulation has become (and remains) a popular approach for modeling and measuring cost uncertainty.

For cost uncertainty analysis, Monte Carlo simulation is typically applied to a WBS to develop an empirical distribution of program cost. The WBS serves as the cost model of the program within which to conduct the simulation. The steps in a Monte Carlo simulation are as follows:

- For each variable in the WBS whose value is uncertain, randomly select one value from the probability distribution that characterizes its uncertainty (e.g., from a uniform or triangle distribution).

- Once a set of possible values for each uncertain variable has been randomly drawn, combine them according to the cost estimation relationships specified in the WBS. This process produces a single randomly generated value for $Cost_{WBS}$, which denotes the work breakdown structure's total cost.

- These steps are repeated thousands of times producing thousands of values of $Cost_{WBS}$. Each value represents one of these thousands of possible outcomes for $Cost_{WBS}$.

- From this step, develop a frequency distribution of the outcomes for $Cost_{WBS}$. This is the empirical probability distribution of $Cost_{WBS}$. It is an approximation to the true (but unknown) underlying distribution of $Cost_{WBS}$, formed by the Monte Carlo simulation running through the WBS.

In cost uncertainty analysis, Monte Carlo simulations are generally static simulations. Static simulations are used to study behavior at a discrete point in time. To illustrate a static simulation, and Monte Carlo sampling, consider the problem of determining the average effort (staff months) to develop a software application. Assume *Effort* is computed by the cost estimation relationship

$$Effort = \frac{Software\ size}{Productivity\ rate} = \frac{S}{P} \tag{8.1}$$

where uncertainties in S and P are given by the uniform distributions shown on the left side of Figure 8.9. In the Monte Carlo method, values of S and P are randomly sampled from their distribution functions. With them, a value for *Effort* is then computed according to Equation 8.1. This process is repeated thousands of times to produce a static simulated distribution of *Effort*. The simulated distribution is an empirical approximation of the exact probability distribution* of *Effort*, shown on the right of Figure 8.9.

In Figure 8.9, ten random samples of S and P are shown. For each S and P pair, the corresponding value for *Effort* is computed according to Equation 8.1. This produces ten random samples (outcomes) of *Effort*, listed on the right side of Figure 8.9. The average of all ten sampled values of *Effort* is 523.14 staff months. This is an empirical approximation of the exact mean of *Effort*, shown in Figure 8.9 as 519.86 staff months. The percentage error between these two values is only 0.63%. Increasing the number of random samples of S and P would further reduce this percentage error. If 100,000 random samples of S and P were drawn, then the percentage error between the

---

* The probability distribution of a combination of random variables can be difficult and, in some cases, intractable to derive in an exact algebraic form (as in Equation 8.2). Monte Carlo simulation always generates an empirical approximation to the exact form of the distribution; in practice, simulation is often the most efficacious approach.

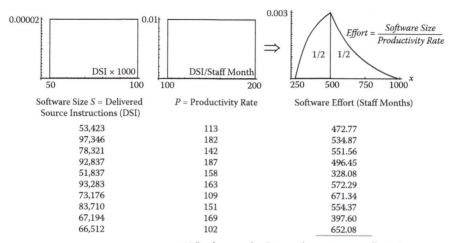

FIGURE 8.9
Monte Carlo sampling: 10 random samples from the distributions of S and P.

simulated mean and the exact mean of *Effort* is reduced to 0.11%. Figure 8.10 shows the negligible difference between the Monte Carlo generated empirical distribution (shown by the dots) and the exact distribution (shown by the solid line). The exact probability distribution of *Effort* in Figure 8.10 is given by Equation 8.2.

$$P(Effort \leq x) = \begin{cases} \left(\dfrac{250}{x} + \dfrac{x}{250}\right) - 2, & \text{if } 250 \leq x < 500 \\ 3 - \left(\dfrac{1000}{x} + \dfrac{x}{1000}\right), & \text{if } 500 \leq x \leq 1000 \end{cases} \quad (8.2)$$

Figure 8.11 illustrates ways to present a probability distribution. Figure 8.11a is the probability density function of *Effort*. Figure 8.11b is the cumulative distribution function for *Effort*, informally called the S-curve. In Figure 8.11, the range of possible values of *Effort* is the interval $250 \leq x \leq 1000$. The PDF or CDF reveal the confidence of not exceeding any value in the range of possible values. For example, the probability that the true software development effort will be less than or equal to $x = 500$ staff months is 50%. In Figure 8.11a, this probability is given by the area under the curve. In Figure 8.11b, this probability is given by $y = 0.5$ along the vertical axis.

Figure 8.11 illustrates a property about the mean, median, and mode of a probability distribution. They are equal when the distribution is normal; however, because the distribution in Figure 8.11 is skewed, the mean,

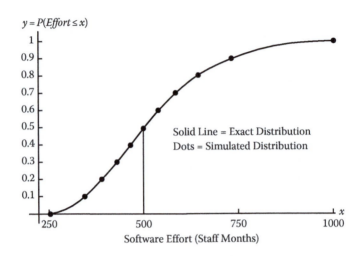

FIGURE 8.10
Comparing a Monte Carlo simulation to the exact distribution of *Effort*.

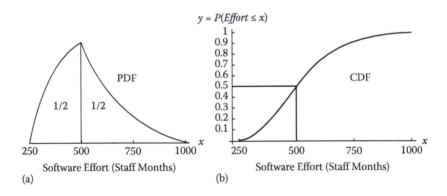

FIGURE 8.11
The exact PDF and CDF of *Effort*.

median, and mode are not equal. Here, the median and the mode are each 500 staff months but the exact mean of this distribution is 519.86 staff months. In this case, the mean is larger than the median or the mode because the probability distribution of *Effort* is skewed to the right.* It is

---

* The mode of a distribution is the value of x at the peak of the distribution. In Figure 8.11, this occurs at $x = 500$ staff months. The median of a distribution is the value of x that occurs at exactly the 50th percentile. In Figure 8.11, the median accounts for $1/2$ the area under the PDF or by the value $y = 0.50$ along the vertical axis of the CDF. In general, the mean, median, and mode of a probability distribution are not always the same.

an exercise for the reader to derive the exact forms of the PDF and CDF in Figure 8.11.

In Monte Carlo simulations, a question often asked is "How many trials (random samples) are needed to have confidence in the outputs of the Monte Carlo simulation?" The statistical literature provides guidelines for determining sample size as a function of the precision desired in the outputs of a simulation. Specifically, the formulas below address the question: "What sample size is needed so that with probability α the true value of an uncertain variable falls between a pair of values generated by the Monte Carlo simulation?"

Morgan–Henrion Guideline (Morgan and Henrion 1990): Define m as the sample size for the Monte Carlo simulation. Let x_p be the p-fractile of the random variable X; that is, $P(X \leq x_p) = p$. Let c satisfy the probability $P(-c \leq Z \leq c) = \alpha$, where $Z \sim N(0,1)$ is the standard normal probability distribution. Then, the pair of fractiles $(\hat{x}_i, \hat{x}_k)$ generated by the Monte Carlo simulation, with

$$i = p - c\sqrt{\frac{p(1-p)}{m}} \tag{8.3}$$

$$k = p + c\sqrt{\frac{p(1-p)}{m}} \tag{8.4}$$

contains x_p with probability α. For different sample sizes, Figure 8.12 shows with probability $\alpha = 0.95$ (which means $c \approx 2$) the values of i and k such that the true median $x_{0.50}$ of the distribution falls between $\hat{x}_i$ and $\hat{x}_k$. The lower and upper curves in Figure 8.12 are generated from Equations 8.3 and 8.4, respectively. As the sample size m increases, the difference between these curves

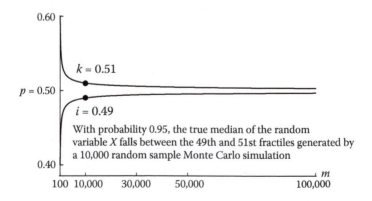

FIGURE 8.12
Sample size for Monte Carlo simulations.

decreases dramatically. With 100 samples, the true median value $x_{0.50}$ of the random variable X falls between the pair of fractiles $\hat{x}_{0.40}$ and $\hat{x}_{0.60}$, generated by the Monte Carlo simulation, with probability 0.95. Increasing that sample size by a factor of 100 brings the same degree of confidence to be within $\hat{x}_{0.49}$ to $\hat{x}_{0.51}$. In practice, 10,000 trials (or random samples) are generally sufficient to meet the precision requirements for Monte Carlo simulations when used for cost uncertainty analysis.

Monte Carlo simulation is commonly applied to a work breakdown structure (WBS) when it is used to derive a probability distribution of a program's total cost. Mentioned in Section 8.2, the WBS is the definitive cost element structure and cost model of a program, where the summation of element costs across WBS levels forms an estimate of a program's total cost. In a similar way, the WBS serves as a cost risk model of the program, where the summation of element cost ranges across WBS levels forms a probability distribution of possible total cost outcomes, one of which is the point estimate. This was illustrated in Figure 8.7. For convenience, this is shown in Figure 8.13. The following illustrates a Monte Carlo simulation applied to a WBS consisting of five cost elements.

Figure 8.14 presents a simple work breakdown structure consisting of five cost elements X_1, X_2, X_3, X_4, and X_5. A point estimate cost for each element is shown, along with an uncertainty distribution around each estimate. For each WBS element, a random value from its cost uncertainty distribution is taken. These randomly selected values are then summed to form one estimate of total cost. This process is repeated thousands of times (e.g., 10,000 times or more) to produce an empirically derived overall probability distribution of total cost. This is the Monte Carlo simulation process, with each circle in Figure 8.14 representing a single thread or single pass through one of the thousands of Monte Carlo samples.

Mentioned earlier, the outcome of this process is a frequency distribution derived from these n sampled values. This distribution is the Monte Carlo simulated distribution of total cost. Figure 8.15 presents the results of 10,000 Monte Carlo samples of the WBS in Figure 8.14. It shows the resultant probability distribution of the WBS total cost, given by Equation 8.5.

$$Cost_{WBS} = X_1 + X_2 + X_3 + X_4 + X_5 \qquad (8.5)$$

In Figure 8.15, the dots are the probability distribution $Cost_{WBS}$ generated by the Monte Carlo simulation of the WBS in Figure 8.14. The simulation produced a total cost mean μ of \$203.3M and a standard deviation σ of \$19M. In Figure 8.15, the solid line is the probability distribution of $Cost_{WBS}$ assuming its possible cost outcomes fall along a normal probability distribution—with mean and variance given by the sums of the means and variances of the WBS cost element distributions in Figure 8.14.

Observe the closeness of the simulated and normal probability distributions in Figure 8.15. There are reasons for this result. One reason is due

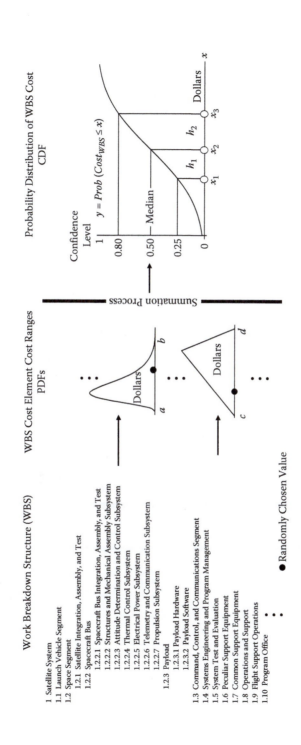

FIGURE 8.13
Monte Carlo simulation applied to a work breakdown structure.

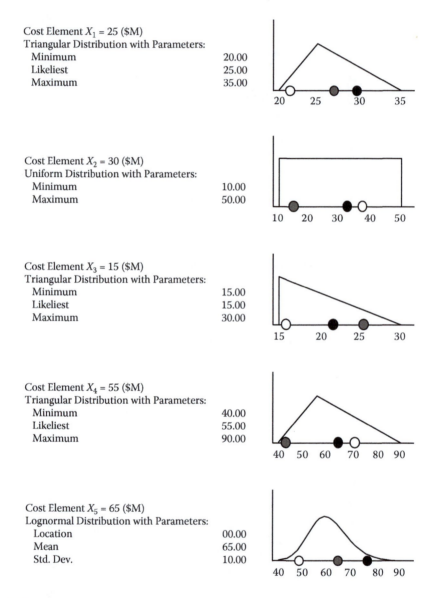

FIGURE 8.14
A WBS for Monte Carlo simulation.

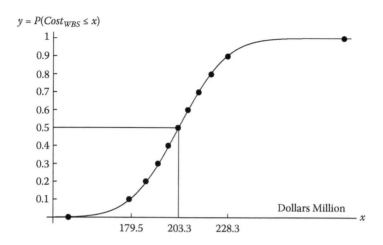

FIGURE 8.15
Probability distribution of $Cost_{WBS}$.

to the assumed mutual independence between WBS element costs X_1, X_2, X_3, X_4, and X_5 given in Figure 8.14. With this, the famous central limit theorem (CLT) enters the picture and ensures the eventual tendency of the simulated distribution to approach a normal distribution.*

Correlation and Monte Carlo Simulations: For years, the cost analysis community has addressed the topic of correlation and how to represent it in a cost uncertainty analysis. Despite a large body of technical work on this topic, the community needs practical guidance consistent with the subtleties of statistical theory. Given this, this section offers a practical approach for modeling correlation and capturing its effects on program cost. The approach is presented in the context of Monte Carlo simulations, but it can easily be used with other cost uncertainty analysis methods. Further in-depth discussions and guidance are presented in Chapter 9; first, some background.

Discussed in Section I of this book, correlation ρ is a statistical measure of the "co-variation" between two random variables. It measures the strength and direction of change in one random variable with change in another random variable. Regarding strength, correlation is a continuous measure

---

* Informally, the CLT establishes the fundamental result that the sum of independent identically distributed random variables with finite variance approaches a normally distributed random variable as their number increases; in particular, if enough random samples (e.g., Monte Carlo samples) are repeatedly drawn from any distribution, the sum of the sample values can be thought of, approximately, as an outcome from a normally distributed random variable. The requirement for identically distributed random variables has been relaxed in modern variants of the original CLT.

whose magnitude ranges between negative one and positive one. Regarding direction, correlation can be positive or negative. *Positive correlations* fall in the interval $0 < \rho \leq 1$. *Negative correlations* fall in the interval $-1 \leq \rho < 0$. Uncorrelated random variables have correlation $\rho = 0$.

From a cost analysis perspective, the strength of correlations between WBS element costs impacts the magnitude of the program's total cost risk— measured by the variance or standard deviation of the program's cost probability distribution.* The positive or negative direction of correlation affects whether total cost risk increases or decreases, respectively. Thus, correlation is a required consideration in modeling the cost uncertainty of a program. Ignoring correlation can be equivalent to setting its value to zero, which can significantly *underestimate* a program's total cost risk. The extent of the potential underestimation is highlighted in Figure 8.16.

Figure 8.16 shows the percent that a program's total cost risk is underestimated if correlation was assumed to be zero between all pairs of WBS element costs, instead of a positive constant value between them. In Figure 8.16, consider the curve shown for a WBS with $n = 10$ cost elements. Suppose $\rho = 0$ was assumed for all cost element pairs in this WBS. If it later became known that the actual correlation was 0.4, then the program's total cost risk has been *underestimated* by 53%. In a 30-element WBS, the underestimation is 72%.

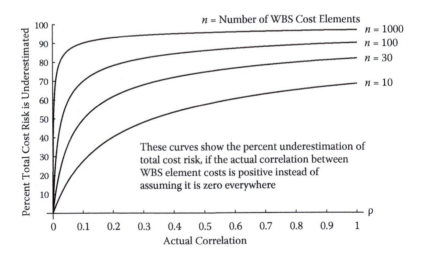

FIGURE 8.16
Potential underestimation of total cost risk. (From Book, S., Why correlation matters in cost estimating, in *32nd Annual DOD Cost Analysis Symposium (DODCAS)*, The Aerospace Corporation, Los Angeles, CA.)

---

* The standard deviation σ is the square root of the variance σ^2 of a random variable.

As shown in Figure 8.16, if the actual ρ is greater than an assumed $\rho = 0$, then the underestimation of a program's total cost risk worsens exponentially with each increase in the number of WBS cost elements. A detailed discussion of this topic is presented in Chapter 9.

The cost of a program derives from a summation process when it is based on a WBS, where a program's total cost is the sum of its work WBS costs. Furthermore, the statistical mean of the program's total cost is the sum of the statistical means of its WBS element costs. However, the statistical variance of the program's total cost is not the sum of the statistical variances of its WBS element costs. To see why, we need only look at the sum of two random variables X and Y. What is the variance of their sum? The answer is given by Equation 8.6.

$$Var(X + Y) = \sigma^2_{X+Y} = Var(X) + Var(Y) + 2\rho_{X,Y}\sigma_X\sigma_Y \qquad (8.6)$$

The variance of $(X + Y)$ is not just the sum of the variance of X plus the variance of Y. The last term in Equation 8.6 is called the covariance (or the "co-variation") between X and Y. In Equation 8.6, $\rho_{X,Y}$ is the Pearson product-moment correlation between X and Y and $\sigma_X\sigma_Y$ is the product of their respective standard deviations. This is technically why and how correlation enters the scene. The Monte Carlo simulation of the work breakdown structure in Figure 8.14 reflected the condition that all pairs of WBS element costs were uncorrelated—a change in the cost of one element was not associated with a change in the cost of another element. However, this condition is not the common case. In WBSs more complex than the one given in Figure 8.14, correlation is often found between many pairs of WBS element costs. If the presence of correlation is such that an increase in the cost of one element is associated with an increase in the cost of another element, then this positive correlation causes the variance of the sum of these WBS element costs to increase. The reason for this is seen in Equation 8.7.

$$Var(X + Y) = \sigma^2_{X+Y} = \begin{cases} Var(X) + Var(Y), & \text{if } \rho_{X,Y} = 0 \\ Var(X) + Var(Y) + \underbrace{2\rho_{X,Y}\sigma_X\sigma_Y}, & \text{otherwise} \end{cases}$$

$$Covariance(X, Y)$$

The variance of $X + Y$ changes by this amount, which

is the covariance between X and Y, when Pearson

product-moment correlation is present between them

$$(8.7)$$

Equation 8.7 is the variance of the sum of just two random variables X and Y if $\rho_{X,Y} = 0$ (X and Y are uncorrelated) or if $\rho_{X,Y} \neq 0$ (X and Y are correlated).

The term $2\rho_{X,Y}\sigma_X\sigma_Y$ is the covariance of X and Y. In Equation 8.7, the covariance becomes larger and larger as more and more random variables with positive correlations are summed. An example is shown by the equation below.

$$Var(X + Y + Z) = \sigma^2_{X+Y+Z}$$

$$= \begin{cases} Var(X) + Var(Y) + Var(Z), & \text{if } \rho_{X,Y}, \rho_{X,Z}, \rho_{Y,Z} = 0 \\ Var(X) + Var(Y) + Var(Z), & \text{otherwise} \\ \quad + \underbrace{2\rho_{X,Y}\sigma_X\sigma_Y + 2\rho_{X,Z}\sigma_X\sigma_Z + 2\rho_{Y,Z}\sigma_Y\sigma_Z} \end{cases}$$

$$Covariance(X, Y, Z)$$

The variance of $X + Y + Z$ changes by this amount, which is the covariance between X, Y, and Z, when Pearson product-moment correlation is mutually present between them

Thus, correlation between WBS element costs in a WBS can have significant effects on the magnitude of cost risk, given by the variance σ^2 or the standard deviation σ of the total cost probability distribution.

Pearson and Spearman Rank Correlation Measures: In Equation 8.6, the correlation coefficient $\rho_{X,Y}$ is the Pearson product-moment correlation. Pearson's correlation measures the strength and direction of the linearity between X and Y. It is the *only technically correct* correlation measure for summing the variances of random variables and, in our context, when summing the variances of a program's WBS element costs.

There are other types of correlation measures in statistics. One is Spearman's rank correlation. Rank correlation is also measured in the interval $-1 \le \rho_{rank} \le 1$. It measures the strength and direction of the monotonicity between two random variables. Monotonicity and linearity can be different behaviors between pairs of random variables. Thus, Spearman's rank correlation and Pearson's product-moment correlation are not guaranteed to produce the same measures. The following illustrates this point.

In Figure 8.17, suppose Y is a random variable that is a function of X. Suppose X is a random variable whose outcomes are uniformly distributed in the interval 0 to 1. In Figure 8.17, the function at the top has a Pearson correlation of $\rho_{X,Y} = 0.8732$ and a rank correlation equal to one ($\rho_{rank} = 1$). The function at the bottom has a Pearson correlation of $\rho_{X,Y} = 0.7861$ and a rank correlation equal to one ($\rho_{rank} = 1$). Why are the Pearson and rank correlations so different? The answer is because Y is a perfectly monotonically increasing function of X over the indicated domains. Rank correlation measures the strength and direction of monotonicity *only*. A function Y can have

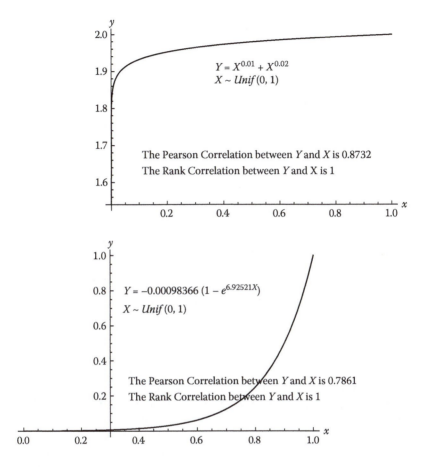

FIGURE 8.17
Comparing Pearson and rank correlation measures.

a perfect rank correlation *but a very different Pearson correlation*. This is because a monotonic function need not be a linear function.

Guidance on Capturing Correlation in Monte Carlo Simulations: As mentioned earlier, correlation cannot be ignored—doing so can be equivalent to setting it equal to zero. Pearson correlation is the *technically correct* correlation measure for summing the variances of random variables. Caution is needed to not double count or mix different types of correlation in the analysis. It can lead to the generation of invalid measures of cost risk or measures that are unrealistically under- or overestimated (refer to Section 9.2).

Chapter 9 presents three ways of valuing correlations between cost elements in a program's work breakdown structure. One approach falls into a class of subjective methods for situations where a lack of time or information may preclude deriving correlations between WBS element costs. This is when correlations might be assigned. The second falls into a class of methods where correlation is mathematically derived from the structures of cost estimation relationships defined within the WBS. The third approach makes use of Monte Carlo simulation to empirically determine any needed correlations.

8.3.2 Method of Moments

The method of moments is a procedure in classical statistics to estimate the unknown mean and variance of a population by random samples from the population. Inferences from the sample mean and sample variance are used to make inferences about the population mean and population variance. A *moment* is a statistical term referring to central tendency measures of a random variable or its probability distribution. The mean is the first moment. The variance is a function of the second moment of a random variable.

In cost uncertainty analysis, the method of moments (MOM) refers to deriving the mean and variance of the cost of a program as functions of the means and variances of the costs of its WBS elements. From these measures, the probability distribution of $Cost_{WBS}$ is formed. The method of moments produces analytically derived measures of cost risk, while Monte Carlo simulation empirically derives them from thousands of random trials or samples.

Method of Moments: Applied to Work Breakdown Structures: The method of moments is commonly applied to a WBS when it is used to derive the probability distribution of $Cost_{WBS}$. Mentioned in Section 8.2, the WBS is the definitive cost element structure and cost model of a program, where the summation of WBS element costs across WBS levels forms an estimate of total program cost. Similarly, the WBS serves as a cost risk model of the program. Here, the summation of WBS element cost ranges across WBS levels forms a probability distribution of possible outcomes of total program cost, one of which is the point estimate. This is shown in Figure 8.18. The following applies a method of moments approach to the same WBS in Figure 8.14, which was used to illustrate the Monte Carlo simulation technique.

Figure 8.19 presents a work breakdown structure consisting of the five cost elements X_1, X_2, X_3, X_4, and X_5. A point estimate cost for each element is shown, along with an uncertainty distribution around each estimate. The cost mean and cost variance of each WBS element is analytically derived instead of empirically generated from random Monte Carlo samples. The cost means and cost variances of the WBS elements are summed and the probability distribution of $Cost_{WBS}$ is formed. In practice, this distribution is usually well approximated by a normal or a lognormal form.* Consider the following example.

$$Cost_{WBS} = X_1 + X_2 + X_3 + X_4 + X_5 \qquad (8.8)$$

---

* The following section will discuss the applicability of normal and lognormal forms in further detail.

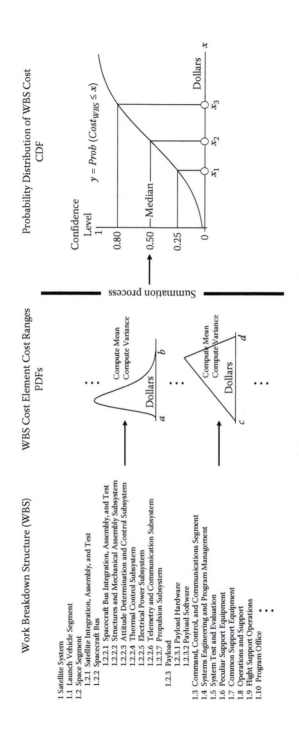

FIGURE 8.18
Method of moments applied to a WBS.

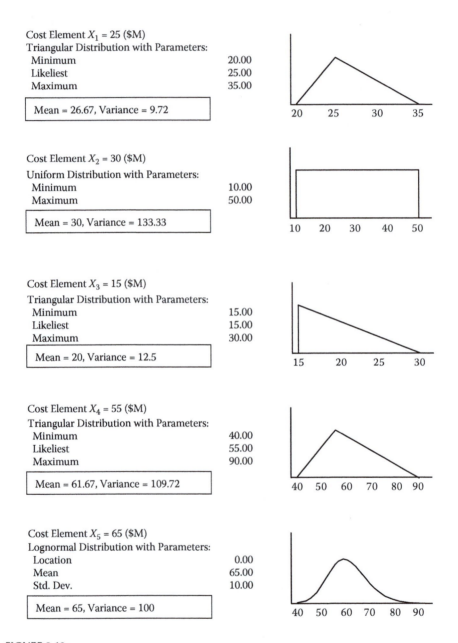

Cost Element X_1 = 25 ($M)
Triangular Distribution with Parameters:
  Minimum   20.00
  Likeliest   25.00
  Maximum   35.00

Mean = 26.67, Variance = 9.72

Cost Element X_2 = 30 ($M)
Uniform Distribution with Parameters:
  Minimum   10.00
  Maximum   50.00

Mean = 30, Variance = 133.33

Cost Element X_3 = 15 ($M)
Triangular Distribution with Parameters:
  Minimum   15.00
  Likeliest   15.00
  Maximum   30.00

Mean = 20, Variance = 12.5

Cost Element X_4 = 55 ($M)
Triangular Distribution with Parameters:
  Minimum   40.00
  Likeliest   55.00
  Maximum   90.00

Mean = 61.67, Variance = 109.72

Cost Element X_5 = 65 ($M)
Lognormal Distribution with Parameters:
  Location   0.00
  Mean   65.00
  Std. Dev.   10.00

Mean = 65, Variance = 100

FIGURE 8.19
A WBS for method of moments.

Thus, the mean and variance of $Cost_{WBS}$ is

$$E(Cost_{WBS}) = E(X_1) + E(X_2) + E(X_3) + E(X_4) + E(X_5) \tag{8.9}$$

$$Var(Cost_{WBS}) = Var(X_1) + Var(X_2) + Var(X_3)$$
$$+ Var(X_4) + Var(X_5) \tag{8.10}$$

From the information in Figure 8.19, it follows that

$$E(Cost_{WBS}) = 26.67 + 30 + 20 + 61.67 + 65 \tag{8.11}$$

$$E(Cost_{WBS}) = 203.34 \ (\$M)$$

$$Var(Cost_{WBS}) = 9.72 + 133.33 + 12.5 + 109.72 + 100$$

$$= 365.27 \ (\$M)^2 \tag{8.12}$$

$$\sigma_{Cost_{WBS}} = \sqrt{Var(Cost_{WBS})} = 19.11 \ (\$M) \tag{8.13}$$

The mean and standard deviation computed by the method of moments, shown by Equations 8.9 through 8.13, are very close to these same statistics derived empirically by the Monte Carlo simulation discussed in Section 8.3.1. Mutual independence between WBS element costs X_1, X_2, X_3, X_4, and X_5 has been assumed in this example. Mutual independence implies these five WBS element costs are mutually uncorrelated (refer to Theorem 5.4). Thus, the variance of the sum of X_1, X_2, X_3, X_4, and X_5 is the sum of their individual variances. Chapter 9 extends this discussion to illustrate how correlations between WBS element costs can be incorporated into $Var(Cost_{WBS})$.

Figure 8.20 presents the probability distribution of $Cost_{WBS}$. The dots depict the probability distribution of $Cost_{WBS}$ generated by a Monte Carlo simulation of the same WBS in Figure 8.14. The simulation produced a mean of $Cost_{WBS}$ equal to \$203.3M and a standard deviation of $Cost_{WBS}$ equal to \$19M. In Figure 8.20, the solid line is the probability distribution of $Cost_{WBS}$ assuming its possible cost outcomes fall along a normal probability distribution—with mean and variance derived by the method of moments. The dashed line is the probability distribution of $Cost_{WBS}$ assuming its possible cost outcomes fall along a lognormal probability distribution—with mean and variance derived by the method of moments.

In Figure 8.20, the simulated, normal, and lognormal probability distributions are almost indistinguishable. Mentioned earlier, a reason for this is the assumed mutual independence between WBS element costs X_1, X_2, X_3, X_4, and X_5 in this example. The central limit theorem (CLT) enters the picture and ensures the eventual tendency of the simulated distribution to approach a

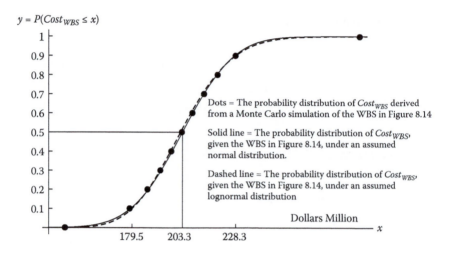

FIGURE 8.20
Probability distribution of $Cost_{WBS}$.

normal distribution. Figure 8.20 visually reveals the lognormal distribution also approximates the normal or simulated distributions of $Cost_{WBS}$. The goodness of these approximations is often seen in cost uncertainty analysis, which greatly supports the use of method of moments approaches. Appendix D illustrates the method of moments applied to a WBS with correlated WBS element costs.

Extending the Method of Moments: The preceding illustrated the method of moments applied to a WBS consisting of five cost elements. This section extends the method of moments to a WBS of n cost elements. Let

$$Cost_{WBS} = X_1 + X_2 + X_3 + \cdots + X_n \tag{8.14}$$

It then follows that the mean or expected value of $Cost_{WBS}$ is

$$E(Cost_{WBS}) = E(X_1) + E(X_2) + E(X_3) + \cdots + E(X_n) \tag{8.15}$$

If the costs of cost elements $X_1, X_2, X_3, \ldots, X_n$ are independent, then the variance of $Cost_{WBS}$ is

$$Var(Cost_{WBS}) = Var(X_1) + Var(X_2) + Var(X_3) + \cdots + Var(X_n) \tag{8.16}$$

If the costs of cost elements $X_1, X_2, X_3, \ldots, X_n$ are *not* independent, then the variance of $Cost_{WBS}$ is

$$Var(Cost_{WBS}) = Var(X_1) + Var(X_2) + Var(X_3) + \cdots + Var(X_n)$$
$$+ 2\rho_{X_1,X_2}\sigma_{X_1}\sigma_{X_2} + 2\rho_{X_1,X_3}\sigma_{X_1}\sigma_{X_3} + \cdots + 2\rho_{X_i,X_j}\sigma_{X_i}\sigma_{X_j}$$
$$(8.17)$$

for all i and j such that $1 \leq i \leq j \leq n$. In Equation 8.17, ρ_{X_i,X_j} is the Pearson product-moment correlation between the costs of WBS element random variables X_i and X_j. Once the values for $E(Cost_{WBS})$ and $Var(Cost_{WBS})$ are computed, they are used in the method of moments to specify a lognormal or normal probability distribution of $Cost_{WBS}$, with mean $E(Cost_{WBS})$ and variance $Var(Cost_{WBS})$. The following illustrates how to create these probability distributions in Excel.

How to Specify a Lognormal Probability Distribution: Given the WBS in Figure 8.19, the method of moments was used to derive the mean and variance of $Cost_{WBS}$; specifically, it was shown that

$$E(Cost_{WBS}) = 203.34 \ (\$M) \quad \text{and} \quad Var(Cost_{WBS}) = 365.27 \ (\$M)^2$$

Microsoft Excel can be used to specify a lognormal distribution with these computed statistics. This is illustrated by the following steps.

Step 1. Transform $E(Cost_{WBS})$ and $Var(Cost_{WBS})$: Transform the mean and variance of $Cost_{WBS}$, computed by the method of moments, into the lognormal parameters μ and σ^2 given by Equations 8.18 and 8.19

$$\mu = \frac{1}{2}\ln\left[\frac{a^4}{a^2 + b}\right] \tag{8.18}$$

$$\sigma^2 = \ln\left[\frac{a^2 + b}{a^2}\right] \tag{8.19}$$

where $a = E(Cost_{WBS})$ and $b = Var(Cost_{WBS})$

Step 2. Compute and Enter Parameters μ and σ^2 into Excel: In this step, we compute the values for μ and σ^2 given by Equations 8.18 and 8.19. Define

$$a = E(Cost_{WBS}) = 203.34(\$M) \tag{8.20}$$

$$b = Var(Cost_{WBS}) = 365.27 \ (\$M)^2 \tag{8.21}$$

From Equations 8.18 and 8.19, it follows that

$$\mu = \frac{1}{2} \ln\left[\frac{a^4}{a^2 + b}\right] = \frac{1}{2} \ln\left[\frac{(203.34)^4}{(203.34)^2 + 365.27}\right] = 5.31048 \tag{8.22}$$

$$\sigma^2 = \ln\left[\frac{a^2 + b}{a^2}\right] = \ln\left[\frac{(203.34)^2 + 365.27}{(203.34)^2}\right] = 0.00879543 \tag{8.23}$$

$$\sigma = \sqrt{\sigma^2} = 0.093784 \tag{8.24}$$

From these equations, the values for a, b, μ, and σ are used to form the lognormal probability distribution of $Cost_{WBS}$. Table 8.1 is one way to use Excel with

TABLE 8.1

Excel Function to Compute Lognormal Probabilities for $Cost_{WBS}$

	A	B	C	D	E	F
1					If	Then
2	Equation				$\alpha =$	$x =$
3	8.20	$a =$	203.34		0.00	151.5
4	8.21	$b =$	365.27		0.10	179.5
5	8.22	$\mu =$	5.31048		0.20	187.1
6	8.24	$\sigma =$	0.093784		0.30	192.7
					0.40	197.7
		Cumulative Lognormal Probability Distribution			0.50	202.4
		of $Cost_{WBS}$			0.60	207.3
					0.70	212.7
					0.80	219.1
					0.90	228.3
					1.00	270.5
					F3 = LOGNORM.INV(E3,C5,C6)	
					F4 = LOGNORM.INV(E4,C5,C6)	
					F5 = LOGNORM.INV(E5,C5,C6)	
					F6 = LOGNORM.INV(E6,C5,C6)	

outputs from the method of moments to derive the lognormal probability distribution of $Cost_{WBS}$. Columns A, B, and C are inputs from Step 2. Column E is the probability α that $Cost_{WBS}$ does not exceed x dollars; that is, $P(Cost_{WBS} \leq x) = \alpha$. In Table 8.1, observe how the values computed by Excel in Column F track to the lognormal probability distribution shown by the dashed line in Figure 8.20.

How to Specify a Normal Probability Distribution: Excel can be used to specify a normal probability distribution of $Cost_{WBS}$, with mean $E(Cost_{WBS})$ and $Var(Cost_{WBS})$ derived from the method of moments. Table 8.2 illustrates how to create this distribution in Excel. Columns A, B, and C are the computed values of $E(Cost_{WBS})$ and $Var(Cost_{WBS})$ derived from the method of moments. Column E is the probability α that $Cost_{WBS}$ does not exceed x dollars; that is $P(Cost_{WBS} \leq x) = \alpha$. In Table 8.2, observe how the values computed by Excel in Column F track to the normal probability distribution shown by the solid line in Figure 8.20.

In the preceding discussion, the method of moments produced an analytically derived measure of the mean and variance of $Cost_{WBS}$. Monte Carlo simulation generates an empirically derived basis for these two measures, as well as an empirically derived probability distribution of $Cost_{WBS}$. To produce the probability distribution of $Cost_{WBS}$ using the method of moments,

TABLE 8.2

Excel Function to Compute Normal Probabilities for $Cost_{WBS}$

	A	B	C	D	E	F
1					If	Then
2	Equation				$\alpha =$	$x =$
3	8.20	$a =$	203.34		0.00	144.3
4	8.21	$b =$	365.27		0.10	178.8
5		$\sqrt{b} =$	19.11		0.20	187.3
6					0.30	193.3
					0.40	198.5
		Cumulative Normal Probability Distribution			0.50	203.3
		of $Cost_{WBS}$			0.60	208.2
					0.70	213.4
					0.80	219.4
					0.90	227.8
					1.00	262.4

$F3 = \text{NORM.INV}(E3,\$C\$5,\$C\$6)$
$F4 = \text{NORM.INV}(E4,\$C\$5,\$C\$6)$
$F5 = \text{NORM.INV}(E5,\$C\$5,\$C\$6)$
$F6 = \text{NORM.INV}(E6,\$C\$5,\$C\$6)$

the form or shape of this distribution must be assumed. Best practice observations, published evidence, and statistical tests indicate the probability distribution of $Cost_{WBS}$ is often approximated by normal or lognormal forms. This is numerically evident in Figure 8.20. Chapter 6 provides an extensive discussion on reasons why these distributional forms are so often seen in cost uncertainty analyses.

In Figure 8.20, the normal and lognormal distributions well approximate the empirically derived probability distribution of $Cost_{WBS}$—shown by the dots generated by the Monte Carlo simulation. There are many technical and empirically observed reasons for this. Primary among them is that a program's total cost is a summation of WBS element costs, including a summation of costs derived from nonlinear cost estimation relationships. Within the WBS, it is typical to have a mixture of independent and correlated element costs. The greater the number of independent WBS element costs, the more it is that the probability distribution of $Cost_{WBS}$ is approximately normal. Why is this? As mentioned in Chapter 6, it is essentially the phenomenon explained by the CLT.

8.4 Summary

The cost of a future system can be significantly affected by uncertainty. The existence of uncertainty implies the existence of a range of possible costs. How can a decision-maker be shown the chance a particular cost in the range of possible costs will be realized? The probability distribution is a recommended approach for providing this insight.

Probability distributions result when independent variables (e.g., weight, power-output, schedule) used to derive a system's cost randomly assume values across ranges of possible values. For instance, the cost of a satellite might be derived using a range of possible weight values, with each value randomly occurring. This approach treats cost as a random variable. It is recognized that values for these variables (such as weight) are not typically known with sufficient precision to predict cost perfectly, *at a time when such predictions are needed*. This point is further articulated by well-respected analyst S. A. Book.*

> The mathematical vehicle for working with a range of possible costs is the probability distribution, with cost itself viewed as a "random variable." Such terminology does not imply, of course, that costs are "random" (though well they may be!) but rather that they are composed of a large number of very small pieces, whose individual contributions to the whole

---

* Book, S. A. 1997. Cost Risk Analysis—A tutorial. In *Risk Management Symposium Proceedings*. Los Angeles, CA: The Aerospace Corporation.

we do not have the ability to investigate in a degree of detail sufficient to calculate the total cost precisely. It is much more efficient for us to recognize that virtually all components of cost are simply "uncertain" and to find some way to assign probabilities to various possible ranges of costs.

An analogue is the situation in coin tossing where, in theory, if we knew all the physics involved and solved all the differential equations, we could predict with certainty whether a coin would fall "heads" or "tails." However, the combination of influences acting on the coin is too complicated to understand in sufficient detail to calculate the physical parameters of the coin's motion. So we do the next best thing: we bet that the uncertainties will probably average out in such a way that the coin will fall "heads" half the time and "tails" the other half. It is much more efficient to consider the deterministic physical process of coin tossing to be a "random" statistical process and to assign probabilities of 0.50 to each of the two possible outcomes, heads or tails.

In summary, cost uncertainty analysis provides decision-makers many benefits and important insights. These include

Establishing a Cost and Schedule Risk Baseline: Baseline probability distributions of program cost and schedule should be developed for a given system configuration (its technical baseline), acquisition strategy, and cost-schedule estimation approach. The baseline provides decision-makers visibility into potentially high-payoff areas for risk reduction initiatives. Baseline distributions assist in determining a program's cost and schedule that simultaneously have a specified probability of not being exceeded. They also provide decision-makers an assessment of the chance of achieving a budgeted (or proposed) cost and schedule, or cost for a given feasible schedule.

Measuring Cost Risk: Cost uncertainty analysis provides a basis for measuring the overall cost risk inherent to a program as a function of its specific uncertainties. This can be measured by the difference between the program point estimate cost and the cost at a predefined confidence level, as set by budgetary decisions or management policies.

Conducting Risk Reduction Trade-off Analyses: Cost uncertainty analysis can be conducted to study the payoff of implementing risk reduction initiatives on lessening a program's cost, schedule, and performance risks. Families of probability distribution functions, as shown in Figure 8.4, can be generated to compare the cost and cost risk impacts of competing design options or acquisition strategies.

Documenting Program Risks and Risk Analysis Inputs: The validity and influence of any cost uncertainty analysis relies on the engineering and cost team's experience, judgment, and knowledge of their program's risks and uncertainties. Documenting the team's insights into these considerations is a critical part of the process. Without it, the veracity of the analysis is easily questioned. Details about the analysis methodology, especially assumptions, are important to document. The methodology must be technically sound, traceable, and offer value-added problem structure and insights otherwise not visible. Decisions that successfully reduce or eliminate risk ultimately rest on human judgment. At best, this is aided by, not directed by, the methods in this book.

Exercises

8.1 Figure 8.8 identified a list of potential areas of program cost estimate uncertainty. Think about areas not shown that should be listed given today's engineering systems and acquisition environment. Include a definition of each new area and give examples, as shown in the second and third columns in Figure 8.8.

8.2 From the information provided in Figure 8.9, derive the probability equation below.

$$P(\textit{Effort} \leq x) = \begin{cases} \left(\dfrac{250}{x} + \dfrac{x}{250}\right) - 2, & \text{if } 250 \leq x < 500 \\[2ex] 3 - \left(\dfrac{1000}{x} + \dfrac{x}{1000}\right), & \text{if } 500 \leq x \leq 1000 \end{cases}$$

8.3 a. For a 30,000 sample run of a Monte Carlo simulation, find the values of i and k such that the true median $x_{0.50}$ of the distribution of X falls between $(\hat{x}_i, \hat{x}_k)$ with probability 0.90, 0.95, and 0.99.

b. For a 50,000 sample run of a Monte Carlo simulation, find the values of i and k such that the true value of the fractile $x_{0.70}$ of the distribution of X falls between $(\hat{x}_i, \hat{x}_k)$ with probability 0.99.

c. In a Monte Carlo simulation, find the number of trials m needed to be within $p = \pm0.01$ of the fractile $x_{0.70}$ of the distribution of X with probability 0.95.

8.4 Build a Monte Carlo simulation in an Excel spreadsheet that generates m trials (random samples) of a lognormal random variable X with mean 203.34 and variance 365.27. Use the number of trials m derived in Exercise 8.3c. For this exercise, use the Excel Function: LOGNORM. INV(RAND( ), μ, σ) and review the discussion associated with Equations 8.18 through 8.24. Do not use a commercial Monte Carlo simulation application or add-in to Excel for this exercise.

8.5 From the m trials of the lognormal random variable X generated from Exercise 8.4, compare and contrast the true values of the fractiles $x_{0.69}, x_{0.70}, x_{0.71}$ with their estimated values $\hat{x}_{0.69}, \hat{x}_{0.70}, \hat{x}_{0.71}$ generated by the Monte Carlo simulation.

References

Dienemann, P. F. 1966. *Estimating Uncertainty Using Monte Carlo Techniques*, RM-4854-PR. Santa Monica, CA: The RAND Corporation.

GAO, 2009. Cost estimating and assessment guide: Best practices for developing and managing capital program costs, GAO-09-3SP, March 2009.

Garvey, P. R., B. J. Flynn, P. Braxton, and R. Lee. December 2012. Enhanced scenario based method for cost risk analysis: Theory, application, and implementation. *Journal of Cost Analysis and Parametrics*, 5(2), 98–142.

Morgan, M. G. and M. Henrion. 1990. *Uncertainty: A Guide to Dealing with Uncertainty in Quantitative Risk and Policy Analysis*. New York: Cambridge University Press.

Rubinstein, R. Y. 1981. *Simulation and the Monte Carlo Method*. New York: John Wiley & Sons, Inc.

Winston, W. L. 1994. *Operations Research: Applications and Algorithms*. Belmont, CA: Duxbury Press.

Additional Reading

Book, S. A. 1995. Fictions we live by. The Aerospace Corporation, Society of Cost Estimating and Analysis, Los Angeles, CA.

Book, S. A. 1999. Why correlation matters in cost estimating. In *32nd Annual DOD Cost Analysis Symposium (DODCAS)*, The Aerospace Corporation, Los Angeles, CA.

Garvey, P. R. 2000. *Probability Methods for Cost Uncertainty Analysis: A Systems Engineering Perspective*. Boca Raton, FL: Chapman-Hall/CRC Press, Taylor & Francis Group.

Lurie, P. M. and M. S. Goldberg. 1998. An approximate method for sampling correlated random variables from partially-specified distributions. *Management Science*, 44(2), 203–218.

Public Law 111–23. 2009. Weapon Systems Acquisition Reform Act of 2009.

Young, P. H. 1995. A generalized probability distribution for cost/schedule uncertainty in risk assessment. In *Proceedings of the 1995 Western Multi-Conference on Business/MIS Simulation Education*, The Society for Computer Simulation, San Diego, CA.

9

Correlation: A Critical Consideration

The importance of correlation as a critical consideration in cost uncertainty analysis cannot be understated. Seen throughout the preceding chapters, correlation can have a significant effect on the measure of a program's cost risk. This chapter presents several approaches to capture and incorporate correlation in cost uncertainty analyses. Guidelines on when one approach is preferred over another are offered. First, a brief refresher on correlation and how it can influence a program's cost uncertainty.

9.1 Introduction

Mentioned earlier, cost estimates are inherently uncertain. A cost estimate is merely one outcome in a probability distribution of possible cost outcomes.* A challenge for the cost analysis community has been which cost to choose, from the distribution of possible costs that is suitably risk-adjusted for budgeting or planning a program. Selecting a risk-adjusted cost also means choosing one with an associated level of confidence it will not be exceeded.

Historically, the median cost from a program's cost probability distribution was chosen. However, by definition, the median cost has only a 50% chance of not being exceeded. Furthermore, it remained unclear what risks the median cost would cover in terms of their cost impacts if they occurred. Even if confidence levels higher than the 50th percentile were chosen, it was becoming clear to practitioners that the spread or range of possible cost outcomes was far too narrow. Was there too much optimism in program cost estimates? Was there a fault somewhere in the mathematical methodology?

The answer is yes to each question. While optimistic assumptions continue to be a factor in cost growth above initial estimates, a key methodology deficiency was inadequate attention to correlations between a program's work breakdown structure (WBS) element costs. Why is correlation so critical? Discussed extensively in Chapters 5 and 6, correlation affects the extent of a program's cost risk, when measured by the variance (or standard deviation) of the sum of its WBS element costs. If a program's total cost is given by

$$Cost = a_1X_1 + a_2X_2 + a_3X_3 + \cdots + a_nX_n \tag{9.1}$$

---

* The same is true for schedule estimates.

where $a_1, a_2, a_3, \ldots, a_n$ are constants and X_1, X_2, $X_3, \ldots, X_n$ are random variables that represent the uncertainties in the WBS element costs, then from Theorems 5.7 and 5.8, the mean and variance of *Cost* are

$$E(Cost) = a_1 E(X_1) + a_2 E(X_2) + a_3 E(X_3) + \cdots + a_n E(X_n) \qquad (9.2)$$

$$Var(Cost) = \sum_{i=1}^{n} a_i^2 Var(X_i) + 2 \sum_{i=1}^{n-1} \sum_{j=i+1}^{n} a_i a_j \rho_{X_i, X_j} \sigma_{X_i} \sigma_{X_j} \qquad (9.3)$$

for all i and j such that $1 \leq i \leq j \leq n$. In Equation 9.3, ρ_{X_i, X_j} is the Pearson product-moment correlation between the costs of WBS element random variables X_i and X_j. Seen in Equation 9.2, the mean of the program's total cost is the sum of the means of its WBS element costs. It is unaffected by correlation. The variance of the program's total cost is the sum of the variances of its WBS element costs *plus* the sum over all covariation between each pair of WBS cost elements. In Equation 9.3, the term

$$2 \sum_{i=1}^{n-1} \sum_{j=i+1}^{n} a_i a_j \rho_{X_i, X_j} \sigma_{X_i} \sigma_{X_j} \qquad (9.4)$$

is the *covariance* of *Cost*, denoted by *Cov(Cost)*, where *Cost* is given by Equation 9.1. It is a function of the correlation between each WBS cost element pair. This is why and how correlation enters a cost uncertainty analysis. From this, it follows that

$$\sigma(Cost) = \sqrt{Var(Cost)} = \sqrt{\sum_{i=1}^{n} a_i^2 Var(X_i) + 2 \sum_{i=1}^{n-1} \sum_{j=i+1}^{n} a_i a_j \rho_{X_i, X_j} \sigma_{X_i} \sigma_{X_j}} \qquad (9.5)$$

Thus, correlation between WBS element costs can have significant effects on the extent of a program's total cost risk. The positive or negative direction of correlation affects whether it increases or decreases cost risk (Equation 9.5), respectively. If $\sigma(Cost)$ is less than it should be, then so is any confidence level derived from this distribution. Consider Figure 9.1.

Suppose Figure 9.1 illustrates two probability distributions of a program's total cost, summed across its WBS cost elements. Suppose the median cost was chosen as the risk-adjusted cost for the program. Suppose distribution A was produced without considering correlation between any pair of WBS element costs. This implicitly means the correlations are zero between all pairs. Suppose distribution B captured the true impacts of correlation on this program's cost and it resulted in doubling the standard deviation of distribution A.

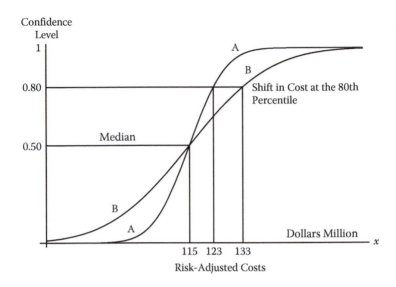

FIGURE 9.1
Potential effects of correlation on a point estimate cost.

Comparing the two distributions in Figure 9.1, the impact of correlation produced a 10 ($M) shift in the program's risk-adjusted cost at the 80th percentile confidence level. The shift in risk-adjusted cost increases monotonically with each unit increase in the confidence level above the 50th percentile. Clearly, correlation can have significant effects on the probability distribution of a program's total cost—chiefly on the upper percentile confidence levels where risk-adjusted costs of a program are often chosen.

From Chapter 5, recall that correlation ρ is a statistical measure of the covariation between two random variables. It measures the strength and direction of change in the value of one random variable with change in the value of another random variable. Regarding strength, ρ is a continuous measure whose value ranges between negative one and positive one. Regarding direction, correlation can be positive or negative. Positive correlations* fall in the interval $0 < \rho \leq 1$. Negative correlations[†] fall in the interval $-1 \leq \rho < 0$. Uncorrelated random variables have correlation $\rho = 0$.

Finally, from Chapter 7, recall that correlation between certain types of random variables have bounds contained within the interval $-1 \leq \rho \leq 1$. For example, the correlation coefficient of the bivariate lognormal distribution is bounded by the interval $-e^{-1} < \rho < 1$. Thus, caution is needed when subjectively assigning correlations to avoid potentially impermissible values.

---

* A cost increase in one area of a program is associated with a cost increase in another area.
[†] A cost increase in one area of a program is associated with a cost decrease in another area.

9.2 Correlation Matters

The preceding discussion showed the importance of properly capturing correlation between a program's WBS element costs, especially positive correlations. Illustrated in Figure 9.1, the potential effects of positive correlations on a program's total cost risk can be dramatic. Consider the following.

Suppose a program's total cost is given by $Cost = X_1 + X_2 + X_3 + \cdots + X_n$, where $X_1, X_2, X_3, \ldots, X_n$ are random variables that represent the uncertainties in the program's WBS element costs. Suppose each WBS element pair has *common* variance σ^2 and nonnegative correlation ρ. From Theorem 5.9

$$\sigma(Cost) = \sqrt{n}\sigma \text{ when } \rho = 0$$

$$\sigma(Cost) = \sqrt{n}\sigma\sqrt{[1 + (n-1)\rho]} \text{ when } 0 < \rho < 1$$

$$\sigma(Cost) = n\sigma \text{ when } \rho = 1$$

Let $\sigma(Cost)$ denote the measure of a program's total cost risk. From this, the percent underestimation of total cost risk when correlation was initially assumed to be equal to 0 ($\rho_0 = 0$) instead of a value in the interval $0 < \rho < 1$ is

$$y = 100\frac{\sqrt{n}\sigma\sqrt{1 + (n-1)\rho} - \sqrt{n}\sigma}{\sqrt{n}\sigma\sqrt{1 + (n-1)\rho}} = 100 - 100\sqrt{\frac{1}{1 + (n-1)\rho}}$$

Figure 9.2 is a graph of y (above) as a function of ρ and the number of WBS cost elements n. Suppose $\rho_0 = 0$ is the initially assumed or implied correlation for all WBS element pairs with $n = 10$ cost elements. If it later became known that the actual correlation ρ was equal to 0.4, then the program's total cost risk has been *underestimated* by 53% if each WBS element pair has a common variance σ^2.

In Figure 9.2, the underestimation worsens exponentially with each increase in the number of WBS cost elements and where $\rho > \rho_0 = 0$. Furthermore, if $\rho_0 = 0$, then a change (increase or decrease) in the cost of one element does not effect change (increase or decrease) in the cost of any other element. This is seldom true. Wrongly assuming or unknowingly setting $\rho_0 = 0$ is the main reason for narrow probability distributions of a program's total cost.

Conversely, the percent overestimation of total cost risk when correlation was initially assumed to be equal to 1 ($\rho_0 = 1$) instead of a value in the interval $0 < \rho < 1$ is

$$y = 100\frac{n\sigma - \sqrt{n}\sigma\sqrt{1 + (n-1)\rho}}{\sqrt{n}\sigma\sqrt{1 + (n-1)\rho}} = 100\sqrt{\frac{n}{1 + (n-1)\rho}} - 100$$

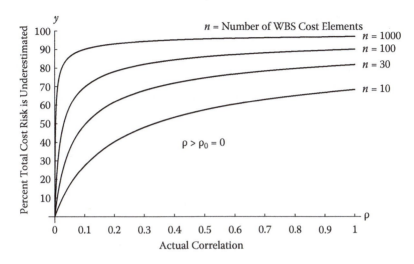

FIGURE 9.2
Underestimation of total cost risk. (From Book, S.A., Why correlation matters in cost estimating, in: *32nd Annual DOD Cost Analysis Symposium (DODCAS)*, Los Angeles, CA, 1999.)

Figure 9.3 is a graph of y (above) as a function of ρ and the number of WBS cost elements n. For example, consider a WBS with $n = 10$ cost elements. Suppose $\rho_0 = 1$ was initially assumed for all WBS cost element pairs. If it later became known that the actual correlation ρ was 0.4, then the program's total cost risk was *overestimated* by 47%. Seen in Figure 9.3, the overestimation

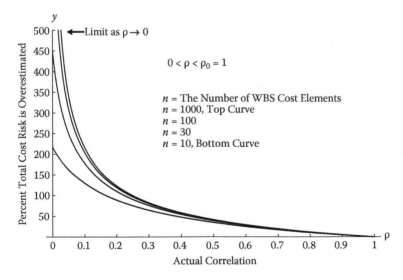

FIGURE 9.3
Overestimation of total cost risk. (From Book, S.A., Why correlation matters in cost estimating, in: *32nd Annual DOD Cost Analysis Symposium (DODCAS)*, Los Angeles, CA, 1999.)

worsens exponentially with each increase in the number of WBS cost elements and where $0 < \rho < \rho_0 = 1$. Furthermore, if $\rho_0 = 1$, then an increase in the cost of one element has a perfect linear increase in the cost of all other elements. This is seldom true and is the main reason for overly elongated probability distributions of a program's total cost. For some random variables, correlations between them are limited to subintervals of $-1 \le \rho \le 1$.

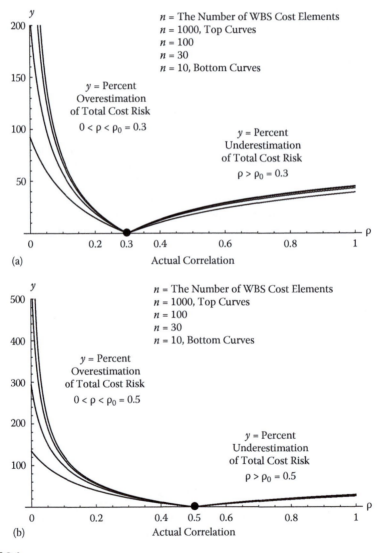

FIGURE 9.4
Under- or overestimation of total cost risk. (From Book, S.A., Why correlation matters in cost estimating, in: *32nd Annual DOD Cost Analysis Symposium (DODCAS)*, Los Angeles, CA, 1999.)

In these cases, assuming $\rho_0 = 1$ would be impermissible (Covert 2013, Garvey and Taub 1992).

Figure 9.4 jointly portrays the extent of overestimating or underestimating a program's total cost risk if assuming an initial correlation ρ_0 in the interval $0 < \rho_0 < 1$ when the actual correlation is ρ, where $0 < \rho < 1$ and $\rho_0 \neq \rho$. If $0 < \rho < \rho_0$, then an overestimation of total cost risk occurs. Conversely, if $\rho > \rho_0$ then an underestimation of total cost risk occurs. Figure 9.4a and b illustrate these outcomes for $\rho_0 = 0.3$ and $\rho_0 = 0.5$, respectively.

9.3 Valuing Correlation

This section presents three ways of valuing correlations between cost elements in a program's WBS. One approach falls into a class of subjective methods for situations where a lack of time or information may preclude deriving correlations between WBS element costs. This is when correlations are assigned. The second falls into a class of methods where correlation is mathematically derived from the structures of cost estimation relationships defined within the WBS. The third approach makes use of Monte Carlo simulation to empirically determine any needed correlations (Gentle 2003).

9.3.1 Assigning Correlations

As mentioned, time or information may be unavailable to assess, assign, or derive correlations between some or all pairs of WBS cost elements that define a program. Moreover, a program with n WBS elements has $n(n-1)/2$ cost element pairs. Thus, a WBS with 10 cost elements has 45 pairs, 50 cost elements have 1225 pairs, and 100 cost elements have 4950 pairs. Clearly, it quickly becomes unwieldy to explicitly assign correlations to all pairs of cost elements; however, ignoring some or all is equivalent to setting them to zero. Given this, what practical options exist such that the important effects of correlation are captured in judicious ways?

One answer to this question is to use Figure 9.2. In Figure 9.2, observe the "knee" in these curves all occur in the interval $0.10 \leq \rho \leq 0.30$. In particular, for $\rho > 0.30$ there is little change in the percent that a program's total cost risk is underestimated by not capturing positive correlation when it is present in the WBS. If the conditions stated are present, then choosing a value for ρ within this interval is a reasonable option in lessening the underestimation of total cost risk. In particular, setting ρ equal to the midpoint of this interval ($\rho = 0.20$) *for all* WBS cost element pairs captures most of the potential underestimation of total cost risk.* Moreover, $\rho = 0.20$ is a permissible

---

* In practice, choosing ρ to lessen the underestimation of cost risk is emphasized. Historically, underestimating the true cost risk of programs is a major reason for cost growth above initial cost estimates.

correlation for the pairs of continuous probability distributions most often used to characterize uncertainties in WBS element costs.

One way to use Figure 9.2 is as follows. Consider a program whose cost is given by

$$Cost = X_1 + X_2 + X_3 + X_4 + X_5 \qquad (9.6)$$

where X_1, X_2, X_3, X_4, and X_5 are independent random variables that denote the program's WBS element costs. If the uncertainty in each element's cost is given by the distribution in Figure 8.19, then it was shown that

$$E(Cost) = E(X_1) + E(X_2) + E(X_3) + E(X_4) + E(X_5)$$
$$= 203.34 \ (\$M) \qquad (9.7)$$

$$Var(Cost) = Var(X_1) + Var(X_2) + Var(X_3) + Var(X_4) + Var(X_5)$$
$$= 365.27 \ (\$M)^2 \qquad (9.8)$$

$$\Rightarrow \sigma = \sigma(Cost) = \sqrt{Var(Cost)} = \sqrt{365.27}$$
$$= 19.11 \ (\$M) \qquad (9.9)$$

Computing *Var(Cost)* by Equation 9.8 implies the correlation between *each pair* of WBS cost elements X_1, X_2, X_3, X_4, and X_5 is zero; that is, the covariance given by Equation 9.4 is zero. In practice, this can mean (1) there is genuinely no correlation between any WBS cost element pair for this program or (2) the analysis did not take correlation into account when it should have been considered or (3) the cost of each WBS element was developed separately such that correlations between them were not possible to determine. If the latter two of these three situations are true, then one option is to adjust *Var(Cost)* with a correlation value from Figure 9.2. The reasoning is this brings *Var(Cost)* closer to its "true" value than the value produced by wrongly assuming $\rho = 0$ between all WBS cost element pairs, as implied by Equation 9.8. The following illustrates how adjusting *Var(Cost)* might be done.

As mentioned earlier, the "knee" in the curves in Figure 9.2 occurs in the interval $0.10 \leq \rho \leq 0.30$. For this case, suppose ρ is set equal to the midpoint ($\rho = 0.2$) of this interval *for all* 10 WBS cost element pairs formed from X_1, X_2, X_3, X_4, and X_5. From Equation 9.4, the covariance of *Cost* is

$$Cov(Cost) = 2 \sum_{i=1}^{n-1} \sum_{j=i+1}^{n} a_i a_j \rho_{X_i,X_j} \sigma_{X_i} \sigma_{X_j} = 2 \sum_{i=1}^{5-1} \sum_{j=i+1}^{5} 0.2 \, \sigma_{X_i} \sigma_{X_j}$$
$$= 226.102 \ (\$M)^2 \qquad (9.10)$$

for $1 \leq i \leq j \leq n = 5$ and where $a_i a_j = 1$ (in this case, as implied by Equation 9.6). The terms in Equation 9.10 derive from the variances of X_1, X_2, X_3, X_4, and X_5 given in Figure 8.19. Thus, from Equation 9.3

$$Var(Cost) = \sum_{i=1}^{5} Var(X_i) + 2\sum_{i=1}^{5-1}\sum_{j=i+1}^{5} 0.2\, \sigma_{X_i}\sigma_{X_j}$$

$$= 365.278 + 226.102 = 591.38\ (\$M)^2 \tag{9.11}$$

$$\Rightarrow \sigma(Cost) = \sqrt{Var(Cost)} = \sqrt{591.38} = 24.32\ (\$M) \tag{9.12}$$

Consequently, incorporating a correlation of $\rho = 0.2$ into the analysis results in a 27% increase in the program's cost risk, when compared to the initial correlation of $\rho_0 = 0$ implied by Equation 9.9. A cost risk equal to 24.31 ($M) for this program is closer to the "true" measure than that produced by Equation 9.9. Shown in Figures 9.2 and 9.4, the underestimation of a program's total cost risk worsens exponentially with each increase in the number of WBS cost elements and when $\rho > \rho_0 \geq 0$.

Choosing a correlation from the interval $0.10 \leq \rho \leq 0.30$ is informally called the *knee-in-the-curve* method. It is intended for situations where a lack of time or information precludes empirically or analytically deriving correlations between some or all pairs of a program's WBS element costs. It is also for situations when correlations between certain WBS element costs are not automatically accounted for through their functional relationships in, for example, a Monte Carlo simulation. Hence, the knee-in-the-curve method should be considered a practical guideline in these circumstances. Mentioned earlier, choosing a value for ρ in this interval generally assures a permissible correlation is used and further lessens the underestimation of a program's total cost risk when $\rho = 0$ is wrongly assumed or implied.

An examination of a program's cost sensitivity to the choice of ρ is recommended when using the knee-in-the-curve method. For example, under an assumed normal distribution, Figure 9.5 shows the sensitivity of cost at the 80th percentile as ρ changes across the values displayed. The shift in the distribution at the 80th percentile level amounts to a 10 million dollar difference between $\rho = 0$ (the left-most curve) and $\rho = 0.5$ (the right-most curve).

In Figure 9.5, observe that the difference between the successive 80th percentile costs is monotonically decreasing for $\rho \geq 0$. Moreover, for $\rho > 0.5$, Table 9.1 reveals that the successive cost differences at this confidence level are increasingly less consequential. This suggests choosing a value for ρ greater than 0.5 is unnecessary, in this case. Furthermore, arbitrarily choosing too large a value for ρ can lead to an impermissible choice if certain types of random variables are being combined in the analysis.

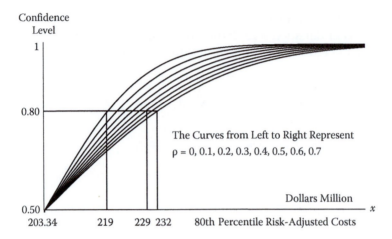

FIGURE 9.5
80th percentile shifts in the probability distribution of *Cost* for varying ρ.*

TABLE 9.1

80th Percentile Costs in Figure 9.5

Correlation ρ	80th Percentile Cost ($M)	Successive 80th Percentile Cost Difference ($M)
0	219.423	—
0.1	221.746	2.323
0.2	223.808	2.062
0.3	225.677	1.869
0.4	227.402	1.725
0.5	229.009	1.607
0.6	230.524	1.515
0.7	231.955	1.431

9.3.2 Deriving Correlations

An alternative to the knee-in-the-curve method is to derive correlations between WBS element costs when relationships between them are specified by mathematical functions. Deriving correlations in such situations was identified by (Garvey and Taub 1992) in their paper on bivariate distributions

---

* In this figure, the probability distribution of *Cost* was assumed to be normal; hence, the 50th percentile cost (median cost) will equal the expected value of *Cost*—which is 203.34 ($M) by Equation 9.7. Although this is always true for the normal distribution, it is not true in general for all univariate probability distributions.

to join cost and schedule uncertainties (refer to Chapter 7). Other writers (Coleman and Gupta 1994) later coined the term *functional correlation* as the correlation among functionally dependent cost estimating relationships.

Case Discussions 6.1 and 6.2 illustrated analytically derived correlations between functionally related random variables of cost estimating relationships. In particular, Theorem 6.1 showed that if $Y = aX + Z$ where a is a real number and X and Z are independent random variables then

$$\rho_{Y,X} = a\frac{\sigma_X}{\sigma_Y} \text{ and } \rho_{Y,Z} = \frac{\sigma_Z}{\sigma_Y}$$

Example 9.1 *Suppose the cost of WBS element Y is a function of the costs of WBS elements X and Z where $Y = 0.5X + Z$. Suppose X and Z are independent random variables with probability density functions given in Figure 9.6. Compute $\rho_{Y,X}$ and $\rho_{Y,Z}$.*

Solution To compute $\rho_{Y,X}$ and $\rho_{Y,Z}$ it is necessary to determine σ_Y, σ_X, and σ_Z. Given the triangular and uniform density functions for X and Z in Figure 9.6, it follows from Theorem 4.3 that $\sigma_X = 3.118$ ($M) and from Theorem 4.2 that $\sigma_Z = 11.547$ ($M). Given that X and Z are independent random variables, from Equation 9.3 (or Theorem 5.8) it follows that

$$Var(Y) = Var(0.5X + Z) = Var(0.5X) + Var(Z) + 2(0.5)(1)\rho_{X,Z}\sigma_X\sigma_Z$$

$$= 0.5^2 Var(X) + 1^2 Var(Z) + 2(0.5)(1)(0)\sigma_X\sigma_Z$$

$$= 0.25(3.118)^2 + (11.547)^2 = 135.763$$

Therefore,

$$\rho_{Y,X} = a\frac{\sigma_X}{\sigma_Y} = 0.5\frac{3.118}{\sqrt{135.763}} = 0.1338 \text{ and } \rho_{Y,Z} = \frac{\sigma_Z}{\sigma_Y} = \frac{11.547}{\sqrt{135.763}} = 0.99$$

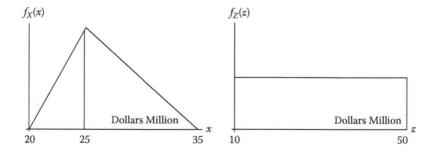

FIGURE 9.6
Probability density functions for X and Z.

These are the derived correlations between Y and X and Y and Z, respectively, for this example. Theorem 6.1 was used to derive these correlations; however, an alternative is to use the definition of correlation given by Equation 5.30. From Equation 5.30, recall that

$$Corr(X, Y) \equiv \rho_{X,Y} = \frac{E(XY) - \mu_X\mu_Y}{\sigma_X\sigma_Y} \tag{9.13}$$

where $\sigma_X > 0$ and $\sigma_Y > 0$. Applying Equation 9.13 to this example, it follows that

$$Corr(X, 0.5X + Z) \equiv \rho_{X,0.5X+Z} = \frac{E(X(0.5X + Z)) - \mu_X\mu_Y}{\sigma_X\sigma_Y}$$

$$= \frac{0.5E(X^2) + E(X)E(Z) - \mu_X\mu_Y}{\sigma_X\sigma_Y}$$

$$= \frac{0.5(Var(X) + (E(X))^2) + E(X)E(Z) - \mu_X\mu_Y}{\sigma_X\sigma_Y} \tag{9.14}$$

where, from Theorem 3.10, $E(X^2) = Var(X) + (E(X))^2$. Thus,

$$\rho_{X,Y} = \rho_{X,0.5X+Z} = \frac{0.5(Var(X) + (E(X))^2) + E(X)E(Z) - \mu_X\mu_Y}{\sigma_X\sigma_Y}$$

$$\rho_{X,Y} = \rho_{X,0.5X+Z}$$
$$= \frac{0.5(3.118^2 + 26.66667^2) + (26.66667)(30) - (26.66667)(43.333335)}{(3.118)(11.652)}$$

where $\mu_Y \equiv E(Y) = E(0.5X + Z) = 0.5E(X) + E(Z) = 43.333335$. Substituting this value into the previous equation

$$\rho_{X,Y} = \rho_{X,0.5X+Z} = 0.1338$$

which, as it should, agrees with the previous finding. It is easily shown that the correlation between two random variables is symmetric; that is, $\rho_{X,Y} = 0.1338 = \rho_{Y,X}$.

Example 9.2 *Suppose data on the cost of an antenna Y was collected and a statistical regression revealed the cost estimation relationship (CER) given by Equation 9.15.*

$$Y = 34.36X_1^{0.5}X_2^{0.8}\varepsilon \tag{9.15}$$

where X_1 is the antenna's aperture diameter (meters), X_2 is the frequency (GHz), and ε is the statistical error associated with the regression. Compute the matrix *of all pairwise correlations of Y with the variables X_1, X_2, and ε given they are independent random variables with probability density functions shown in Figure 9.7.*

Solution In this case, the cost of the antenna Y is a product of three random variables X_1, X_2, and ε; furthermore, X_1 and X_2 are nonlinear random variables. Given this, the Mellin transform (Section 5.5) is a convenient way to derive their moments. They are needed to compute the pairwise correlations of Y with the variables X_1, X_2, and ε.

The correlation pairs of Y with X_1, X_2, and ε are, respectively

$$\text{Corr}(Y, X_1) \equiv \rho_{Y,X_1} = \frac{E(YX_1) - \mu_Y\mu_{X_1}}{\sigma_Y\sigma_{X_1}} \tag{9.16}$$

$$\text{Corr}(Y, X_2) \equiv \rho_{Y,X_2} = \frac{E(YX_2) - \mu_Y\mu_{X_2}}{\sigma_Y\sigma_{X_2}} \tag{9.17}$$

$$\text{Corr}(Y, \varepsilon) \equiv \rho_{Y,\varepsilon} = \frac{E(Y\varepsilon) - \mu_Y\mu_\varepsilon}{\sigma_Y\sigma_\varepsilon} \tag{9.18}$$

Since X_1, X_2, and ε were given to be independent random variables, it follows from Theorem 5.4 that the correlations between all of their pairs are zero. So, the only nonzero correlations in the expression for Y are those given by Equations 9.16 through 9.18. The following shows the derivation of ρ_{Y,X_1} given by Equation 9.16. Deriving ρ_{Y,X_2} and $\rho_{Y,\varepsilon}$ is an exercise for the reader.

Derivation of $E(YX_1)$: Given $Y = 34.36X_1^{0.5}X_2^{0.8}\varepsilon$, it follows that

$$E(YX_1) = E\left(\left(34.36X_1^{0.5}X_2^{0.8}\varepsilon\right)X_1\right) = E\left(34.36X_1^{1.5}X_2^{0.8}\varepsilon\right)$$

$$= 34.36E\left(X_1^{1.5}\right)E\left(X_2^{0.8}\right)E(\varepsilon) \tag{9.19}$$

The terms $E\left(X_1^{1.5}\right)$ and $E\left(X_2^{0.8}\right)$ are easily computed by Mellin transforms; specifically,

$$E(YX_1) = 34.36E\left(X_1^{1.5}\right)E\left(X_2^{0.8}\right)E(\varepsilon)$$

$$= 34.36M_{X_1}(2.5)M_{X_2}(1.8)E(\varepsilon) \tag{9.20}$$

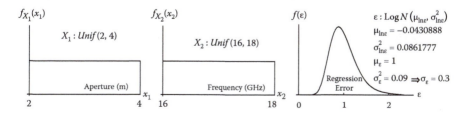

FIGURE 9.7
Probability density functions for X_1, X_2, and ε.

where $M_X(s) = E(X^{s-1})$. From Equation 5.108, the Mellin transforms $M_{X_1}(2.5)$ and $M_{X_2}(1.8)$ can be computed by Equation 9.21

$$M_X(s) = \frac{1}{s(b-a)} \left(b^s - a^s \right) \tag{9.21}$$

since X_1 and X_2 are each uniformly distributed, as shown in Figure 9.7; therefore, it follows that

$$M_{X_1}(2.5) = \frac{1}{2.5(4-2)} \left(4^{2.5} - 2^{2.5} \right) = 5.26862915 \tag{9.22}$$

$$M_{X_1}(1.8) = \frac{1}{1.8(18-16)} \left(18^{1.8} - 16^{1.8} \right) = 9.645373367 \tag{9.23}$$

Substituting these values into Equation 9.20, and recognizing that $E(\varepsilon) = 1$ (why?) it follows that

$$E(YX_1) = 34.36(5.26862915)(9.645373367)(1) = 1746.102882 \tag{9.24}$$

Derivation of μ_Y: Given $Y = 34.36X_1^{0.5}X_2^{0.8}\varepsilon$, it follows that

$$\mu_Y \equiv E(Y) = E\left(34.36X_1^{0.5}X_2^{0.8}\varepsilon \right) = 34.36E\left(X_1^{0.5} \right)E\left(X_2^{0.8} \right)E(\varepsilon) \tag{9.25}$$

$$= 34.36M_{X_1}(1.5)M_{X_2}(1.8)E(\varepsilon) \tag{9.26}$$

$$= 34.36(1.723857265)(9.645373367)(1) \tag{9.27}$$

$$= 571.3123246 \tag{9.28}$$

Thus far, we have

$$\rho_{Y,X_1} = \frac{1746.102882 - (571.3123246)\mu_{X_1}}{\sigma_Y \sigma_{X_1}}$$

$$= \frac{1746.102882 - (571.3123246)(3)}{\sigma_Y \sqrt{1/3}} \tag{9.29}$$

where $\mu_{X_1} = 3$ and $\sigma_{X_1} = \sqrt{1/3}$ are the mean and standard deviation of $X_1 \sim Unif\,(2,4)$. So, it remains to compute σ_Y in the denominator of Equation 9.29.

Derivation of σ_Y: Given $Y = 34.36X_1^{0.5}X_2^{0.8}\varepsilon$, it follows that

$$Var(Y) = E\left(Y^2 \right) - (E(Y))^2 = E\left(Y^2 \right) - (\mu_Y)^2 = E\left(Y^2 \right) - (571.3123246)^2 \tag{9.30}$$

where $\sigma_Y = \sqrt{Var(Y)}$. To determine σ_Y, it remains to compute $E\left(Y^2 \right)$ in Equation 9.30. This too can be done with the aid of Mellin transforms, as follows:

Given $Y = 34.36X_1^{0.5}X_2^{0.8}\varepsilon$, it follows that $Y^2 = (34.36)^2 X_1 X_2^{1.6}\varepsilon^2$. From this, we can write

$$E\left(Y^2\right) = E\left((34.36)^2 X_1 X_2^{1.6}\varepsilon^2\right) = (34.36)^2 E\left(X_1\right) E\left(X_2^{1.6}\right) E\left(\varepsilon^2\right)$$

$$= (34.36)^2 (3) E\left(X_2^{1.6}\right) E\left(\varepsilon^2\right)$$

$$= (34.36)^2 (3) M_{X_2}(2.6)\left(Var(\varepsilon) + (E(\varepsilon))^2\right)$$

$$(9.31)$$

since $E(X_1) = \mu_{X_1} = 3$ and by definition $Var(\varepsilon) = E(\varepsilon^2) - (E(\varepsilon))^2 \Rightarrow E(\varepsilon^2) = Var(\varepsilon) + (E(\varepsilon))^2$. From Figure 9.7, since $X_2 \sim Unif(16, 18)$ and $\varepsilon \sim LogN(1, 0.3)$, it follows that

$$M_{X_2}(2.6) = \frac{1}{2.6(18-16)}\left(18^{2.6} - 16^{2.6}\right) = 93.10192718$$

$$E\left(\varepsilon^2\right) = Var(\varepsilon) + (E(\varepsilon))^2 = 0.3^2 + (1)^2 = 1.09$$

$$E\left(Y^2\right) = (34.36)^2 (3)(93.10192718)\left(0.3^2 + (1)^2\right) = 359,428.6848$$

We can now complete the computation of Equation 9.30:

$$Var(Y) = E\left(Y^2\right) - (571.3123246)^2 = 359,428.6848 - (571.3123246)^2$$

$$= 33,030.91261 \qquad (9.32)$$

Therefore, $\sigma_Y = \sqrt{33,030.91261} = 181.7440855$ and it follows that

$$\rho_{Y,X_1} = \frac{1746.102882 - (571.3123246)(3)}{(181.7440855)\sqrt{1/3}} = 0.306546357 \qquad (9.33)$$

As mentioned earlier, deriving ρ_{Y,X_2} and $\rho_{Y,\varepsilon}$ is an exercise for the reader. Given that, the matrix of pairwise Pearson correlations for the antenna cost estimating relationship $Y = 34.36X_1^{0.5}X_2^{0.8}\varepsilon$ is

$$
\begin{array}{c c c c c c}
 & Y & X_1 & X_2 & \varepsilon \\
\begin{array}{c} Y \\ A = X_1 \\ X_2 \\ \varepsilon \end{array} &
\left(\begin{array}{cccc}
1 & 0.3065 & 0.0854 & 0.9430 \\
0.3065 & 1 & 0 & 0 \\
0.0854 & 0 & 1 & 0 \\
0.9430 & 0 & 0 & 1
\end{array}\right)
\end{array}
$$

Matrix A is called the *correlation matrix* of Y. The zero entries in this matrix are due to the given independence of the random variables X_1, X_2, and ε (refer to Theorem 5.4). The entries along the diagonal of this matrix will always equal one. Why? Showing this is an exercise for the reader. This completes the discussion and solution to Example 9.2.

A correlation matrix has a number of important properties. One property is that correlations contained in the matrix must be mathematically feasible; that is, there are limits on the amount of correlation that can exist between certain types of random variables. From Chapter 7, recall that the correlation coefficient of the bivariate lognormal distribution is bounded by the interval $-e^{-1} < \rho < 1$.

Another property of a correlation matrix is that all pairwise correlations must be internally consistent. For example, suppose X, Y, and Z are random variables with $\rho_{X,Y} = 0.9$ and $\rho_{Y,Z} = 0.9$. Given this, it would be inconsistent to specify $\rho_{X,Z} = 0$ since it is not possible for X and Z to be wholly uncorrelated when $\rho_{X,Y} = 0.9$ and $\rho_{Y,Z} = 0.9$. In this case, it can be shown that the smallest possible correlation between X and Z is 0.621. Thus, the set of correlations $\rho_{X,Y} = 0.9$, $\rho_{Y,Z} = 0.9$, and $\rho_{X,Z} = 0.621$ are consistent. From matrix algebra, a correlation matrix is internally consistent if it is nonnegative definite.[*]

This concludes the discussion on analytically deriving correlations using the algebra of random variables. Shown in Example 9.2, the Mellin transform is a convenient technique when it is necessary to compute moments of random variables raised to integer and noninteger powers. Refer to Covert (2013) for further examples of analytically derived correlation coefficients for cost estimating relationships and the use of Mellin transforms for this purpose.

9.3.3 Using Monte Carlo Simulation

Monte Carlo simulation is another way to tease out correlations between random variables of interest. It is a convenient method to employ when WBS cost element equations are too complex, or intricate, in their functional relationships to analytically derive correlations of interest. The antenna cost estimating relationship in Example 9.2 is used to show the application of Monte Carlo simulation for this purpose.

Let $Y = 34.36 X_1^{0.5} X_2^{0.8} \varepsilon$ as defined by Equation 9.13. Suppose Y was programmed into an Excel spreadsheet and a Monte Carlo simulation was run within that application. Suppose the simulation generated 10,000 sample values of Y, for each randomly drawn value of X_1, X_2, and ε from their probability distributions in Figure 9.7. Figure 9.8 shows the simulation results.

In Figure 9.8, the table on the left lists outcomes of the first 20 trials from a 10,000 trial Monte Carlo simulation. Correlations between variables can then be derived from the 10,000 sample values of Y, X_1, X_2, and ε. These correlations are empirically determined rather than being analytically derived, as in

---

[*] This is a technical concept in matrix theory. In general, any real symmetric matrix is nonnegative definite if and only if all of its eigenvalues are nonnegative. Eigenvalues c_i are roots to the determinant equation $|\mathbf{A} - c\mathbf{I}| = 0$, where $\mathbf{A}$ is an n by n matrix, $\mathbf{I}$ is the identity matrix, and c is a scalar. Refer to Gentle (2007) for a further discussion on matrix algebra and its applications.

First 20 trials from a 10,000 trial Monte Carlo simulation of Y

Trial	A Y	B X_1	C X_2	D ε
1	716.0635	3.8597438	16.037734	1.152138609
2	665.89094	2.1035162	16.840955	1.395672944
3	471.22698	3.1512404	16.34593	0.826434629
4	352.30958	3.0802721	17.879527	0.581690456
5	719.12834	3.2249405	17.356313	1.188300871
6	844.77407	2.1880223	16.413226	1.772174089
7	745.41913	3.7210749	17.391141	1.144855604
8	698.75886	3.4622632	16.519926	1.159278187
9	702.53114	2.4212408	17.549093	1.327974382
10	666.79891	2.79625	16.941987	1.206374776
11	566.71725	2.7154908	16.236625	1.07644663
12	400.41521	3.431783	16.664714	0.662611851
13	591.91529	3.5340655	16.386318	0.978326911
14	582.83037	2.527662	17.18258	1.096626695
15	491.23833	3.1256018	17.877357	0.80524793
16	338.61037	2.1701524	16.80498	0.69992514
17	677.73691	2.2611489	17.685819	1.317477663
18	669.83565	3.6147271	17.355055	1.045530127
19	672.19718	2.5063492	17.97116	1.225354782
20	1029.2042	3.5760607	17.343243	1.616001296
⋮ 10,000	⋮	⋮	⋮	⋮

$$Y = 34.36X_1^{0.5} X_2^{0.8}\varepsilon$$

Pearson Correlations Derived from the Monte Carlo Simulation

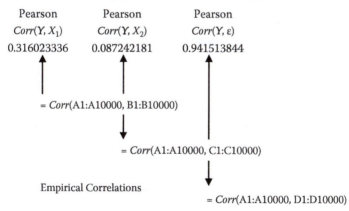

Pearson $Corr(Y, X_1)$	Pearson $Corr(Y, X_2)$	Pearson $Corr(Y, \varepsilon)$
0.316023336	0.087242181	0.941513844

$= Corr(\text{A1:A10000, B1:B10000})$

$= Corr(\text{A1:A10000, C1:C10000})$

Empirical Correlations

$= Corr(\text{A1:A10000, D1:D10000})$

FIGURE 9.8
Monte Carlo simulation of $Y = 34.36X_1^{0.5}X_2^{0.8}\varepsilon$.

the preceding section. The right half of Figure 9.8 shows the empirical correlations between Y and its pairings with X_1, X_2, and ε. They were computed by the Excel *Correl* function. Observe how the empirical correlations compare favorably with their analytically derived values, as summarized by matrix A in Section 9.3.2. Differences between them are due to random errors endemic in all simulations. From a practical view, such differences are negligible.

9.4 Summary

The importance of correlation as a critical consideration in cost uncertainty analysis cannot be understated. Seen herein, and throughout the preceding chapters, correlation can have a significant effect on the amount of cost risk in programs. This chapter presented several approaches for capturing and incorporating correlation in cost uncertainty analyses. Guidelines when one approach is preferred over another were also offered. This chapter concludes with key practice points as they relate to this topic.

Practice Point: Do not introduce rank correlation into cost uncertainty analysis. Rank correlation is not the technically correct correlation in the variance of the sum of WBS element cost random variables. Furthermore, Pearson product-moment correlations will already be captured in a Monte Carlo simulation if functional relationships between WBS element costs are defined. Mixing Pearson product-moment correlations with rank correlations leads to (1) double counting the effects of correlation on a program's total cost risk and (2) a simulation that produces results whose interpretation is unknown.

Practice Point: Monte Carlo simulations will automatically capture Pearson product-moment correlations that may be present between WBS element costs by virtue of the way analysts define their equations (or relationships) in a cost model. In practice, it is recommended that analysts express associations within the WBS through functional relationships (cost equations), as illustrated throughout this book. This allows the Pearson correlations *implied by these relationships* to be routinely captured in the overall analysis. Pearson correlations that originate from logically defined functional relationships in a WBS are more likely to be accepted in cost reviews than debating the merits of those assigned by subjective assessments.

Practice Point: If analysts need to assign values to correlations not automatically captured in a Monte Carlo simulation, then do so in accordance with the guidelines presented throughout this chapter. Realize that certain types of probability distributions cannot be positively correlated at the maximum value of 1 or negatively correlated at the minimum value of -1 in the correlation interval $-1 \leq \rho \leq 1$. Therefore, caution is needed to avoid assigning an impermissible correlation between the costs of WBS elements. For instance, recall that the correlation coefficient of the bivariate lognormal distribution is bounded by the interval $-e^{-1} < \rho < 1$. In addition, all pairwise correlations in an analysis must be internally consistent. From matrix algebra, a correlation matrix is internally consistent if it is nonnegative definite.

Exercises

9.1 In Example 9.2, the cost of an antenna Y was given by the cost estimating relationship $Y = 34.36 X_1^{0.5} X_2^{0.8} \varepsilon$. (a) Given that X_1, X_2, and ε are independent random variables with probability density functions shown in Figure 9.7, derive ρ_{Y,X_2} and $\rho_{Y,\varepsilon}$. (b) Compare these analytically derived results with their values determined empirically by the Monte Carlo simulation shown in Figure 9.8.

9.2 Section 9.3.3 demonstrated the efficacy of using Monte Carlo simulation as a way to empirically determine correlations between random variables, rather than deriving them through an analytical approach as shown by the Mellin transform. From a practical perspective, time may be such that empirically derived correlations will suffice in a given situation. (a) Recognizing this, build your own Monte Carlo simulation of the problem in Example 9.2 and use it to empirically determine ρ_{Y,X_1}, ρ_{Y,X_2}, and $\rho_{Y,\varepsilon}$. (b) Compare your results with those shown in Figure 9.8.

9.3 In Example 9.2, the *correlation matrix* of Y is given by matrix A as follows. Show that the entries along the diagonal of this matrix will always equal one.

$$A = \begin{array}{c c} & \begin{array}{c c c c} Y & X_1 & X_2 & \varepsilon \end{array} \\ \begin{array}{c} Y \\ X_1 \\ X_2 \\ \varepsilon \end{array} & \left(\begin{array}{c c c c} 1 & 0.3065 & 0.0854 & 0.9430 \\ 0.3065 & 1 & 0 & 0 \\ 0.0854 & 0 & 1 & 0 \\ 0.9430 & 0 & 0 & 1 \end{array} \right) \end{array}$$

9.4 Given that X_1, X_2, and ε are independent random variables with probability density functions shown in Figure 9.7, show that the cost estimating relationship $Y = X_1 X_2 \varepsilon$ has the following correlation matrix A.

$$A = \begin{array}{c c} & \begin{array}{c c c c} Y & X_1 & X_2 & \varepsilon \end{array} \\ \begin{array}{c} Y \\ X_1 \\ X_2 \\ \varepsilon \end{array} & \left(\begin{array}{c c c c} 1 & 0.5303 & 0.0936 & 0.8267 \\ 0.5303 & 1 & 0 & 0 \\ 0.0936 & 0 & 1 & 0 \\ 0.8267 & 0 & 0 & 1 \end{array} \right) \end{array}$$

9.5 Figure 9.4 jointly portrays the extent of overestimating or underestimating a program's total cost risk, if assuming an initial correlation ρ_0 in the interval $0 < \rho_0 < 1$ when the actual correlation is ρ, where $0 < \rho < 1$ and $\rho_0 \neq \rho$. Figure 9.4a and b illustrated these outcomes for $\rho_0 = 0.3$ and $\rho_0 = 0.5$, respectively.

a. Build a table of numerical values associated with the curves shown in Figure 9.4a and b.

b. Extend this table to include $\rho_0 = 0.1, 0.2, 0.4, 0.6, 0.7, 0.8, 0.9$.

References

Book, S. A. 1999. Why correlation matters in cost estimating. In *32nd Annual DOD Cost Analysis Symposium (DODCAS)*, Los Angeles, CA.

Coleman, R. J. and S. S. Gupta. 1994. An overview of correlation and functional dependencies in cost risk and uncertainty analysis. In *28th Annual Department of Defense Cost Analysis Symposium*, Williamsburg, VA.

Covert, R. P. 2013. Analytic method for probabilistic cost and schedule risk analysis. National Aeronautics and Space Administration (NASA), Office of Program Analysis and Evaluation (PA&E), Cost Analysis Division (CAD). Prepared for NASA under subcontract to Galorath Incorporated, El Segundo, CA.

Garvey, P. R. and A. E. Taub. 1992. A joint probability model for cost and schedule uncertainties. In *26th Annual DOD Cost Analysis Symposium (ADODCAS)*, September, Washington, DC. *The Journal of Cost Analysis*, Spring 1997.

Gentle, J. E. 2003. *Random Number Generation and Monte Carlo Methods*, 2nd edn. New York: Springer Science+Business Media, Springer-Verlag.

Gentle, J. E. 2007. *Matrix Algebra: Theory, Computations, and Applications in Statistics*. New York: Springer Science+Business Media, LLC.

10

Building Statistical Cost Estimating Models

All models are wrong, but some are useful.

George E.P. Box, Statistician (1919–2013)

The cost estimating and analysis community has a variety of models and methods for generating point estimates of a program's cost from its work breakdown structure (WBS) elements. They include engineering build-up techniques,* cost estimation by analogy, and parametric models. This chapter focuses on parametric models and the use of statistical regression methods to develop them.

10.1 Introduction

Statistical regression is a process for discovering relationships between a dependent random variable Y and one or more independent random variables $X_1, X_2, X_3, \ldots, X_n$. Its purpose is to statistically predict or forecast a value of Y from data observed or collected on values of $X_1, X_2, X_3, \ldots, X_n$. In the cost estimating and analysis community, models derived by this process are considered parametric in that predictions about Y are made as a function of one or more observations about $X_1, X_2, X_3, \ldots, X_n$.

Advances in computing technologies and mathematical methods have made regression a desirable approach for building statistical cost estimating models. They are often used to generate statistically-based point estimates of various WBS cost elements and probability distributions around them. This chapter describes classical statistical regression techniques and presents the general error regression method (GERM)—a major advance in practice and technique.

The most important aspect of a statistical cost estimation model is the historical (observed) data upon which it is built. Knowing the pedigree of these data is of paramount importance. There are many considerations in this part of the model building process; they include the following:

Cost Data Collection: The first type of data to collect is historical cost data, which is easier said than done. Ideally, historical costs are available

---

* Also known as "grass-roots" or "bottom-up" estimation methods.

within government or industry cost reports such as earned value and cost performance reports. Examples of historical cost data collected on past U.S. Department of Defense programs are described in Section 12.4.

Once data sources are located, it must then be decided what "costs" will be collected. The most useful data to collect are the dollars charged to a project. Materials and hours are useful data to collect, but this type of data must also be accompanied by material prices and billing rates. Unit cost data are very important to collect, especially in models designed to estimate costs of more than one item, because they provide useful resources for calculating learning, or cost improvement effects. If unit data cannot be collected, then lot cost or average unit cost data and the range of units (e.g., the first and last unit numbers) to which they pertain should be collected.

Cost Data Normalization: The data used to create the cost model must be normalized to reflect consistent WBS definitions and scope, the appropriate fiscal year, and correct procurement quantity. The three broad areas of normalization are quantity, inflation, and content.

Normalization for quantity ensures comparing the same type of cost, whether it is at the total, lot, or unit level. The most common method of normalizing for quantity is through the use of cost improvement or learning curves. The classical technique is to normalize all data points to the theoretical first unit cost using either unit theory or cumulative average theory learning curves (Book and Burgess 1996). This is discussed in Chapter 11.

Normalization for inflation is needed to express all collected cost data in consistent dollar terms, called base year dollars (BY$). Historical cost data comes in many forms. Sometimes they are reported in constant fiscal year dollars (FY$) or across differing fiscal years. Cost data can be given in terms of the sum of expenditures. Nonetheless, they must all be transformed into consistent dollar terms before they are used to build a statistical cost estimation model. A recommended practice is to normalize cost data to constant fiscal year (FY) dollars. For example, when each data point is normalized to FYXX$, then results predicted by the cost model will also be in FYXX$.

Normalization for content requires a mapping of different WBS cost element definitions into a consistent WBS dictionary for the model. This step is required to ensure that different programs with different WBS cost element definitions are represented correctly in the model.

Cost Driver Determination: Collecting data on technical and programmatic independent variables (the input parameters to a model) that drive cost is as important as collecting the cost data itself. For example, cost drivers for a solar array might be the array material, power generating efficiency, and the solar array area. Other examples of cost drivers are weight of a payload, the amount of software to develop, and schedule. Care is needed when specifying or identifying too many cost drivers for a model. Doing so affects the number of degrees of freedom* for a model, which can affect the accuracy of its forecasts or predictions.

Correlations: Plotting each cost driver (each independent variable) against cost (the dependent or prediction variable) and computing their correlation aids in identifying which driver (or drivers) will most influence the predicted cost. However, a strong correlation does not alone indicate the superiority of a cost driver. A cost driver that is strongly correlated with cost could be a proxy for a better driver. If the dataset is small, then strong correlations could give false positives concerning the predictive merit of the cost driver.

Correlation considerations should be investigated before decisions are made on the choice of cost drivers for a model. An important factor to consider is the correlation between candidate cost drivers. When modeling cost as a dependent variable, the cost drivers are the independent variables. This ideally means the cost drivers must satisfy the rules of statistical independence (Chapter 5).

10.2 Classical Statistical Regression

This section introduces classical regression methods and how they are used to build cost estimating relationships (CERs).[†] In the cost estimating and analysis community, CERs are expressed in various algebraic forms as illustrated here:

- Single variable linear: $f(x_1) = a + bx_1, f(x_1) = bx_1$
- Single variable nonlinear: $f(x_1) = a + bx_1^c, f(x_1) = bx_1^c, f(x_1) = ab^{x_1}$
- Multivariable linear and nonlinear: $f(x_1, x_2) = ax_1 + bx_2, f(x_1, x_2, x_3) = ax_1^b x_2^c x_3^d$

These expressions use the following notation: f is the estimated cost of an item or WBS element; x_1, x_2, and x_3 are independent variables that drive the estimated cost; a, b, c, and d are constants derived from a regression analysis. CERs can be a function of multiple independent variables, as shown by the last algebraic form. Building statistical cost estimating models with multiple independent variables (i.e., multiple cost drivers) is also discussed herein.

---

* This term refers to the number of independent variables minus the number of coefficients used to specify a statistical model.

[†] In this chapter, the terms *cost estimating relationship* and *statistical cost estimation model* are synonymous.

To begin, the following presents the classical ordinary least squares (OLS) regression method.

10.2.1 Ordinary Least Squares Regression

Dating back centuries, the method of least squares was originally developed to find a function that best fits a set of observations. The phrase "best fit" refers to the function that minimizes the sum of the squared differences between the data observed and the data estimated by the function.

There are a variety of least squares methods that go by an assortment of names, many of which mean the same. There are linear and nonlinear least squares methods. Linear least squares methods find the equation of a *line* that best fits a set of observations. The simplest form of this method is called ordinary least squares (OLS). Nonlinear least squares methods find the equation of a *curve* that best fits a set of observations.

There is a great deal of scholarship on the mathematics of least squares methods, with numerous books on the topic available to readers.* Given this, the purpose of this section is to provide a light introduction to linear OLS and how it applies in building statistical CERs. This provides the needed background and context for the GERM, the main theme of this chapter.

We begin with the following. Define n as the number of observations contained in a dataset, with x_i as the value of the ith independent variable (the cost driver parameter) for each observation ($i = 1, \ldots, n$) and y_i as the value of the dependent variable (its observed value) associated with each x_i. OLS linear regression involves finding constants a and b such that best fits a dataset of observations (x_i, y_i) for $i = 1, \ldots, n$. OLS linear regression assumes an additive error form, where error ε_i is the difference between the observed value y_i and its estimated value produced by $f(x_i)$. This is given by Equation 10.1.

$$\varepsilon_i = f(x_i) - y_i \quad \text{for } (i = 1, 2, \ldots, n) \tag{10.1}$$

where $f(x_i) = a + bx_i$. The objective is to find numerical values for a and b that minimize the standard error of the estimate *SEE* given by Equation 10.2.[†]

$$SEE = \sqrt{\frac{1}{n-2} \sum_{i=1}^{n} (f(x_i) - y_i)^2} \tag{10.2}$$

The solution to this minimization problem is given by Theorem 10.1.

---

* See Larsen and Marx (2001).
† Equation 10.2 is called the adjusted *SEE*. It applies when using linear OLS to fit $f(x) = a + bx$ to a dataset of observations.

Theorem 10.1 *If a dataset contains n points (x_i, y_i) for $i = 1, \ldots, n$, then the line $f(x) = a + bx$ that minimizes Equation 10.2 is defined by*

$$a = \frac{\sum y_i - b \sum x_i}{n}$$

$$b = \frac{n \sum x_i y_i - (\sum x_i)(\sum y_i)}{n \sum x_i^2 - (\sum x_i)^2}$$

(10.3)

The proof of this theorem is widely available* and left for the reader to study. How well the equation of the line determined from Theorem 10.1 fits a dataset can be assessed by the following quality measures.

Standard Error: Equation 10.2 is the adjusted *SEE* associated with finding *a* and *b* such that the regression model $f(x) = a + bx$ is the optimal linear fit to a dataset of observations. The *SEE* is the root-mean-square of all additive errors when estimating the observed data with the regression model.

Bias: Bias is a measure of all positive and negative errors made in estimating the observed data with the regression model. It reflects how well the model fits the observations and whether it has a tendency to overestimate or underestimate these data. Bias is measured by Equation 10.4. In this form, it is called the *sample bias* of the regression model $f(x_i)$ with respect to the observed values y_i.

$$Bias = \frac{1}{n} \sum_{i=1}^{n} (f(x_i) - y_i)$$

(10.4)

Pearson's Correlation Squared: R^2 is a measure that expresses the goodness of fit between the observed values in a dataset and the estimated values of these data from the linear regression model. R^2 can be computed by Equation 10.5

$$R^2 = 1 - \frac{\sum_{i=1}^{n} \varepsilon_i^2}{\sum_{i=1}^{n} (y_i - \bar{y})^2}$$

(10.5)

where $\bar{y}$ is the arithmetic average of the observed values y_i. Equation 10.5 is sometimes referred to as the coefficient of determination. As a goodness of fit measure, it represents the proportion of the total variability in the dataset that is explained by the linear OLS model. For linear models of the form $f(x) = a + bx$, R^2 is the same as the square of the correlation between the observed values of y_i and its estimated value produced by $f(x_i)$.

---

* See Larsen and Marx (2001).

R^2 is a measure that falls between 0 and 1, inclusively. If $\varepsilon_i = 0$ for all $i = 1, \ldots, n$ then the OLS linear model explains 100% ($R^2 = 1$) of the total variability in the dataset. Here, the linear model is a perfect fit to the data. If $R^2 = 0$, then the observed values of y_i cannot be predicted by the OLS linear model—the model is unsuited as a statistical prediction tool. Thus, the closer R^2 is to 1, the better the linear model's fit to the data.

Example 10.1 *Use OLS to find the parameters a and b of the CER given by* $f(x) = a + bx$ *that best fits the dataset in Table 10.1.*

Solution Theorem 10.1 is used to find the equation of the line $f(x) = a + bx$ that minimizes the sum of squared errors given by Equation 10.2. The dataset contains $i = 1, \ldots, n = 7$ observations from programs that tracked cost y (in dollar units) as a function of the number of staff x on a team. In this case, x_i is the value of the ith independent variable and y_i is the value of the dependent variable associated with each x_i.

From Theorem 10.1, the parameters a and b that best fit $f(x) = a + bx$ to these data are computed by

$$a = \frac{\sum y_i - b \sum x_i}{n} \text{ and } b = \frac{n \sum x_i y_i - \left(\sum x_i\right)\left(\sum y_i\right)}{n \sum x_i^2 - \left(\sum x_i\right)^2}$$

Substituting the values for x_i and y_i from the dataset into the formulas for a and b yields $a = -3.207$ and $b = 0.9692$. Thus, the CER given by $f(x) = a + bx$ that best fits the dataset is

$$f(x) = -3.207 + 0.9692x \tag{10.6}$$

Figure 10.1 is a scatterplot of the dataset and the CER given by Equation 10.6. The observed data is shown by the dark circles in Figure 10.1a and b. Estimates of the observed data generated by the CER are shown by the open circles in Figure 10.1b. The solid line is a plot of the CER through those points.

TABLE 10.1

Example 10.1 Dataset

Program Number	Number of Staff, x	Observed Cost, y
1	7.9	1.380
2	8.2	3.395
3	9.8	7.201
4	11.5	10.900
5	16.4	15.434
6	19.7	16.074
7	23.6	17.274

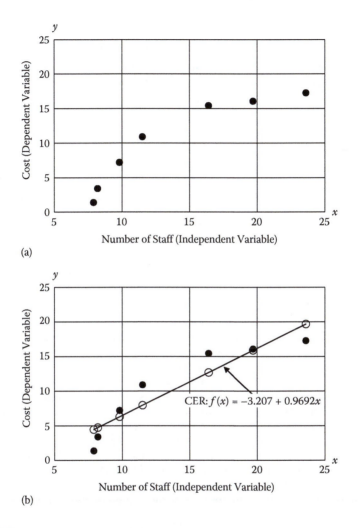

FIGURE 10.1
Example 10.1 scatterplot of observations and linear model fit. (a) Observed data. (b) Linear Model Fit.

A number of observations and statistical measures can be made about the goodness of the linear model's fit to this dataset. For instance, the scatterplot in Figure 10.1a reveals a nonlinear trend instead of a linear trend in the data. Although the OLS linear regression found the equation of the line that minimizes the sum of squared errors between these observations and the model's predicted values, from Equation 10.5, the R^2 measure is only 0.86. This means the model given by Equation 10.6 explains only 86% of the total variability in the dataset, with the remaining 14% unexplained. Table 10.2 summarizes these and other derived statistical measures.

TABLE 10.2

Example 10.1 CER $f(x)$ and its Quality Measures

Program Number	Number of Staff, x	Observed Cost, y	Estimated Cost, Regression Model	Additive Error
1	7.9	1.380	4.449	3.069
2	8.2	3.395	4.740	1.345
3	9.8	7.201	6.291	−0.910
4	11.5	10.900	7.938	−2.962
5	16.4	15.434	12.688	−2.746
6	19.7	16.074	15.886	−0.188
7	23.6	17.274	19.666	2.392

Theorem 10.1: OLS Derived Values	Model Quality Measures
$a = -3.207$	$SEE = 2.613$
$b = 0.9692$	$Bias = 0.000$
	$R^2 = 0.861$

Shown in Table 10.2, the SEE (from Equation 10.2) is equal to 2.613 (dollar units). In linear OLS, the SEE is an additive error bound on the regression model. From Equation 10.4, the bias measure is exactly zero. This reflects a mathematical property of linear OLS; that is, when fitting the function $f(x) = a + bx$ to a dataset by OLS, the sum of the additive errors (also called the sum of the residuals) will always equal zero. This implies the bias quality measure, when derived by the linear OLS method, will always equal zero.

Figure 10.2 shows dotted lines above and below the solid line given by the CER $f(x)$. They represent one standard error (± 2.613 dollar units) above and below the values of y predicted by $f(x)$. The diamond shown is the coordinate $(\bar{x}, \bar{y})$. This point is the average of the observed values of x and y (in Table 10.1). In linear OLS, the regression model $f(x) = a + bx$ will always pass through this point.

This concludes the Example 10.1 solution discussion; however, the dataset in this example will be revisited in Example 10.3. A closer look at the nonlinear shape of the scatterplot in Figure 10.1 will be presented and the GERM will find a nonlinear CER model of this dataset.

Some Algebraic Properties of Linear OLS: In general, the sum of the additive errors (the sum of the residuals) is equal to zero when linear OLS is used to build models of the form $f(x) = a + bx$. As mentioned earlier, this means the bias measure will equal zero as demonstrated in Table 10.2. The average of the observed values of y will equal the average of their estimated values produced by the linear regression model; for example, in Table 10.2 the averages of the data in columns three and four (from the left) are equal. Hence, the linear regression model formed by OLS will always pass through the mean of

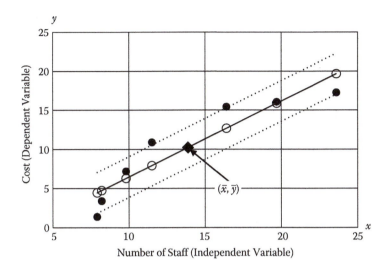

FIGURE 10.2
Example 10.1 dataset and scatterplot of observations.

the dataset, indicated by the coordinate $(\bar{x}, \bar{y})$. These are just a few of the many interesting and useful properties of linear OLS. The reader is encouraged to study Larsen and Marx (2001) for a further discussion about the statistical and algebraic properties of this regression method.

10.2.2 Nonlinear Ordinary Least Squares Regression

As mentioned earlier, there are linear and nonlinear least squares methods. The preceding discussion introduced OLS as a method for finding the equation of a *line* that best fits a set of observations. However, a line might not truly characterize the trend of the observations when visualized in a graphical way. Regression errors minimized to a line can still be large relative to regression errors minimized to a curve. In many cases, a curve is a better choice with which to build a regression model. To address these circumstances, nonlinear OLS methods are available. A popular nonlinear OLS technique is logarithmic OLS (LOLS). The following introduces LOLS and presents considerations when applying it to the development of statistical CERs.

The idea behind LOLS is to logarithmically transform a nonlinear equation to a linear form, from which linear OLS methods are applied to find the parameters of the regression model. For example, consider the dataset of observations in Table 10.1. The scatterplot of these data, shown in Figure 10.1a, suggests fitting them with a nonlinear CER such as $f(x) = ax^b$.

To apply LOLS, the log transform of $f(x) = ax^b$ is determined. This results in the expression

$$\log f(x) = \log(ax^b) = A + b \log x \tag{10.7}$$

where $A = \log a$ or equivalently $a = 10^A$. Equation 10.7 is now a linear equation defined in logarithmic space or "log-space" $(\log x, \log y)$ instead to being in arithmetic space (x, y). In LOLS, the objective is to find numerical values for A and b that minimizes the log-space standard error of the estimate $LSEE$, where

$$LSEE = \sqrt{\frac{1}{n-2} \sum_{i=1}^{n} (\log f(x_i) - \log y_i)^2} \tag{10.8}$$

The solution to this minimization problem is given by Theorem 10.2, which is a variant of Theorem 10.1.

Theorem 10.2   *Suppose a dataset contains n points (x_i, y_i) for $i = 1, \ldots, n$. The log-linear equation $\log f(x) = A + b \log x$ that minimizes Equation 10.8 is defined by*

$$A = \frac{\sum \log y_i - b \sum \log x_i}{n} \tag{10.9}$$

$$b = \frac{n \sum \log x_i \log y_i - \left(\sum \log x_i\right)\left(\sum \log y_i\right)}{n \sum (\log x_i)^2 - \left(\sum \log x_i\right)^2} \tag{10.10}$$

When Theorem 10.2 is applied to the dataset in Table 10.1, it follows that $A = -1.218$ and $b = 1.901$. Therefore, from Equation 10.7

$$\log f(x) = -1.218 + (1.901) \log x \tag{10.11}$$

in log-space, or equivalently

$$f(x) = 0.0605 x^{1.901} \tag{10.12}$$

in arithmetic space. Equation 10.11 is the log-linear CER regression model. Equation 10.12 is the nonlinear CER regression model.

Figure 10.3 shows a scatterplot of the dataset of observations in Table 10.1, along with the CER in log-space (Equation 10.11) and in arithmetic space (Equation 10.12). In Figure 10.3a and b, the observed data are shown by the dark circles. Estimates of the observed data generated by the CER are shown by the open circles. In Figure 10.3a, the line is a graph of the estimates of the observed data generated by the CER in log-space. In Figure 10.3b, the curve is a graph of the estimates of the observed data generated by the CER in arithmetic-space. The scatterplot suggests these data follow a nonlinear

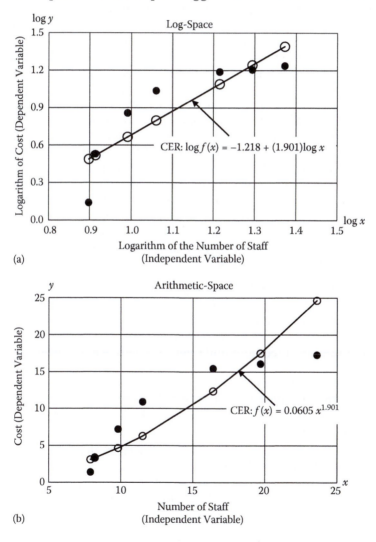

FIGURE 10.3
Log linear OLS regression of $f(x) = ax^b$ on the Table 10.1 dataset. (a) Log space. (b) Arithmetic space.

model; however, the CER in Figure 10.3a or b does not appear to be a good fit in log-space or in arithmetic space.

In Figure 10.3a, the CER is a linear model, in log-space, of observations that follow a nonlinear trend in log-space. It can be shown that the square of the correlation R^2 between the observed values and their estimated values produced by the CER, in log-space, is 0.756. The CER has a log-space standard error *LSEE* equal to 0.2236, which, with respect to the dataset in Table 10.1, is in *logarithmic* dollar units.

In Figure 10.3b, the CER is a nonlinear model, in arithmetic-space, of observations that follow a nonlinear trend in arithmetic-space. It can be shown that the square of the correlation R^2 between the observed values and their estimated values produced by the CER, in arithmetic-space, is 0.776. Moreover, in Figure 10.3b, the CER is a convex function that is opposite to the concave trend of these data.* The CER has an arithmetic-space standard error *SEE* equal to 4.4, which, with respect to the dataset in Table 10.1, is in dollar units.

Despite the nonlinearity of the observations in Figure 10.3, the CER $f(x) = -3.207 + 0.9692x$ (Equation 10.6) developed by linear OLS has better goodness of fit statistics than the CER developed by logarithmic OLS shown in Figure 10.3a and b. Still, the regression errors minimized to the CER given by the line $f(x) = -3.207 + 0.9692x$ might be large relative to regression errors minimized to a curve. The question is "what curve" has regression errors less than those of a line? Unfortunately, as shown earlier, LOLS will not always find a curve that satisfies this question. Moreover, LOLS is restricted to nonlinear forms where the properties of logarithms enables their nonlinearity to be transformed into log-linearity.

Issues with Logarithmic Ordinary Least Squares: In general, building statistical cost estimating models by the method of LOLS is an undesirable approach. There are mathematical and practical considerations that make LOLS problematic as a tool for regression analysis of cost data. The following discusses these considerations and offers alternative approaches.

- A log-space CER developed by LOLS does not estimate cost—it estimates the *logarithm* of cost. The LOLS process minimizes error if the goal was to estimate log cost, but the logarithm of cost is not a meaningful measure.

- In LOLS, the error term ε of the nonlinear model $f(x) = ax^b$ must be multiplicative and not additive; that is, it must be of the form $f(x) = ax^b \varepsilon$ and not $f(x) = ax^b + \varepsilon$. In the latter, the additive error prohibits transforming $f(x)$ into a log-linear form where linear OLS

---

* A real-valued function $f(x)$ that is differentiable and defined on an interval I is convex when any line segment joining two points on its graph lies above the graph. It is concave when any line segment joining two points on its graph lies below the graph.

could be used to find the parameters a and b that minimizes the log-space standard error.

Figure 10.4 is a scatterplot of observations that exhibit additive error and multiplicative error trends. In practice, it is always best to visually examine a dataset (when possible) for its error profile before considering the regression approach and candidate linear or nonlinear model forms.

- Nonlinear CERs derived by LOLS must have standard errors expressible as a percentage of the estimate, while linear CERs derived by OLS must have standard errors expressed as plus/minus dollar values. Thus, LOLS cannot handle situations where a cost analyst needs a linear CER with a multiplicative error or a nonlinear CER with an additive error.

- Any CER developed using LOLS regression must pass through the origin of the graph. That is, even if the data collected does not appear to have a best fit line that passes through the point $(0,0)$ on a graph, it is constrained as such by the properties of the log-linear model. Furthermore, nonlinear CERs whose coefficients are derived by LOLS must have fixed-cost equal to zero, while linear CERs whose coefficients are derived by OLS are permitted to have nonzero fixed-cost terms.*

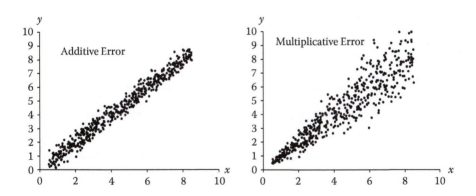

FIGURE 10.4

Observations exhibiting additive or multiplicative error trends. (From Eskew, H.L. and Lawler, K.S., *J. Cost Anal.*, Spring, 105, 1994.)

---

* In a CER, a fixed cost term is a constant that captures the cost incurred regardless of the numerical value of the independent variable(s) that drive cost (e.g., power, weight). Fixed cost is often associated with the start-up cost of development or production.

In the eighteenth and early nineteenth centuries, LOLS regression was considered the best (and, in fact, the only) method for fitting nonlinear algebraic relationships to a dataset of observations. No other option was available 200 years ago. Today, advances in computing technology and statistical optimization provide alternatives that are free of the shortcomings of LOLS. The latest advance in statistical regression techniques is called the *general error regression method* (GERM). GERM eliminates all the mentioned concerns with LOLS regression. GERM derives functional relationships having optimal (unbiased and minimum possible) error of estimation, while allowing the analyst to choose to minimize additive error or multiplicative error regardless of whether the functional relationship is best expressed by a linear or a nonlinear equation.

10.3 General Error Regression Method

Ordinary least squares regression, either linear or nonlinear, has been applied in the past to historical cost data to derive CERs. A fundamental assumption of OLS regression is that the error model upon which it is based be additive. More precisely, each observed value of cost is assumed to be a function of cost driving parameters (independent variables) plus a random error term that does not depend on the parameters. Unfortunately, this assumption is often invalid.

A case in point is where the values of observed ("actual") costs in a dataset change by an order of magnitude or more as a function of the cost driving parameters. Here, the random error is more realistically considered to be proportional to the magnitude of the cost, thereby effectively depending on the parameters. In this situation, it is often more realistic to assume a multiplicative error model than an additive error model. In past, this has been dealt with by taking logarithms of both sides of the model form and then applying additive-error linear regression. However, there are a number of difficulties in working with logarithmic transformations, in particular, the meaning of log-dollars when building a statistical CER.

The logarithmic transformation procedure also unnecessarily binds one to a specific class of regression equation forms. It is far from clear that the appropriate forecasting error is the one that is being minimized. Furthermore, use of OLS regression sometimes produces a curve fit that favors data points with large observed values. This is because an additive error model attempts to minimize the sum of squared deviations from all observed data points. This can give data points with large values the opportunity to perhaps unduly influence the determination of the "best fitting" curve. Use of the multiplicative error model will reduce the influence of large data values in a dataset.

GERM* is a new, general least squares approach that can treat additive and multiplicative error models. By examining the historical data, GERM allows the analyst to decide whether an additive or multiplicative error regression model is best. This flexibility allows one to select an appropriate functional form for the regression, independently of the choice of error model. In the past, linear OLS regression involved an additive error model, while non-linear forms (if they could be handled at all) required the assumption of a multiplicative error model. GERM allows the functional form to be chosen independently of the error probability distribution. Furthermore, the choice of the functional form of a CER developed through GERM is essentially unrestricted.

GERM derives functional relationships having optimal unbiased and minimum possible error of estimation, while allowing the analyst to choose to minimize additive error or multiplicative error regardless of whether the functional relationship is best expressed by a linear or a nonlinear equation. Thus, previously unavailable functional relationships can now be fit to the observed or collected data, such as $f(x) = a + bx^c$ or $f(x) = a + bx^c + d^x$ functions.

In GERM, n is defined as the number of observations contained in the dataset, x_i as the value of the ith independent variable (the cost driver parameter) for each observation ($i = 1, \ldots, n$), y_i as the value of the dependent variable (its observed value) associated with each x_i, and m as the number of constants† specified in the form of the CER or regression model (with $m < n$). Define the degrees of freedom for the CER or regression model as $n - m$.

GERM refers to the regression method where estimates of parameter values of a functional relationship are derived through constrained optimization. GERM derives these parameters without relying on assumptions about the nature of the error distribution. There are two varieties of GERM models—those with additive errors and those with multiplicative errors. The following discusses each of these models.

10.3.1 Additive Error Form

This section presents the additive error form of GERM. The three measures of CER statistical quality reported are as follows:

---

* GERM was developed during the period 1994–1998 by S.A. Book, P.H. Young, and N.Y. Lao. The reader is directed to the following references for further information: Book and Lao 1998, Book and Young 1997, Young 1999.
† This is the number of constant coefficients in the form of the CER plus the number of constants that appear in any exponents.

Standard Error: The standard error of the estimate is the root-mean-square of all *additive* errors when estimating the observed data with the regression model. This is the minimization objective in the additive form of GERM.

Bias: Bias is the algebraic sum of the errors (positive and negative values) made in estimating the observed data with the regression model, averaged over the number of observations. In the additive error form of GERM, the bias is constrained to be zero.

Pearson's Correlation Squared: In GERM, this measure is the square of the correlation between the *observed values* of the *dependent variable* and the *estimated values* of the *dependent variable* generated by the regression model. In GERM, the closer this measure is to 1, the better the fit of the regression model to these values. In GERM, Pearson's correlation squared is computed this way for linear or nonlinear regression model forms.

In the additive form of GERM, error ε_i is the difference between the observed value y_i and its estimated value $f(x_i)$ computed from the derived regression function. This is given by Equation 10.13.

$$\varepsilon_i = f(x_i) - y_i \quad \text{for } (i = 1, 2, \ldots, n) \tag{10.13}$$

The objective is to find values of the constants that define the algebraic form chosen for $f(x)$ that minimizes the standard error of the estimate *SEE*, given by Equation 10.14* for the additive GERM,

$$SEE = \sqrt{\frac{1}{n-m} \sum_{i=1}^{n} \varepsilon_i^2} = \sqrt{\frac{1}{n-m} \sum_{i=1}^{n} (f(x_i) - y_i)^2} \tag{10.14}$$

subject to the constraint that the bias is zero, as given by Equation 10.15

$$Bias = \frac{1}{n} \sum_{i=1}^{n} (f(x_i) - y_i) = 0 \tag{10.15}$$

In the additive error form of GERM, this procedure is called minimum-error regression under a zero-bias constraint or the ZME ("zimmy") technique (Book and Lao 1998). ZME was developed to yield CERs *guaranteed* to have the minimum possible error among all unbiased CERs with the chosen algebraic form. ZME CERs are derived using an operations research method known as *constrained optimization*, a numerical analysis method which searches for the constants of the CER that minimizes the standard error (Equation 10.14) under the zero bias constraint (Equation 10.15). Today's

---

* Equation 10.14 is called the adjusted *SEE* for GERM in its additive error form.

advanced computing technology* has greatly facilitated solutions to this problem and the subsequent use of GERM by the cost estimating and analysis community.

The additive error form of GERM has probabilistic structure $Y_i = f(X_i) + \varepsilon_i$ where Y_i and X_i are random variables and ε_i is a random error such that $E(\varepsilon_i) = 0$ and $Var(\varepsilon_i)$ is a constant additive-error dispersion around $E(\varepsilon_i)$. In GERM, the nature of the underlying error probability distribution is unimportant—unlike least squares regression where the error must be assumed to have a normal probability distribution.

Example 10.2 *Consider the dataset in Table 10.3. Use the GERM ZME technique to find a pair of values for a and b such that $f(x) = a + bx$ has minimum-error under a zero bias constraint.*

Solution GERM ZME implies using the additive error form. Table 10.3 shows the dataset of $n = 10$ observations. An Excel model is built and the Excel *Solver* add-in is used to run the ZME optimization (minimization in GERM). Table 10.4 shows the setup for this analysis. Table 10.4 contains the dataset and the degrees of freedom for this model, which was given as $f(x) = a + bx$. The columns labeled "Estimated Hours, Regression Model," "Additive Error," and "Model Quality Measures" are computed from the starter values shown for a and b.

Starter values are the initial conditions chosen by the analyst to stimulate the optimization routine in Excel *Solver*. The lower right corner of

TABLE 10.3

Example 10.2 Dataset

Program Number	Number of Requirements, x	Observed Hours, y
1	73	1445
2	128	2448
3	92	2052
4	329	7121
5	217	3929
6	64	1683
7	192	3705
8	201	4226
9	256	5141
10	171	3156

Source: Larsen, R. and Marx, M. 2001. *An Introduction to Mathematical Statistics and Its Applications*, 3rd edn. Englewood Cliffs, NJ: Prentice-Hall, pp. 338–344.

---

* For example, Microsoft's Excel *Solver* can be used to solve the optimization problem formulated when applying GERM.

TABLE 10.4

Example 10.2 Dataset and Analysis Setup

	A	B	C	D	E
1	Number of Program Data Points		$n = 10$	**Regression Model Form**	
2	Number of Regression Constants		$m = 2$	$f(x) = a + bx$	
3	Degrees of Freedom		$n - m = 8$		
4	Program Number	Number of Requirements, x	Observed Hours, y	Estimated Hours, Regression Model	Additive Error
5	1	73	1445	148.0	−1297.0
6	2	128	2448	258.0	−2190.0
7	3	92	2052	186.0	−1866.0
8	4	329	7121	660.0	−6461.0
9	5	217	3929	436.0	−3493.0
10	6	64	1683	130.0	−1553.0
11	7	192	3705	386.0	−3319.0
12	8	201	4226	404.0	−3822.0
13	9	256	5141	514.0	−4627.0
14	10	171	3156	344.0	−2812.0
15		Starter Values		Model Quality Measures	
16		$a = 2 \equiv B16$		$SEE = 3893.2 \equiv E16$	
17		$b = 2 \equiv B17$		$Bias = -3144.0 \equiv E17$	
18				$R^2 = 0.9712$	

Table 10.4 shows the derived statistics for the three quality measures for the additive form of GERM. In Table 10.4, they are computed *from the starter values* for a and b. Figure 10.5 shows the Excel *Solver* window and the features selected to run the ZME optimization for the data in Table 10.4. The optimization algorithm begins with the starter values shown for a and b and iterates numerically until the search algorithm converges to a solution. The optimization results are shown in Table 10.5.

In this example, *Solver* found the minimum-error under a zero bias constraint occurs for $a = -2.056$ and $b = 20.2708$, as shown in Table 10.5. *Solver* generated this by the GRG algorithm.* Thus, the regression model with the statistical quality measures shown in Table 10.5 is given by Equation 10.16.

$$f(x) = a + bx = -2.056 + 20.2708x \qquad (10.16)$$

The model derived in Example 10.2 reflects a minimum-error regression under a zero-bias constraint. Figure 10.6 plots this regression model versus the observed data. The observed data is shown by the dark circles. Estimates

---

* The optimization method shown is the generalized reduced gradient algorithm, developed by Leon Lasdon of the University of Texas at Austin and Allan Waren of Cleveland State University.

FIGURE 10.5
Example 10.2 Excel *Solver* setup.

of the observed data generated by the regression model are shown by the open circles, with the solid line being a plot of $f(x) = a + bx = -2.056 + 20.2708x$ through those points. The quality measures for this model, shown in Table 10.5, suggest Equation 10.16 is a good fit to this dataset.

In this example, ZME produced through an optimization process, the same results that would be obtained by a linear OLS regression of $f(x) = a + bx$ to the dataset in Table 10.3. Showing this is an exercise for the reader.

10.3.2 Multiplicative Error Form

This section presents the multiplicative error form of GERM. Three measures of CER statistical quality are reported. They are as follows.

Percentage Standard Error: The multiplicative error form of GERM expresses the standard error of the estimate in *percentage* terms instead of expressing it in dollar values. This provides practical benefits to cost analysts. The first is that expressing cost estimating error in percentage terms offers stability of meaning across a wide range of programs, time periods, and estimating situations.

A percentage error of, say, 30%, retains its meaning whether a $10,000 component or a $10,000,000,000 program is being estimated. A standard error expressed in dollars, say, $59,425, is a large error when estimating a $100,000 component, but is less significant when estimating a $10,000,000,000

TABLE 10.5

Example 10.2 *Solver* Solution for *a* and *b*

	A	B	C	D	E
1	Number of Program Data Points		$n = 10$	**Regression Model Form**	
2	Number of Regression Constants		$m = 2$	$f(x) = a + bx$	
3	Degrees of Freedom		$n - m = 8$		
4	Program Number	Number of Requirements, x	Observed Hours, y	Estimated Hours, Regression Model	Additive Error
5	1	73	1445	1477.7	32.7
6	2	128	2448	2592.6	144.6
7	3	92	2052	1862.9	−189.1
8	4	329	7121	6667.0	−454.0
9	5	217	3929	4396.7	467.7
10	6	64	1683	1295.3	−387.7
11	7	192	3705	3889.9	184.9
12	8	201	4226	4072.4	−153.6
13	9	256	5141	5187.3	46.3
14	10	171	3156	3464.2	308.2
15	Solver Derived Values			Model Quality Measures	
16	$a = -2.056$			$SEE = 313.8$	
17	$b = 20.2708$			$Bias = 0.0$	
18				$R^2 = 0.9712$	

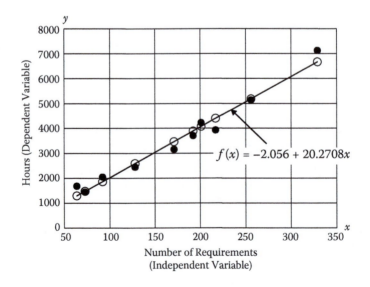

FIGURE 10.6
Example 10.2 regression model versus the observed data.

program. Regardless, a standard error expressed in dollars can make a CER unusable at the low end of its data range, where relative magnitudes of the estimate and its standard error can be inconsistent. Describing a CER in terms of percentage standard error offers the additional benefit that linear and nonlinear forms can be compared on the basis of quality.

In the early history of CER development, nonlinear CERs were derived by taking logarithms of both sides of the equation $f(x_1) = ax_1^b$. Classical linear regression methods were then applied to the form $\log f(x_1) = \log a + b \log x_1$. Cost models developed during that age typically contained a mix of CERs, some linear and some nonlinear, whose standard errors were as incommensurable as apples and oranges. Standard errors of linear CERs were universally expressed in dollars ("apples"), while standard errors of nonlinear CERs were reported as percentages ("oranges"). It was difficult to compare the precision of different candidate CER forms, let alone to combine the apples and oranges to determine the error of the cost estimate of a total program. Using percentage standard error to measure the precision of all CER forms resolves this inconsistency.

Percentage Bias: Percentage bias is the algebraic sum of the percentage errors (positive and negative values) made in estimating the observed data with the regression model, averaged over the number of observations. Thus, percentage bias measures the net percentage error. In the multiplicative error form of GERM, percentage bias is constrained to be zero.

Pearson's Correlation Squared: In GERM, this measure is the square of the correlation between the *observed values* of the *dependent variable* and the *estimated values* of the *dependent variable* generated by the regression model. In GERM, the closer this measure is to 1, the better the fit of the regression model to these values. In GERM, Pearson's correlation squared is computed this way for linear or nonlinear regression model forms.

In the multiplicative form of GERM, error ε_i is the difference between the estimated value $f(x_i)$ and the observed value y_i divided by its estimated value $f(x_i)$, where $f(x_i)$ is computed from the derived regression function. This is given by Equation 10.17.

$$\varepsilon_i = \frac{f(x_i) - y_i}{f(x_i)} \quad \text{for } (i = 1, 2, ..., n) \tag{10.17}$$

The objective is to find values of the constants that define the algebraic form chosen for $f(x)$ that minimizes the standard error of the estimate *SEE*, given by the following* for the multiplicative GERM:

$$SEE = \sqrt{\frac{1}{n-m} \sum_{i=1}^{n} \varepsilon_i^2} = \sqrt{\frac{1}{n-m} \sum_{i=1}^{n} \left(\frac{f(x_i) - y_i}{f(x_i)} \right)^2} \tag{10.18}$$

---

* Equation 10.18 is called the adjusted *SEE* for GERM in its multiplicative error form.

which is subjected to the constraint that the bias is zero, as given by Equation 10.19

$$Percent\ Bias = \frac{1}{n} \sum_{i=1}^{n} \left(\frac{f(x_i) - y_i}{f(x_i)} \right) = 0 \qquad (10.19)$$

In the multiplicative error form of GERM, this procedure is informally called minimum-percentage error regression under a zero percentage bias constraint or the ZMPE ("zimpy") technique (Book and Lao 1998). ZMPE was developed to yield CERs _guaranteed_ to have the minimum possible percentage error among all unbiased CERs with the chosen algebraic form. ZMPE CERs are derived using an operations research method known as _constrained optimization_, a numerical analysis method that searches for the constants of the CER that minimizes the percentage standard error (Equation 10.18) under the zero percentage bias constraint (Equation 10.19).

The multiplicative error form of GERM has a probabilistic structure $Y_i = f(X_i)\varepsilon_i$, where Y_i and X_i are random variables and ε_i is a random error such that $E(\varepsilon_i) = 1$ and $Var(\varepsilon_i)$ is a constant multiplicative-error dispersion around $E(\varepsilon_i) = 1$. In GERM, the nature of the underlying error probability distribution is unimportant—unlike least squares regression where the error must be assumed to have a normal probability distribution.

Example 10.3 _Consider the scatterplot in Figure 10.1. Use GERM ZMPE to find a set of values for a, b, and c such that the nonlinear function $f(x) = a + bx^c$ has minimum-percentage error under a zero percentage bias constraint._

Solution This example revisits Example 10.1, where the dataset evidenced a nonlinear trend in the observations. This example illustrates how GERM ZMPE can be used to find a nonlinear regression model to this dataset— one that will minimize the percentage error under a zero percentage bias constraint.

GERM ZMPE implies using the multiplicative error form. Table 10.1 shows the dataset of $n = 7$ observations that correspond to the points in Figure 10.1. An Excel model is built and its _Solver_ feature is used to run the ZMPE optimization (minimization in GERM). Table 10.6 shows the setup for this analysis.

Table 10.6 contains the dataset from Example 10.1 and the degrees of freedom for this model, which is given as $f(x) = a + bx^c$. The columns labeled "Estimated Cost, Regression Model," "Multiplicative Error," and "Model Quality Measures" are computed from the three starter values shown for _a_, _b_, and _c_. As mentioned earlier, starter values are the initial conditions chosen by the analyst to stimulate the optimization routine in Excel _Solver_.

The lower right corner of Table 10.6 shows the derived statistics for the three quality measures for the multiplicative form of GERM. In Table 10.6, they are computed _from the starter values_ for _a_, _b_, and _c_. Thus, they do not represent the results from the optimization, which is shown in Table 10.7.

TABLE 10.6

Example 10.3 Dataset and Analysis Setup

	A	B	C	D	E
1	Number of Program Data Points		$n = 7$	**Regression Model Form**	
2	Number of Regression Constants		$m = 3$	$f(x) = a + bx^c$	
3	Degrees of Freedom		$n - m = 4$		
4	Program Number	Number of Staff, x	Observed Cost, y	Estimated Cost, Regression Model	Multiplicative Error
5	1	7.9	1.380	381.069	0.9964
6	2	8.2	3.395	386.356	0.9912
7	3	9.8	7.201	413.050	0.9826
8	4	11.5	10.900	439.116	0.9752
9	5	16.4	15.434	504.969	0.9694
10	6	19.7	16.074	543.847	0.9704
11	7	23.6	17.274	585.798	0.9705
12	Starter Values			Model Quality Measures	
13	$a = 100 \equiv B13$			$SEE = 1.2957 \equiv E13$	
14	$b = 100 \equiv B14$			$Bias = 0.9794 \equiv E14$	
15	$c = 0.50 \equiv B15$			$R^2 = 0.9026$	

TABLE 10.7

Example 10.3 *Solver* Solution for a and b

	A	B	C	D	E
1	Number of Program Data Points		$n = 7$	**Regression Model Form**	
2	Number of Regression Constants		$m = 3$	$f(x) = a + bx^c$	
3	Degrees of Freedom		$n - m = 4$		
4	Program Number	Number of Staff, x	Observed Cost, y	Estimated Cost, Regression Model	Multiplicative Error
5	1	7.9	1.380	1.4767	0.0655
6	2	8.2	3.395	2.9241	−0.1610
7	3	9.8	7.201	8.2420	0.1263
8	4	11.5	10.900	11.3528	0.0399
9	5	16.4	15.434	15.0558	−0.0251
10	6	19.7	16.074	16.0030	−0.0044
11	7	23.6	17.274	16.5926	−0.0411
12	Solver Derived Values			Model Quality Measures	
13	$a = 17.6149$			$SEE = 0.1119$	
14	$b = -2957.696$			$Bias = 0.0000$	
15	$c = -2.5212$			$R^2 = 0.9919$	

FIGURE 10.7
Example 10.3 Excel *Solver* setup.

Figure 10.7 shows the Excel *Solver* window and the features selected to run the ZMPE optimization for the data in Table 10.6. The optimization algorithm begins with the starter values shown for a, b, and c and iterates numerically until the search algorithm converges to a solution.

Figure 10.8 is a plot of this regression model versus the observed data. The observed data is shown by the dark circles. Estimates of the observed data generated by the CER $f(x) = 17.6149 - (2957.696)x^{-2.5212}$ are shown by the open circles. A visual inspection suggests this nonlinear CER is a markedly improved fit to the dataset than the linear model $f(x) = -3.207 + 0.9692x$ derived in Example 10.1.

In addition, the quality measures in Table 10.7 quantifies the goodness of the CER's fit to this dataset. Derived under a zero bias constraint, the standard error of the estimate for $f(x) = 17.6149 - (2957.696)x^{-2.5212}$ comes to within $\pm 11.2\%$ of the observed costs in Table 10.7. This is accompanied by a very high R^2 of 0.992, measured as the square of the correlation between the *observed* costs and the *estimated costs* (shown in Table 10.7) generated by the CER $f(x) = 17.6149 - (2957.696)x^{-2.5212}$.

This example demonstrates the flexibility of GERM over OLS and LOLS. The model form $f(x) = a + bx^c$ is not amenable to OLS because $f(x)$ is a nonlinear function, which requires a multiplicative (not additive) error. Recall that OLS requires an additive error model. The model form $f(x) = a + bx^c$ is not

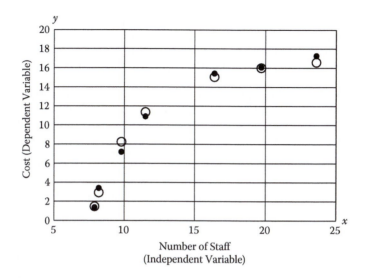

FIGURE 10.8
Example 10.3 regression model versus the observed data.

amenable to LOLS because the log-transform of $f(x)$ cannot be written in a log-linear form; that is, $\log f(x) = \log(a + bx^c)$ cannot be expressed as a linear sum of logarithms—*a* necessary condition for LOLS. However, the method of general error regression found a set of values for parameters a, b, and c of the model $f(x) = a + bx^c$ that has minimum-percentage error under a zero percentage bias constraint.

Example 10.4 *Consider the antenna cost dataset in Table 10.8. Use the GERM ZMPE technique to find a set of values for a, b, and c such that the CER* $f(x_1, x_2) = a + bx_1^c x_2^d$ *has minimum-percentage error under a zero percentage bias constraint.*

TABLE 10.8

Example 10.4 Antenna Historical Cost Dataset

Program Number	Observed Diameter (meters, m) x_1	Observed Frequency (GHz) x_2	Observed Cost ($K)
1	3	2	44.68
2	3	2	85.85
3	5	2	77.55
4	15	22	11,881.17
5	20	22	15,666.08
6	4	10	1242.96
7	3	12	2088.57
8	5	10	2438.03

Solution GERM ZMPE implies using the multiplicative error form. Table 10.8 shows the dataset of $n = 8$ antenna program cost histories. An Excel model is built and the Excel *Solver* add-in is used to run the ZMPE optimization (minimization in GERM). Table 10.9 shows the setup for this analysis.

In Table 10.9, the optimization algorithm begins with the starter values shown for a, b, c, and d and iterates numerically until the search algorithm converges to a solution. Table 10.9 contains the dataset from Example 10.4 and the degrees of freedom for this model.

The columns labeled "Estimated Cost, Regression Model," "Multiplicative Error," and "Model Quality Measures" are computed from the starter values shown for a, b, c, and d. As mentioned earlier, starter values are the initial conditions chosen by the user to stimulate the optimization routine in Excel *Solver*.

The lower right corner of Table 10.9 shows the derived statistics for the three quality measures for the multiplicative form of GERM. In Table 10.9, they are computed *from the starter values* for a and b. Thus, they do not represent the results from the optimization, which is shown in Table 10.10. Figure 10.9 shows a three-dimensional plot of this nonlinear CER.

TABLE 10.9

Example 10.4 Dataset and Analysis Setup

	A	B	C	D	E	F
1	Number of Program Data Points		$n = 8$		**Regression Model Form**	
2	Number of Regression Constants		$m = 4$		$f(x_1, x_2) = a + bx_1^c x_2^d$	
3	Degrees of Freedom		$n - m = 4$			
					Estimated	
			Observed		Cost,	
4	Program	Observed	Frequency	Observed	Regression	Multiplicative
	Number	Diameter (meters, m) x_1	(GHz) x_2	Cost ($K)	Model	Error
5	1	3	2	44.68	74.00	0.3962
6	2	3	2	85.85	74.00	−0.1601
7	3	5	2	77.55	202.00	0.6161
8	4	15	22	11,881.17	217,802.00	0.9454
9	5	20	22	15,666.08	387,202.00	0.9595
10	6	4	10	1242.96	3202.00	0.6118
11	7	3	12	2088.57	2594.00	0.1948
12	8	5	10	2438.03	5002.00	0.5126
13		Starter Values			Model Quality Measures	
14		$a = 2$			$SEE = 0.8735$	
15		$b = 2$			$Bias = 0.5096$	
16		$c = 2$			$R^2 = 0.9598$	
17		$d = 2$				

TABLE 10.10

Example 10.4 *Solver* Solution for a, b, c, and d

	A	B	C	D	E	F
1	Number of Program Data Points		$n = 8$		**Regression Model Form**	
2	Number of Regression Constants		$m = 4$		$f(x_1, x_2) = a + bx_1^c x_2^d$	
3	Degrees of Freedom		$n - m = 4$			
4	Program Number	Observed Diameter (meters, m) x_1	Observed Frequency (GHz) x_2	Observed Cost ($K)	Estimated Cost, Regression Model	Multiplicative Error
5	1	3	2	44.68	65.70	0.3200
6	2	3	2	85.85	65.70	−0.3066
7	3	5	2	77.55	83.78	0.0744
8	4	15	22	11,881.17	12,618.54	0.0584
9	5	20	22	15,666.08	14,213.96	−0.1022
10	6	4	10	1242.96	1689.07	0.2641
11	7	3	12	2088.57	2103.98	0.0073
12	8	5	10	2438.03	1853.41	−0.3154
13	Solver Derived Values				Model Quality Measures	
14	$a = -11.1490$				$SEE = 0.3103$	
15	$b = 13.5340$				$Bias = 0.0000$	
16	$c = 0.4135$				$R^2 = 0.9891$	
17	$d = 1.8501$					

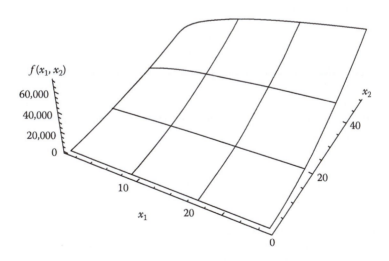

FIGURE 10.9

A three-dimensional plot of $f(x_1, x_2) = -11.149 + 13.534 x_1^{0.413503} x_2^{1.8501}$.

10.4 Summary

This chapter presented a light introduction to ordinary least squares as a method for developing CERs. In keeping with the aim of Section II of this book, the emphasis has been on practical applications of ordinary least squares (OLS), and its variants, rather than an in-depth theoretical treatment of statistical regression.

This chapter also introduced an advance in methodology called general error regression. As illustrated, the general error regression method (GERM) is a robust alternative to OLS approaches—especially logarithmic ordinary least squares (LOLS). It is a recommended protocol for building CERs by statistical regression. As shown, GERM separates the problem of whether estimating error should be additive or multiplicative from the problem of whether the form of a regression model should be linear or nonlinear. This offers analysts the choice of minimizing additive or multiplicative errors, regardless of the functional form of the relationship.

It is possible that constrained optimization methods, like the GRG algorithm used in the GERM examples, will produce locally optimal rather than globally optimal results. Thus, it is best to use GERM with a variety of regression constant starter values and choose those with the best-quality metrics. At present, a statistical significance test (such as the t-statistic in OLS) has not been established to assess a GERM ZMPE regression's goodness of fit to a dataset. This is an open research question that is worthy of further exploration and study by members of the statistical sciences communities.

A well-known aphorism asserts that "a man cannot serve two masters." The cost analysis version of this adage is that a CER cannot be optimized with respect to two different criteria. Because GERM requires not two but three criteria (standard error, bias, and Pearson's correlation squared), it follows that GERM CERs cannot be optimal with respect to all three criteria. For example, a CER optimized with respect to standard error will not, in general, have zero bias and maximum possible correlation. A CER optimized to have maximum possible correlation will not have minimum possible standard error and zero bias.

The choice of which CER to use is a trade-off that must be made by the cost analyst. Cost should be regressed against single cost drivers and combinations of multiple cost drivers. Finding the "best" statistical cost estimating model is a judgment that can be guided by the GERM quality metrics. In the case of ZMPE regression, the R^2 value and the percent standard error are very useful comparison tools. As a general rule, if one model form is a stronger fit to the observed data than another model form, then it will have a higher R^2 and a lower percent standard error (Anderson 2002).

The coefficients for all cost drivers should also be analyzed. If the resulting relationship between cost and a cost driver is different than would be expected from engineering judgment, then this model is most likely not one that should be used to predict future cost. Correlations between cost drivers should also be calculated. Often two or more cost drivers could provide very similar relationships with cost. Including correlated drivers together in a cost model could yield inconsistent results in a regression model. In statistics, this is known as multicollinearity—and the cost modeler should look for this potential problem.

GERM is a significant advance in the ability to create CERs. It allows building them for virtually any type of linear or nonlinear form, with additive or multiplicative errors. Furthermore, measures of statistical quality for GERM-derived CERs are free from a priori statistical requirements or assumptions such as normally distributed error probabilities. GERM's flexibility and ease of use allows for the derivation of a variety of potential CERs with zero bias and minimum additive error (ZME) or zero percent bias and minimum percent error (ZMPE).

Another general error regression approach is iteratively reweighted least squares (IRLS), introduced in Appendix F. IRLS is a least squares procedure where the constants of a regression function are determined by iterating through a minimization of a weighted sum of additive squared errors. Today, the widespread availability of advanced computing technologies greatly facilitates the ease with which general error regression methods can be used throughout the cost and greater analytic communities.

Exercises

10.1 a. Find the linear OLS solution to the dataset given in Example 10.2 and confirm the quality measures of the linear regression model shown in Table 10.5.

 b. Verify the LOLS solution given in Section 10.2.2 to the dataset in Table 10.1.

10.2 Consider the scatterplot in Figure 10.1. Use GERM ZMPE to find a set of values for a and b such that

 a. the nonlinear function $f(x) = \sqrt{a + bx}$ has minimum-percentage error under a zero percentage bias constraint.

 b. the nonlinear function $f(x) = ab^{\ln x}$ has minimum-percentage error under a zero percentage bias constraint. Use the starter values $a = 2$ and $b = 1$ in both cases.

References

Anderson, T. P. 2002. A tutorial on CER development. Washington Chapter SCEA, March 23, 2002.

Book, S. and E. Burgess. 1996. The learning rate's overpowering impact on cost estimates and how to diminish it. *Journal of Parametrics*, 16, 33–57.

Book, S. and N. Lao. 1998. Minimum-percentage error regression under zero-bias constraints. In *Proceedings of the Fourth Annual U.S. Army Conference on Applied Statistics*, U.S. Army Research Laboratory, Adelphi, MD, pp. 47–56.

Book, S. and P. Young. 1997. General-error regression for deriving cost-estimating relationships. *The Journal of Cost Analysis*, Fall 1997, 1–28.

Eskew, H. L. and K. S. Lawler. 1994. Correct and incorrect error specifications in statistical cost models. *Journal of Cost Analysis*, Spring, 105–123.

Larsen, R. and Marx, M. 2001. *An Introduction to Mathematical Statistics and Its Applications*, 3rd edn. Englewood Cliffs, NJ: Prentice-Hall, pp. 338–344.

Young, P. H. 1999. The meaning of the standard error of the estimate in logarithmic space. *The Journal of Cost Analysis and Management*, Winter 1999, 59–65.

Additional Reading

Fylstra, D. L. Lasdon, J. Watson, and A. Waren. 1998. Design and use of the Microsoft Excel solver. *Interfaces*, 28(5), 29–55.

Stump, E. 2010. Regression for cost analysts, Galorath Associates. Accessed May 12, 2014 from www.galorath.com/images/uploads/stump_regr_for_cost_analysis.pdf.

11

Mathematics of Cost Improvement Curves

This chapter introduces a phenomenon associated with producing an item over and over again. The phenomenon is called cost improvement (CI)*. It refers to a lessening in the cost of an item produced in large quantities in the same way over a period of time. An item's CI is often attributed to a lessening in time for its production due to learning from the repetition of tasks in its manufacture.

This chapter presents two main topics associated with CI in an item's recurring production cost. The first describes the phenomenon of CI and methods to measure and mitigate its effects on the uncertainty in production cost estimates. The second illustrates how the general error regression method (GERM†) can be applied to build cost estimating relationships (CERs) of an item's recurring production costs, in the presence of CI effects.

11.1 Introduction

Cost improvement (CI) is used to describe the phenomenon of the general improvement (reduction) of costs between two successive items, efforts, or programs. CI can be experienced between similar successive programs or between successive production units and repeated tasks. Cost reductions due to CI make it an attractive and even expected phenomenon that must be considered in cost estimates.

For the purposes of this chapter, define *recurring costs* as the costs associated with *repeating* efforts. A rule of thumb to determine whether costs of a first production unit are recurring or nonrecurring is to examine the differences between the cost of the first production unit and the cost of the second production unit. Costs that are repeatedly incurred for both production units are considered *recurring costs*, whereas unique, nonrepeated costs incurred for the first production unit are considered *nonrecurring costs*.

---

* The terms "cost progress" and "learning" are often used as synonyms to "cost improvement" (Godberg and Touw 2003).
† Chapter 10 presents a discussion of the GERM.

The literature on CI is focused on the recurring cost perspective,* but nonrecurring CI can also exist (Boehm 1981, Stump 1988). Recurring CI can occur between successive units within a program and when the same units are successively produced for different programs. Nonrecurring CI can occur from reuse of existing designs, materials, equipment, effort, or products.

Nonrecurring CIs between successive programs are accomplished through the partial (or full) reuse of designs, development articles, software, and other nonrecurring engineering effort. Savings from nonrecurring costs (presumably at the expense of the first program) are from the subsequent program to capitalize on reuse of artifacts from the first project. For example, design requirements might be achieved in a less costly manner than if the predecessor program had not existed.

An accurate measurement of CI is difficult because it is influenced by many factors (Covert 2014). A major one is cost accounting structures, which are often different between organizations. This affects how labor hours and costs are accumulated and tracked, which makes comparisons across organizations difficult. Other factors include the following:

- Learning or gained experience in value added effort.[†]
- Skill mix changes that increase or decrease cost.
- Process "shortcuts" that eliminate effort or expenses.
- Yield improvements that reduce cost.
- Production rate increases allowing for amortization of pooled costs and greater cost efficiency.
- Technological advances allowing greater yield and process efficiency.
- Material price discounts based on larger quantity purchases.
- Inflation adjustments for multiyear programs to account for the time varying costs.

The actual observation of CI implies it covers the same work content between successive units. It should be independent of other recurring and nonrecurring CIs such as skill mix changes, process shortcuts, yield improvements, production rate increases, material discounts, and inflation adjustments.

The theories behind CI factors require them to be treated independently from each other and modeled using different mathematical equations. As a practical consideration, it is difficult to segregate the effects of CI due to learning from other CI factors in historical cost data. Thus, all observed CI between successively produced units is treated as learning and is assumed

---

* See Yelle (1979) and Covert (2014) for a comprehensive survey.
[†] An example of a value-added effort is the assembly of parts to produce a finished product. An example of a non-value-added effort is financial accounting to track the labor hours to assemble those parts.

to follow the familiar "power" form attributable to learning curve theory, as shown by Equation 11.1. For this reason, the word "learning" in this chapter is meant to include all effects that result in lower (improved) costs of successively produced units.* Although the computations expressed herein are in dollar units, the original work on learning curve theory focused on the reduction in labor hours between successive units in a production process (Hudgins 1966).

11.2 Learning Curve Theories

Theories of learning in mass production of aircraft began to be develop in the first half of the twentieth century, with two predominating the original literature.[†] One is the cumulative average learning curve theory (Wright 1936). The other is unit learning curve theory (Crawford 1944, Crawford and Strauss 1947). Both theories state that as production quantities double, costs reduce at a constant percentage rate called the learning rate or the learning curve slope. In both theories, the cost Y_x of successively produced units x is expressed mathematically by the form $Y_x = Ax^B$. Although they share a similar form, the distinction between cumulative average and unit learning curve theories is in their interpretation of Y_x. The following introduces both theories and illustrates their similarities and differences.

Cumulative Average Learning Curve Theory: The cumulative average learning curve theory states as quantities double, the cumulative average costs (or hours) decline by a constant percentage.[‡] Cumulative average cost

---

* Other treatments of CI are described by Badiru (1992); Bierman and Dyckman (1971); Nussbaum (1994); and Stump (1988). More general formulations are discussed by Levine et al. (1989). These latter authors make a distinction between "learning" and a more inclusive concept they refer to as "cost progress."

[†] One of the first investigations into airframe production data, which led to the formulation of the learning curve theory, was conducted by Major Leslie MacDill, the commanding officer at McCook Field (nearby Wright-Patterson Air Force Base) in 1925. McCook Field was home to airplane research, development, and production engineering. T.P. Wright, who began his research in the early 1920s, is credited with the first publication of learning curve theory in 1936 while working for the Curtiss Aeroplane Company. In the *Journal of the Aeronautical Sciences,* "Factors Affecting the Cost of Airplanes," he showed that as the number of aircraft produced increases, the cumulative average cost to produce the aircraft decreases at a constant rate. This became known as the cumulative average learning curve theory or the Wright curve. After World War II, while working for Lockheed Corporation, J.R. Crawford proposed that as the number of aircraft produced increases, the unit cost to produce those aircraft decreases at a constant rate. This became known as the unit learning curve theory or the Crawford curve (This historical note is excerpted from "Fundamentals of Cost Analysis," BCF106, U.S. Defense Acquisition University, August 2008).

[‡] From this point forward, the term cost refers to the number of dollars or hours to produce a unit or perform an activity.

is the sum of the costs to produce the first x units divided by the cumulative number of units produced x. Equation 11.1 is the general expression for cumulative average learning curve theory.

$$Y_x = Ax^B \tag{11.1}$$

where

Y_x is the cumulative average unit cost (CAUC) to produce units 1 through x
A is the first unit value expressed as cost, commonly denoted by T1*
B is the cumulative average learning exponent $= \ln(LR_{CA})/\ln(2)$
LR_{CA} is the cumulative average learning rate, where $0 \le LR_{CA} \le 1$

If values for A and either B or LR_{CA} are known, then the CAUC of the first x units can be computed. For instance, if an item's first unit cost is $1000 and the cumulative average learning rate (theoretical or estimated) is 90% then, from Equation 11.1, the CAUC of the first $x = 8$ units is

$$CAUC = Y_x = Ax^B = 1000(8)^{\ln(0.90)/\ln(2)} = 1000(8)^{-0.1520} = \$729$$

Thus, the cumulative cost (CC) of all 8 units is the CAUC times the number of units ($x = 8$)

$$CC = xAx^B = Ax^{B+1} = 1000(8)^{1-0.1520} = 1000(8)^{0.8480} = \$5832$$

Let the theoretical unit cost of any unit x be denoted by Tx. This measure can be found by subtracting the CAUC of all previous $(x-1)$ units from the CAUC of x units. Specifically,

$$Tx = Ax^{B+1} - A(x-1)^{B+1} = A[x^{B+1} - (x-1)^{B+1}]$$

Continuing from the discussion above, the theoretical unit cost of unit 8 is

$$T8 = 1000(8)^{0.8480} - 1000(8-1)^{0.8480} = 5832 - 5207.63 = \$624.37$$

Suppose the first 8 units were built in two lots. Suppose Lot 1 consists of unit 1 through unit 4 and Lot 2 consists of unit 5 through unit 8. Given this, the cost of Lot 1 is the cumulative cost of all 4 units; that is,

$$T_{Lot1} = Ax^{B+1} = 1000(4)^{0.8480} = \$3240$$

To determine the cost of Lot 2, let F and L denote the first unit and last unit in Lot 2, respectively. The theoretical unit cost of Lot 2 is then determined by

---

* The term A is known as the theoretical first unit cost (T1). This can be an estimated value or a theoretical value.

subtracting the total cost of Lot 1 (i.e., $F - 1 = 5 - 1 = 4$ units) from the total cost of all 8 units. This yields the following:

$$T_{Lot2} = AL^{B+1} - A(F - 1)^{B+1} = A[L^{B+1} - (F - 1)^{B+1}]$$
$$= 1000(8)^{0.8480} - 1000(4)^{0.8480} = 5832 - 3240 = \$2592$$

Unit Learning Curve Theory: Unit learning curve theory states that as quantities double, the cost of individual units (effort or hours) decline by a constant percentage. Equation 11.2 is the general expression for unit learning curve theory.

$$Y_x = Ax^B \tag{11.2}$$

Although Equation 11.2 looks the same as Equation 11.1, the terms in Equation 11.2 mean the following:

Y_x is the unit cost to produce unit number x, also denoted by T_x
A is the first unit value expressed as cost, commonly denoted by T1*
x is the unit number
B is the unit learning exponent $= \ln(LR_U)/\ln(2)$
LR_U is the unit learning rate

If values for A and either B or LR_U are known, then the theoretical unit cost T_x of any unit x can be computed. For instance, if an item's first unit cost (theoretical or estimated) is \$1000 and the unit learning rate is 90% then, from Equation 11.2, the unit costs (dollars) of the first $x = 8$ units are

$$x = 1 \Rightarrow Y_1 = A(1)^B = A(1)^{-0.1520} = A = T1 = 1000$$
$$x = 2 \Rightarrow Y_2 = A(2)^B = A(2)^{-0.1520} = T2 = 900$$
$$x = 3 \Rightarrow Y_3 = A(3)^B = A(3)^{-0.1520} = T3 = 846.21$$
$$x = 4 \Rightarrow Y_4 = A(4)^B = A(4)^{-0.1520} = T4 = 810$$
$$x = 5 \Rightarrow Y_5 = A(5)^B = A(5)^{-0.1520} = T5 = 782.99$$
$$x = 6 \Rightarrow Y_6 = A(6)^B = A(6)^{-0.1520} = T6 = 761.59$$
$$x = 7 \Rightarrow Y_7 = A(7)^B = A(7)^{-0.1520} = T7 = 743.95$$
$$x = 8 \Rightarrow Y_8 = A(8)^B = A(8)^{-0.1520} = T8 = 729$$

---

* The term A is known as the theoretical first unit cost (T1). It can be an estimated or a theoretical value.

Thus, the cumulative unit cost (CUC) of all 8 units is*

$$CUC = A \sum_{i=1}^{x=8} i^B = A(1)^B + A(2)^B + \cdots + A(8)^B = T1 + T2$$

$$+ \cdots + T8 = \$6573.74$$

In unit learning curve theory, the CAUC at unit number x can also be determined by dividing the CUC by x. Continuing from the earlier discussion

$$CAUC = \frac{A}{x} \sum_{i=1}^{x=8} i^B = \frac{CUC}{x} = \frac{6573.74}{8} = \$821.72$$

As before, suppose the first eight units were built in two lots with Lot 1 consisting of unit 1 through unit 4 and Lot 2 consisting of unit 5 through unit 8. Then the cost of Lot 1 and Lot 2 are found by summing the costs of the units contained in each lot. In this case, it follows that

$$T_{Lot1} = T1 + T2 + T3 + T4 = \$3556.21$$

$$T_{Lot2} = T5 + T6 + T7 + T8 = \$3017.53$$

11.2.1 Similarities and Differences

The unit and cumulative average learning theories given by Equations 11.1 and 11.2 share the power form $Y_x = Ax^B$ and the first unit cost A. However, they each express their independent variable x and their dependent variable Y_x in different units. Hence, their learning exponent B and learning rates (LR_U and LR_{CA}) are also different and cannot be interchanged. An estimate using cumulative average learning curve theory must use the CA learning rate LR_{CA} to determine the AUC or the costs of different quantities (such as the first 10 units). This same distinction applies to unit theory. The following presents a convenient summary of the governing equations for these two learning curve theories.

Equations 11.1 and 11.2 are versatile and can be used to compute the cost of any particular unit (such as unit x), the cost of the first x units, the AUC of the first x units, the cost of each unit in a lot that leads to a total lot cost, and the AUC of units in a lot. These costs can be determined if the first unit cost A, the learning curve slope B, and the first F and last L units in a lot are known. The following is a summary of the equations for the cumulative average (Equations 11.3 through 11.7) and unit learning curve (Equations 11.8 through 11.12) theories.

---

* The index i is used in the summand to represent the unit numbers 1 to x.

Summary Equations for Cumulative Average and Unit Learning: $A = T1$, i = index indicating the unit number (e.g., i = 2 is unit 2), F = first unit number, L = last unit number.

Using Cumulative Average Learning Theory

$$\text{Cost of each unit } x = A\left(x^{B+1} - (x-1)^{B+1}\right) \tag{11.3}$$

$$\text{Cost of the first } x \text{ units} = Ax^{B+1} \tag{11.4}$$

$$\text{Average Unit Cost of } x \text{ units } (AUC_x) = Ax^B \tag{11.5}$$

$$\text{Lot cost}^* = A\left(L^{B+1} - (F-1)^{B+1}\right) \tag{11.6}$$

$$\text{Lot AUC}^\dagger (AUC_{lot}) = \frac{A}{L-F+1}\left(L^{B+1} - (F-1)^{B+1}\right) \tag{11.7}$$

Using Unit Learning Theory

$$\text{Cost of each unit } x = Ax^B \tag{11.8}$$

$$\text{Cost of the first } x \text{ units} = A\sum_{i=1}^{x}(i^B) \tag{11.9}$$

$$\text{Average Unit Cost of } x \text{ Units } (AUC_x) = \frac{A}{x}\sum_{i=1}^{x}(i^B) \tag{11.10}$$

$$\text{Lot cost}^\ddagger = A\sum_{i=F}^{L}(i^B) \tag{11.11}$$

$$\text{Lot AUC}^\S (AUC_{lot}) = \frac{A}{L-F+1}\sum_{i=F}^{L}(i^B) \tag{11.12}$$

Example 11.1 *Find the unit cost and AUC for each of the first eight units in a lot using the unit and cumulative average learning curve theories. Assume $A = 1000$, $LR_U = 0.90$, and $LR_{CA} = 0.90$.*

Solution Using unit learning theory, the unit costs of the first eight units is computed by Equation 11.8, where $B = \ln(0.9)/\ln(2) = -0.1520$. It follows that $Y_1 = Ax^B = 1000(1)^{-0.1520} = 1000$, $Y_2 = 1000(2)^{-0.1520} = 900$, $Y_3 = 1000(3)^{-0.1520} = 846.21$. The remaining unit costs are computed in a similar way, with the results $Y_4 = 810$, $Y_5 = 782.99$, $Y_6 = 761.59$, $Y_7 = 743.95$, and $Y_8 = 729$.

In unit learning theory, the AUC of each unit x is computed from Equation 11.10. In particular, $AUC_1 = 1000$, $AUC_2 = (1000 + 900)/2 = 950$, $AUC_3 = (1000 + 900 + 846.21)/3 = 915.40$. The remaining AUCs are

---

* Use Equation 11.4 if the lot begins with unit 1.
† Use Equation 11.5 if the lot begins with unit 1.
‡ Use Equation 11.9 if the lot begins with unit 1.
§ Use Equation 11.10 if the lot begins with unit 1.

TABLE 11.1

Unit and AUCs using Unit and CA Theory

Unit Number	Unit Theory, $LR_U = 0.90$		CA Theory, $LR_{CA} = 0.90$	
	Unit Cost	AUC	Unit Cost	AUC
1	1000.00	1000.00	1000.00	1000.00
2	900.00	950.00	800.00	900.00
3	846.21	815.40	738.62	846.21
4	810.00	889.05	701.38	810.00
5	782.99	867.84	674.93	782.99
6	761.59	850.13	654.58	761.59
7	743.95	834.96	638.12	743.95
8	729.00	821.72	624.37	729.00

computed in a similar way, with the results $AUC_4 = 889.05$, $AUC_5 = 867.84$, $AUC_6 = 850.13$, $AUC_7 = 834.96$, and $AUC_8 = 821.72$.

In cumulative average theory, the unit costs of the first eight units is computed by Equation 11.3, where $B = \ln(0.9)/\ln(2) = -0.1520$. From this, it follows that $Y_1 = 1000$, $Y_2 = 1000[(2)^{1-0.1520} - 1000] = 1000[(2)^{0.8480} - 1] = 800$, $Y_3 = 1000[(3)^{0.8480} - (2)^{0.8480}] = 738.62$. The remaining unit costs are computed in a similar way, with the results $Y_4 = 701.38$, $Y_5 = 674.93$, $Y_6 = 654.58$, $Y_7 = 638.12$, and $Y_8 = 624.37$.

In cumulative average theory, the AUC for each unit x is computed using Equation 11.5. Here, $AUC_1 = 1000$, $AUC_2 = 1000(2)^{-0.1520} = 900$ and, $AUC_3 = 1000(3)^{-0.1520} = 846.21$. The remaining AUCs are computed in a similar way, with the results $AUC_4 = 810$, $AUC_5 = 782.99$, $AUC_6 = 761.59$, $AUC_7 = 743.95$, and $AUC_8 = 729$.

Table 11.1 summarizes and compares these computations. Although the learning rates in this example were equal, the unit and AUCs in cumulative average learning theory are less than their values in unit learning theory. This will always be true.

11.2.2 Limitations and Considerations

There are limitations and considerations with the learning curve theories. Among them are (1) the effect of learning is not endless and can come to an abrupt stop, (2) the adoption of new processes will mean the old learning curve will be replaced by a new one, (3) breaks in production over periods of time sufficient to warrant a loss of learning can occur, and (4) incorrect valuation of the learning rate has a significant impact on the cost model developed and any estimates that are produced.

In addition, there is a practical limit of 50% for the cumulative average learning curve slope LR_{CA}. For example, if the first unit cost T1 = \$1000, and $LR_{CA} = 0.50$ then, from Equation 11.5, the CAUC of two units $x = 2$ is equal to $Y_2 = 1000(2)^B = 1000(2)^{-1} = \$500.$* This means that the second unit[†] must be free! Thus, under cumulative average theory, the learning rate must always be greater than 50%. In the case of unit learning theory, the learning rate must be greater than 0%.

Another limitation in the use of learning curve theory is the lack of credible historical data that are analogous, applicable, or consistent. This situation makes it difficult or even impossible to determine the "correct" learning rate for a particular cost estimating context.[‡] Nevertheless, the estimator's choice of learning rate exerts a major, perhaps dominant, impact on the estimate of the total spending profile of a large production program. Even if nonrecurring and first-unit production costs are estimated precisely, small variations in the learning rate substantially outweigh all other contributions to the uncertainty in the estimate. This is especially true in large-quantity procurements, such as aircraft or missile programs.

As mentioned above, learning rates are sensitive to the particular cost estimating context. With this, it is advisable to understand the impact of learning rate uncertainty upon the uncertainty in estimated costs. Therefore, in preparing an estimate of the cost of a production run of N units, the following example illustrates comparing AUC estimates at each of two CA learning rates.

Example 11.2 *Compare the AUC estimates for production run sizes $N = 10$, 20, 50, 100, 200, 500, 1000, 2000, and 5000 at CA learning rates $LR_{CA} = 0.90$ and $LR_{CA} = 0.95$.*

Solution With Equation 11.5, the ratio of 95% learning to 90% learning is as follows:

$$\frac{N^{\ln(0.95)/\ln(2)}}{N^{\ln(0.90)/\ln(2)}} = N^{0.7024}$$

Table 11.2 lists values of that ratio for the N production-run sizes given here. These values can be interpreted in the following way: If we assume a CA learning rate of 95% when estimating the cost of a production run of

TABLE 11.2

AUC Ratios for 95% CA Learning versus 90% CA Learning

Production Run Size N	AUC Ratio	Production Run Size N	AUC Ratio	Production Run Size N	AUC Ratio
10	1.20	100	1.43	1000	1.71
20	1.26	200	1.51	2000	1.81
50	1.36	500	1.62	5000	1.94

---

[*] In this case, $B = \ln(0.5)/\ln 2 = -1$.
[†] The second unit must be free because the cumulative average of the first two units is $500. This means the cumulative cost is $2 \times \$500 = \$1000$. Given the first unit is $1000, then the second unit's cost must be $0.
[‡] See Bierman and Dyckman (1971, pp. 88–89); and Nanda and Alder (1982, pp. 133–138).

200 units, our cost estimate will be 51% higher than if we had assumed a CA learning rate of 90%. Worded another way, if we assume a CA learning rate of 90% for our estimate, but the actual CA learning rate turns out to be 95%, we will experience a 51% overrun in production cost. The possibility of learning rate impacts of this magnitude on cost estimates tend to be overlooked and are clearly important considerations in production cost estimating.

11.2.3 Learning Rate Impacts on T1 Costs

Suppose one is interested in comparing average unit and total production costs at learning rates $x\%$ and $y\%$, where $x > y$. The ratio of AUC at $x\%$ learning ($LR_{CA} = x/100$) to AUC at $y\%$ learning ($LR_{CA} = y/100$) can be calculated as follows:

$$\frac{AUC_N \text{ at } x\%}{AUC_N \text{ at } y\%} = \frac{T1\left(N^{(\ln(x/100)/\ln 2)}\right)}{T1\left(N^{\ln(y/100)/\ln 2}\right)} = N^{(\ln(x/y)/\ln 2)} \qquad (11.13)$$

Table 11.3 lists values of this ratio for typical production-run sizes and learning rates:

Table 11.3 can be used to calculate other ratios. For example, the AUC ratio for 95% to 85% CA learning for a 10-unit production run is $1.197 \times 1.209 = 1.45$. In Table 11.3, this is the product of the ratios for 95%–90% and 90%–85%, respectively.

So far, differences in learning rate applied to T1 estimates and how they affect average unit and total program cost estimates have been discussed, but that is only half the story. Developing T1-based cost models can involve conjecture. First, historical cost data are gathered and normalized to reflect T1 costs in constant-year dollars. Moreover, for multiple unit programs only total costs, lot costs, or AUCs are typically available. Given this, analysts must assume a learning rate to derive T1 costs. Unsurprisingly, the magnitude of a program's cost estimate can depend heavily on the learning rate.

TABLE 11.3

AUC Ratios for Typical Production Situations

CA Learning Rates $x\%$–$y\%$	Number of Units Produced N								
	10	20	50	100	200	500	1000	2000	5000
100%–95%	1.186	1.248	1.336	1.406	1.480	1.584	1.667	1.755	1.878
95%–90%	1.197	1.263	1.357	1.432	1.512	1.624	1.714	1.809	1.943
90%–85%	1.209	1.280	1.381	1.462	1.548	1.669	1.768	1.872	2.018
85%–80%	1.223	1.300	1.408	1.496	1.589	1.722	1.830	1.944	2.106
80%–75%	1.239	1.322	1.439	1.535	1.638	1.784	1.903	2.029	2.210
75%–70%	1.258	1.347	1.476	1.582	1.694	1.856	1.989	2.131	2.334

Example 11.3 *Suppose the AUC of x = 10 units is $1; that is, AUC$_{10}$ = 1.00. Using CA learning theory, compute the implied T1 costs given learning rates LR$_{CA}$ = 0.95 and LR$_{CA}$ = 0.85.*

Solution Under CA learning curve theory, from Equation 11.5

$$AUC_x = Ax^B = T1x^B \tag{11.14}$$

$$\Rightarrow T1 = AUC_x(x^{-B}) \tag{11.15}$$

With AUC$_{10}$ = 1.00 and LR$_{CA}$ = 0.95

$$T1 = AUC_{10}\left(10^{-[\ln(LR_{CA})/\ln 2]}\right) = 1.00\left(10^{-[\ln(0.95)/\ln 2]}\right) = 1.1857 \approx 1.19$$

With AUC$_{10}$ = 1.00 and LR$_{CA}$ = 0.85

$$T1 = AUC_{10}\left(10^{-[\ln(LR_{CA})/\ln 2]}\right) = 1.00\left(10^{-[\ln(0.85)/\ln 2]}\right) = 1.7158 \approx 1.72$$

Depending on the learning rate, the T1 cost increases by 45% when an 85% CA learning rate is used instead of a 95% CA learning rate. Figure 11.1 illustrates this result.

11.2.4 Historically Derived Learning Rates and Cost Models

As discussed in Chapter 10, the most important aspect of constructing a statistical cost model is the historical (observed) data upon which it is built. This includes knowing how the data was normalized to reflect consistent WBS

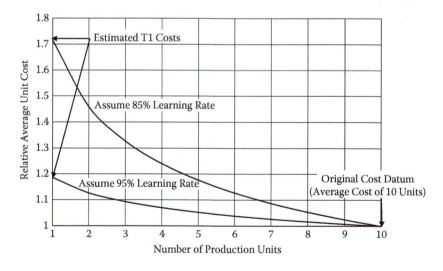

FIGURE 11.1
Effects of assumed learning rates on estimating the T1 costs.

definitions and scope, the appropriate fiscal year, and, from a learning curve perspective, the correct procurement quantity. For quantity normalization, actual lot cost data from multiple programs must be normalized (adjusted) to provide x and y variables that fit the forms shown in Equations 11.3 (for CA learning theory) and 11.8 (for unit learning theory).

Normalization for quantity ensures comparing the same type of cost, whether it is at the total, lot, or unit level. The most common method of normalizing for quantity is through the use of CI or learning curves. The classical technique is to normalize all data points to the theoretical first unit cost using either unit theory or cumulative average theory learning curves (Book and Burgess 1996). For each program in a historical cost database, a T1 cost and learning rate can be derived that is characteristic of that program. Book and Burgess (2003) demonstrated this technique using the CA learning theory with the sample lot cost and quantity data presented in Table G.1. This is known in the community as the two-step process. The following discusses this process and the issues that arise.

Consider the dataset of production program cost histories in Table 11.4. Assume these costs have been normalized to reflect consistent WBS definitions and scope, inflation, and fiscal year.

First step: From these histories, the first step is to compute the AUC of each program. This is the ratio of the program's total cost to its total units. For each program, the resultant ratio is shown in Table 11.4 by the entries in the column titled Average Unit Cost (AUC, $).

Next, suppose analysts determined that cumulative average learning was the appropriate theory for these data and a learning rate of 84% was found by pooled regression.* From Equation 11.15

$$T1 = AUC_x(x^{-B}) \tag{11.16}$$

where
  $B = \ln(0.84)/\ln(2) = -0.2515$
  x is each program's production run total units, given in Table 11.3

With this, the estimated T1 cost of each program is computed by Equation 11.16, with the results (rounded) shown in the last column of Table 11.4.

Second step: The second step involves building a regression model to derive estimates of each program's T1 cost as a function of its observed weight. The objective is to produce a CER for statistically generating T1 costs for programs similar to those represented in Table 11.4. The following illustrates building this CER by GERM ZMPE regression.† Table 11.5 is the dataset and analysis

---

* Pooled regression refers to a single regression of the data from each of the programs using indicator or dummy variables representing the individual T1 costs of each program. See Covert (2014) for information on pooled regression.
† ZMPE regression is fully discussed in Chapter 10.

TABLE 11.4

Dataset: Production Program Cost Histories

Program Name	Weight (lb)	Production Run Total Units	Production Run Total Cost	Average Unit Cost (AUC, $)	Estimated T1 Cost ($)
A	985	2457	1,421,090	578	4117
B	985	2084	1,669,203	801	5474
C	510	12,458	2,245,059	180	1929
D	190	12,808	390,110	30	324
E	190	21,968	746,836	34	420
F	190	5407	215,409	40	347
G	190	8813	383,814	44	432
H	510	6451	1,172,213	182	1653

setup for this regression. The data in columns A, B, and C are from the program cost histories in Table 11.4. The data in columns D and E are for the setup of the GERM ZMPE regression, which will fit a model to the form $f(x) = ax^b$.

Discussed in Chapter 10, GERM ZMPE implies using the multiplicative error form. Table 11.5 shows the dataset of $n = 8$ observations that correspond

TABLE 11.5

Dataset and Analysis Setup

	A	B	C	D	E
1	Number of Program Data Points		$n = 8$	**Regression Model form**	
2	Number of Regression Constants		$m = 2$	$f(x) = ax^b$	
3	Degrees of Freedom		$n - m = 6$		
		Observed	Estimated	Estimated	
4	Program Name	Weight (lbs), x	T1 Cost, y	T1 Cost, Regression Model	Multiplicative Error
5	A	985	4117	985	−3.1797
6	B	985	5474	985	−4.5574
7	C	510	1929	510	−2.7824
8	D	190	324	190	−0.7053
9	E	190	420	190	−1.2105
10	F	190	347	190	−0.8263
11	G	190	432	190	−1.2737
12	H	510	1653	510	−2.2412
13		Starter Values		Model Quality Measures	
14		$a = 1 \equiv$ B14		$SEE = 2.8258 \equiv$ E14	
15		$b = 1 \equiv$ B15		$Bias = -2.0970 \equiv$ E15	
16				$R^2 = 0.9567$	

to the program histories in Table 11.4. An Excel model is built and its *Solver* feature is used to run the ZMPE optimization (minimization in GERM). The columns labeled "Estimated T1 Cost, Regression Model," "Multiplicative Error," and "Model Quality Measures" are computed from the two starter values shown for *a* and *b*. As mentioned previously, starter values are the initial conditions chosen by the user to stimulate the optimization routine in Excel *Solver*.

Figure 11.2 shows the Excel *Solver* window and the features selected to run the ZMPE optimization for the dataset in Table 11.5. The optimization algorithm begins with the starter values shown for *a* and *b* and iterates numerically until the search algorithm converges to a solution. The optimization results are shown in columns D and E in Table 11.6, along with quality measures about the regression model formed.

Figure 11.3a is a plot of the observed data versus the data generated by the regression model built by GERM ZMPE. The dark circles are the

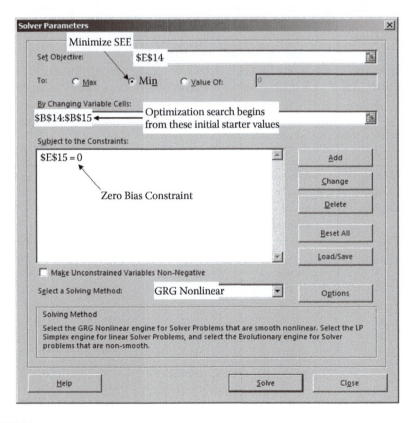

FIGURE 11.2
Excel *Solver* setup for the dataset in Table 11.5.

TABLE 11.6

Solver Solution for a and b

	A	B	C	D	E
1	Number of Program Data Points		$n = 8$	**Regression Model form**	
2	Number of Regression Constants		$m = 2$	$f(x) = ax^b$	
3	Degrees of Freedom		$n - m = 6$		
		Observed	Estimated	Estimate	
4	Program	Weight	T1	T1 Cost,	Multiplicative
	Name	(lbs), x	Cost, y	Regression Model	Error
5	A	985	4117	4859	0.1527
6	B	985	5474	4859	−0.1266
7	C	510	1929	1757	−0.0980
8	D	190	324	382	0.1517
9	E	190	420	382	−0.0996
10	F	190	347	382	0.0915
11	G	190	432	382	−0.1310
12	H	510	1653	1757	0.0591
13		Solver Values		Model Quality Measures	
14		$a = 0.1149$		$SEE = 0.1360$	
15		$b = 1.5455$		$Bias = 0.0000$	
16				$R^2 = 0.9640$	

observed weights and estimated T1 costs from columns B and C of Table 11.6. The open circles are the estimated T1 costs generated from the regression model

$$f(x) = 0.1149x^{1.5455} \tag{11.17}$$

where x is the observed weight. Equation 11.17 is a GERM ZMPE CER for estimating the T1 cost of a program with characteristics similar to the program histories in Table 11.4. To conduct this analysis, it was necessary to assume a learning theory and a learning rate for the dataset presented. In this case, a cumulative average learning theory and a learning rate of 0.84 was assumed. Clearly, these assumptions can have significant effects on the magnitudes of estimated T1 costs that are generated by statistically derived CERs. An example is shown in Figure 11.3b. Figure 11.3b shows the results of the same analysis summarized in Figure 11.3a when the assumed learning rate in Step 1 is 0.74 instead of 0.84.

Production cost data histories have to be normalized to a specific number of units, usually the theoretical first unit by applying assumptions on

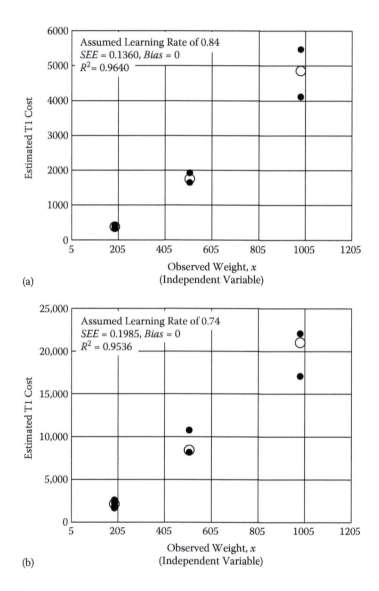

FIGURE 11.3
T1 cost estimating relationship versus the observed data.

learning theory and learning rates—none of which are known absolutely from the data. Uncertainties in learning rates affects production cost model development to the same degree they affect building development CERs. The two-step method illustrated above is prone to the effects of uncertainty. Recognizing this, Book and Burgess (2003) and Covert and Wright (2010) created single-step methods for developing production CERs that do not require learning assumptions—and the effects that uncertainty brings on the quality of production CERs.

11.3 Production Cost Models Built by Single-Step Regression

This section presents the single-step regression approach for developing production CERs. It avoids the problem of guessing learning rates associated with production cost data as seen in the two-step model development process. Two forms of the single-step regression approach are described.

The first form is based on cumulative average learning curve theory, which uses quantity as an independent variable (QAIV) in the regression of lot cost data. The second form is based on unit learning curve theory, which uses the unit number as an independent variable (UAIV) in the regression of lot cost data. Production cost models built by single-step regression provides superior quality measures as compared to those obtained from their respective two-step model development processes.

11.3.1 Quantity as an Independent Variable (QAIV)

QAIV was introduced by Book and Burgess (2003) as a single step technique to develop production CERs from lot data using cumulative average theory. Their approach uses a multivariate CER with weight, lot size, and prior quantity as independent variables. Single-step QAIV regression operates on dependent and independent variables based entirely on known historical facts, no learning-rate assumption is needed for either cost data normalization or T1 cost estimating. The following example illustrates this characteristic of QAIV.

> **Example 11.4** *Appendix G, Table G.1, provides a dataset of lot costs and quantity histories from a collection of production programs. From these data, use GERM ZMPE to derive a QAIV CER of the generic form $y = aw^b x^c$.*
>
> *Solution* Table G.1 provides data on the weight of each unit w, the unit numbers of the first F and last L units produced in a lot, and the AUC of each lot. The generic QAIV CER form $y = aw^b x^c$ allows for estimating the cumulative average cost of the first x units and is versatile in its form like the generic cumulative average learning equation, shown by Equation 11.1. In Equations 11.3 through 11.7, the x term and exponent B can be manipulated to allow for estimating a variety of costs, among those is the cost of any particular lot. The lot cost variant of this equation is given by Equation 11.6, which replaces the term x^B with the terms $(L^{B+1} - (F-1)^{B+1})$. This allows for estimating the cost of any lot given the unit numbers of the first F and last L units in any particular lot.

Thus, the generic QAIV CER can be modified in a similar fashion to estimate the cost of any particular lot. With GERM ZMPE, the modified QAIV CER is given by Equation 11.18. This equation does the following: (1) estimates the observed lot costs from the collected data; (2) provides the regression constants a, b, and c; and (3) has a minimum SEE with zero bias. To achieve these outcomes using cumulative average learning theory, Equation 11.6 is tailored as follows:

$$y(w) = \begin{cases} aw^b(L^{c+1} - (F-1)^{c+1}), & \text{if } F > 1 \\ aw^bL^{c+1}, & \text{if } F = 1 \end{cases} \qquad (11.18)$$

where
  a, b, and c are constants
  w is weight (lb)
  F is the first unit in the lot
  L is the last unit in the lot

The values of F and L are calculated from the production cost data. Table 11.7 is the dataset and analysis setup for the GERM ZMPE regression $y = aw^b x^c$.

Equation 11.18 is analogous to cumulative average learning theory. The term aw^b is equivalent to the T1 cost and c is the cumulative average learning exponent. Since $c = \ln(LR_{CA})/\ln(2)$, once its value is determined from the GERM ZMPE regression, one can derive a single cumulative average learning rate $LR_{CA} = 2^c$ that is representative of the dataset. With this, the uncertainty in estimating the learning rate shown in the two-step process has been removed.

Table 11.7 shows the dataset and analysis setup to execute GERM ZMPE regression on Equation 11.18. Table 11.8 shows the *Solver* solution results for the constants a, b, and c. The resulting QAIV CER is $y = 0.422w^{1.453}x^{-0.328}$ with model quality measures $SEE = 0.2637$, $R^2 = 0.8612$, and $Bias = 0$. The results from this regression can be used to derive the learning rate implied by the dataset, which in this case is $LR_{CA} = 2^c = 2^{-0.328} = 0.79664$. For the set of production program cost histories in Table G.1, the overall learning rate under cumulative average theory is approximately 80%. The capability to derive the learning rate demonstrates the advantage of QAIV over the traditional two-step process, where learning rates must be assumed along with the uncertainties those assumptions bring.

11.3.2 Unit as an Independent Variable (UAIV)

Covert and Wright (2012) developed the unit theory equivalent of QAIV called unit as an independent variable. UAIV is a single-step technique using GERM ZMPE to develop CERs from lot data using unit theory (Covert 2010). Like QAIV, the UAIV approach uses a multivariate CER with weight, lot size, and prior quantity as independent variables. The existence of the

TABLE 11.7
Example 11.4 Dataset and Analysis Setup

	A	B	C	D	E	F	G
1	Number of Program Data Points			n = 69		**Regression Model Form**	
2	Number of Regression Constants			m = 3		$y(w) = aw^b(L^{c+1} - (F-1)^{c+1})$	
3	Degrees of Freedom			n − m = 66			
						Estimated Lost	
4	Lot Num., i	Lot ID	Weight, lbs, w	First Unit Number, F	Last Unit Number, L	Observed Lot Cost, y	Cost, Regression Model	Multiplicative Error
5	1	A1	985	1	37	108,675.81	36,445.00	1.9819
6	2	A2	985	38	63	58,512.61	25,610.00	1.2848
⋮	⋮	⋮	⋮	⋮	⋮	⋮	⋮	⋮
72	68	H9	510	4408	5307	114,181.3	459,000.00	−0.7512
73	69	H10	510	5308	6451	127,010.39	583,440.00	−0.7823
74		Starter Values				Model Quality Measures		
75		a = 1				SEE = 0.7528		
76		b = 1				Bias = −0.3659		
77		c = 0				$R^2 = 0.4913$		

TABLE 11.8

Example 11.4 *Solver* Solution for a, b, and c

	A	B	C	D	E	F	G
1	Number of Program Data Points			$n = 69$			
2	Number of Regression Constants			$m = 3$			
3	Degrees of Freedom			$n - m = 66$			**Regression Model Form** $y(w) = aw^b (L^{c+1} - (F - 1)^{c+1})$	
4	Lot Num., i	Lot ID	Weight, lbs, w	First Unit Number, F	Last Unit Number, L	Observed Lot Cost, y	Estimated Lost Cost, Regression Model	Multiplicative Error
5	1	A1	985	1	37	108,675.81	106,756.75	0.0180
6	2	A2	985	38	63	58,512.61	45,925.41	0.2741
⋮	⋮	⋮	⋮	⋮	⋮	⋮	⋮	⋮
72	68	H9	510	4408	5307	114,181.3	135,767.29	−0.1590
73	69	H10	510	5308	6451	127,010.39	162,116.83	−0.2166
74	Solver Values				Model Quality Measures			
75	$a = 0.422$						$SEE = 0.2637$	
76	$b = 1.453$						$Bias = 0.0000$	
77	$c = -0.328$						$R^2 = 0.8612$	

combination of QAIV (based on cumulative average learning theory) and UAIV (based on unit learning curve theory) regression techniques allows analysts to create single-step production cost estimating relationships based entirely on known historical facts and to completely forego the learning-rate assumptions and uncertainties required in the two-step model development processes.* The following example illustrates this characteristic of UAIV.

Example 11.5 *Appendix G, Table G.1, provides a dataset of lot costs and quantity histories from a collection of production programs. From these data, use GERM ZMPE to derive a UAIV CER of the generic form $y = aw^b x^c$.*

Solution Table G.1 provides data on the weight w of each unit, the unit numbers of the first F and last L units produced in a lot, and the cost of each lot. The generic UAIV CER form $y = aw^b x^c$ allows for estimating the cost of unit x and is versatile in its form like the generic unit learning equation, shown by Equation 11.2. In Equations 11.8 through 11.12, the x term and exponent B can be manipulated to allow for estimating a variety of costs, among those is the cost of any particular lot. The lot cost variant of this equation is given by Equation 11.11 which replaces the term x^B with the summation $\sum_{1-F}^{L}(i^B)$. This allows for estimating the cost of any lot given the unit numbers of the first F and last L units in any particular lot.

The generic UAIV CER can be modified in a similar fashion to estimate the cost of any particular lot. With GERM ZMPE, the modified UAIV CER is given by Equation 11.19. This equation does the following: (1) estimates the observed lot costs from the collected data; (2) provides the regression constants a, b, and c; and (3) has a minimum *SEE* with zero bias. To achieve these outcomes using unit learning theory, Equation 11.11 is tailored as follows:

$$y(w) = aw^b \sum_{i=F}^{L} (i^c) \tag{11.19}$$

where
 a, b, and c are constants
 w is weight (lb)
 F is the first unit in the lot
 L is the last unit in the lot

The values of F and L are calculated from the production cost data. Table 11.9 is the dataset and analysis setup for the GERM ZMPE regression $y = aw^b x^c$.

Equation 11.19 is analogous to unit learning theory. The term aw^b is equivalent to the T1 cost and c is the unit learning exponent. Since

---

* Covert (2014) provides a detailed examination of the assumptions and uncertainties used in two-step model development processes using unit and cumulative average learning theories.

TABLE 11.9

Example 11.5 Dataset and Analysis Setup

	A	B	C	D	E	F	G
1	Number of Program Data Points			$n = 69$		**Regression Model Form**	
2	Number of Regression Constants			$m = 3$		$y(w) = aw^b \sum\limits_{i=F}^{L} (i^c)$	
3	Degrees of Freedom			$n - m = 66$			
4	Lot Num., i	Lot ID	Weight, lbs, w	First unit Number, F	Last unit Number, L	Observed Lot Cost, y	Estimated lost cost, Regression Model	Multiplicative Error
5	1	A1	985	1	37	108,675.81	36,445.00	1.9819
6	2	A2	985	38	63	58,512.61	25,610.00	1.2848
⋯	⋯	⋯	⋯	⋯	⋯	⋯	⋯	⋯
72	68	H9	510	4408	5307	114,181.30	459,000.00	−0.7512
73	69	H10	510	5308	6451	127,010.39	583,440.00	−0.7823
74	Starter Values					Model Quality Measures	
75	$a = 1$					$SEE = 0.7471$	
76	$b = 1$					$Bias = -0.3659$	
77	$c = 0$					$R^2 = 0.4913$	

$c = \ln(LR_U)/\ln(2)$, once its value is determined from the GERM ZMPE regression, one can derive a single unit learning rate $LR_U = 2^c$ that is representative of the dataset. With this, the uncertainty in estimating the learning rate shown in the two-step process has been removed.

Table 11.9 shows the dataset and analysis setup to execute GERM ZMPE regression on Equation 11.19. Table 11.10 shows the *Solver* solution results for the constants a, b, and c. The resulting UAIV CER is $y = 0.302w^{1.447}x^{-0.331}$, with model quality measures $SEE = 0.2608$, $R^2 = 0.8603$, and $Bias = 0$. As mentioned, the results from this regression can be used to derive the unit learning rate implied by the dataset, which in this case is $LR_U = 2^c = 2^{-0.331} = 0.79481$. For the set of production program cost histories in Table G.1, the overall learning rate under unit theory is approximately 80%. The capability to derive the unit learning rate demonstrates the advantage of UAIV over the traditional two-step process, where learning rates must be assumed along with the uncertainties those assumptions bring.

11.4 Summary

In this chapter, we introduced the phenomenon of CI. CI is the cost reduction associated with repeated, successive effort. Learning is one form of recurring CI and is governed by two learning curve theories, the cumulative average and unit theories. These theories are used extensively in data normalization process of creating cost models and in estimating the costs of multiple, successive units.

One practical consideration of using these theories is what learning rate to use in any particular situation. Errors in estimating the exact value of the learning rate translate into large errors of the cost estimates of large-quantity production scenarios. There is also an important uncertainty consideration when using learning curve theory to produce quantity-normalized first unit cost (i.e., T1) data in the first step of the cost model development process—the normalized T1 cost data will have uncertainties that are not modeled (i.e., neglected) in the second step of the cost model development process. To overcome these issues, we introduced two types of innovative, single step-regressions: QAIV and UAIV.

QAIV and UAIV regressions use single-step processes that produce cost models with better *SEE* and *Bias* quality measures than traditional two-step-derived, T1-based cost models. QAIV and UAIV do not rely on assumptions that neglect the uncertainty in the two-step process, so their *SEEs* provide a more accurate portrayal of their predictive abilities than do their equivalent T1-based cost models developed using a two-step process.

In this chapter, we demonstrated the development of QAIV and UAIV cost models using ZMPE regressions and further showed the flexibility and

TABLE 11.10

Example 11.5 *Solver* Solution for a, b, and c

		A	B	C	D	E	F	G
1		Number of Program Data Points			$n = 69$			
2		Number of Regression Constants			$m = 3$			
3		Degrees of Freedom			$n - m = 66$		**Regression Model Form**	
							$y(w) = aw^b \sum\limits_{i=F}^{L} (i^c)$	
4	Lot Num., i	Lot ID	Weight, lbs, w	First Unit Number, F	Last Unit Number, L	Observed Lot Cost, y	Estimated Lost Cost, Regression Model	Multiplicative Error
5	1	A1	985	1	37	108,675.81	103,209.03	0.0530
6	2	A2	985	38	63	58,512.61	46,227.18	0.2658
⋮	⋮	⋮	⋮	⋮	⋮	⋮	⋮	⋮
72	68	H9	510	4408	5307	114,181.30	135,389.10	−0.1566
73	69	H10	510	5308	6451	127,010.39	161,555.14	−0.2138
74		Solver Values					Model Quality Measures	
75		$a = 0.302$					$SEE = 0.2608$	
76		$b = 1.447$					$Bias = 0.0000$	
77		$c = -0.331$					$R^2 = 0.8603$	

efficiency of using GERM (Chapter 10). Without the development and use of GERM to solve these types of regression problems, it is unlikely that innovative regression forms such as QAIV or UAIV would have been discovered or the neglected uncertainty modeling issues with the traditional two-step methods of cost model development would have been mitigated.

Exercises

11.1 Create the unit learning curve theory equivalent of Table 11.3 using unit learning curve slopes.

11.2 Assume a particular learning rate $LR = 0.90$ is used to calculate the total cost of a production lot. Since this cost is dependent on whether the cumulative average or unit learning theory is used, which theory produces the greater error if the true learning curve slope $LR = 0.95$.

11.3 Which CER has better quality metrics, the QAIV CER developed in Example 11.4 or the UAIV CER developed in Example 11.5? Explain the potential sources of the differences in these metrics and discuss circumstances when one of them is preferred over the other.

References

Badiru, A. B. 1992. Computational Survey of univariate and multivariate learning curve models. *IEEE Transactions on Engineering Management*, 39(2), 176–188.

Bierman, H. and T. R. Dyckman 1971. *Managerial Cost Accounting*. New York: Macmillan.

Boehm, B. 1981. *Software Engineering Economics*. Englewood Cliffs, NJ: Prentice-Hall.

Book, S. and E. Burgess 1996. The learning rate's overpowering impact on cost estimates and how to diminish it. *Journal of Parametrics*, 16, 33–57.

Book, S. and E. Burgess. 2003. A way out of the learning-rate morass: Quantity as an Independent Variable (QAIV). In *36th Annual DoD Cost Analysis Symposium*, Williamsburg, VA.

Covert, R. 2010. ZMPE implementation of unit curve analysis. In *PRICE Regional Meeting*, Washington, DC.

Covert, R. 2014. *Cost Improvement*. Palm Beach Gardens, FL: Covarus, LLC.

Covert, R. and N. Wright. 2012. Estimating relationship development spreadsheet and unit-as-an-independent variable regressions. In *2012 Society for Cost Estimating and Analysis (SCEA)/International Society of Parametric Analysts (ISPA) Conference*, Orlando, FL.

Crawford, J. R. 1944. Statistical accounting procedures in aircraft production. *Aeronautical Digest*, 44, 78.

Crawford, J. R. and E. Strauss 1947. *Crawford-Strauss Study*. Dayton, OH: Air Material Command.

Goldberg, M. S. and A. E. Touw. 2003. *Statistical Methods for Learning Curves and Cost Analysis*. Linthicum, MD: Institute for Operations Research and Management Sciences (INFORMS).

Hudgins, E. R. 1966. Learning curve fundamentals. U.S. Army Weapons Command, Defense Documentation Center.

Levine, D. B., S. J. Balut, and B. R. Harmon. 1989. Technology and cost progress. The *Journal of Cost Analysis*, Fall, 79–95.

Nanda, R. and G. L. Alder. 1982. *Learning Curves—Theory and Application*. Norcross, GA: Institute of Industrial Engineers.

Stump, E. J. 1988. Parametric tools of the trade: Learning curve analysis. In *ISPA Workshop*.

Wright, T. P. 1936. Factors affecting the cost of airplanes. *Journal of the Aeronautical Sciences*, 3(2), 122–128.

Yelle, L. E. 1979. The learning curve: Historical review and comprehensive survey. *Decision Sciences*, 10(2), 302–328.

Additional Reading

Anderlohr, G. 1969. What production breaks cost. *Industrial Engineering*, 1(9), 34–36.

Asher, H. 1956. *Cost-Quantity Relationships in the Airframe Industry*. Santa Monica, CA: RAND.

Birkler, J., J. Large, G. Smith, and F. Timson. 1993. *Reconstituting a Production Capability: Past Experience, Restart Criteria, and Suggested Policies*. Santa Monica, CA: RAND.

Book, S. and N. Lao. 1998. Minimum-percentage error regression under zero-bias constraints. In *Proceedings of the Fourth Annual U.S. Army Conference on Applied Statistics*, U.S. Army Research Laboratory, pp. 47–56.

Covert, R. 2005. Cost improvement curves: Modeling reuse, learning, amortization and yield. Association of Cost Engineering (ACostE) Learning Event. Royal Heritage Motor Museum, Filton, U.K.

Covert, R. 2006. Errors-in-variables regression. In *Joint Space Systems Cost Analysis Group (SSCAG)/European Aerospace Working Group on Cost Engineering (EACE)/Society for Cost Analysis and Forecasting (SCAF) Meeting*, London, U.K.

Covert, R. 2007. Advances in CER development: Errors-in-variables regression. In *EACE*, Frascati, Italy.

Valerdi, R. 2005. The constructive systems engineering cost model (COSYSMO). PhD dissertation, University of Southern California, Los Angeles, CA.

12

Enhanced Scenario-Based Method

This chapter presents the last formal method for cost uncertainty analysis discussed in this book. Called the enhanced scenario-based method (eSBM), it was developed from a need in the cost analysis community to simplify aspects of probability-based approaches. This chapter describes eSBM, identifies key features that distinguish it from other methods, and provides illustrative examples.

12.1 Introduction

The enhanced scenario-based method (eSBM) was created as an alternative to the use of advanced statistical methods for generating measures of cost risk.* A central feature of eSBM is its emphasis on defining and costing the impacts of scenarios as the basis for deriving a range of possible program costs and assessing cost estimate confidence. Scenarios are written narratives about potential risk events that, if they occur, increase program cost beyond what was estimated or planned. The aim is to protect the program from the realization of these scenarios to avoid the potential cost risks they pose.

Defining scenarios that identify risk events a program may face is an ideal practice. It builds the rationale and arguments to justify contingencies that may be needed to protect program cost from the realization of these events.

---

* eSBM was first published in 2008 and it was called the scenario-based method (SBM). SBM offered a simpler alternative to advanced statistical methods for generating measures of cost risk. Since 2008, enhancements to SBM continued. They included integrating historical cost growth data into its cost risk analysis algorithms that measure cost estimate confidence. Collectively, these improvements have led to the enhanced scenario-based method—the name used henceforth. eSBM has appeared in the following U.S. government cost risk analysis handbooks and guides:
  - United States Air Force, Air Force Cost Analysis Agency, 2007. *Cost Risk and Uncertainty Analysis Handbook (CRUH)*.
  - NASA, 2008. *Cost Estimating Handbook*.
  - GAO, 2009. *Cost Estimating and Assessment Guide*, GAO-09-3SP.
  - Missile Defense Agency (MDA), 2012. *Cost Estimating and Analysis Handbook*.
  - United States Air Force, Air Force Cost Analysis Agency, 2012. *Joint Cost-Schedule Risk and Uncertainty Handbook*.

The clear expression of scenarios, and the description of the risk events they contain, is a critical part of the analysis process that is often lacking in practice. Scenarios can lead to cost reserve decisions made with an understanding of which risk events, if they occur, these reserves can cover.

Another feature of eSBM is the simplified analytical methods it uses to derive a distribution of possible program costs and measures of cost estimate confidence. Seen throughout this book, the intricacies in the mathematics of Monte Carlo simulation and in method of moments approaches are often very subtle. These subtleties must be understood if errors are to be avoided. Recognizing this, eSBM arose in response to many in the community who asked "Can a valid cost risk analysis, one that is traceable and defensible, be conducted with minimal (to no) reliance on Monte Carlo simulation or other advanced statistical methods?" To address this, eSBM was created and developed for use in either a nonstatistical or statistical form.

The nonstatistical form of eSBM produces deterministic, nonprobabilistic, measures of cost risk as a function of the risk events identified in one or more scenarios (mentioned above). The statistical form of eSBM provides the additional capability to generate probabilistic measures of cost estimate confidence. The simplified analytics of both forms eases the mathematical burden on analysts. It allows them time to focus instead on crafting well-written scenarios and then analyze the cost risks they potentially pose to the program.

12.2 Nonstatistical eSBM

Figure 12.1 illustrates the process flow of the nonstatistical implementation of eSBM. The first step is input to the process. It is the program's point estimate cost (PE). Mentioned throughout this book, the point estimate is the cost that does not include allowances for uncertainty. The PE is the sum of the

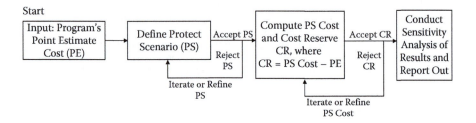

FIGURE 12.1
The nonstatistical eSBM process.

WBS element costs across the program's work breakdown structure *without adjustments for uncertainty.*

The next step in Figure 12.1 is defining and documenting the scenario aspect of eSBM, as explained in Section 12.1. This will now be referred to as the protect scenario. The protect scenario identifies the major known risks to the program—those events the program must monitor and guard against occurring. The protect scenario is not arbitrary, nor should it reflect extreme worst-case events. It should reflect a set of possible risks that, if realized, would cause the program's PE to be higher than planned. In practice, it is envisioned that management will converge on an "official" protect scenario after deliberations on the one initially defined. The objective is to ensure all parties reach a consensus understanding of the program's risks and how they are best described by the protect scenario.

Once the protect scenario is established, the program's cost is estimated by supposing all the risk events contained in the protect scenario occur. This estimate is called the protect scenario cost and is denoted by PS. The amount of cost reserve dollars (CR) needed to protect the program's cost from the occurrence of these risks is the *cost difference between the PS and the PE.* Shown in Figure 12.1, there may be additional refinements to the cost estimated for the protect scenario, based on management reviews and other considerations. The process may be iterated until the reasonableness of the amount of the cost reserve dollars to plan or budget for the program is accepted by management.

The final step in Figure 12.1, is a sensitivity analysis to identify critical cost drivers and assumptions associated with the protect scenario cost and the program's point estimate cost. It is recommended that the analysis measure the sensitivity of CR with respect to variations in the parameters associated with these drivers and assumptions. Defining and evaluating alternative protect scenarios is encouraged since numerous variations and excursions on them could be conjectured. Considering this, various trade-offs between a set of proposed protect scenarios can be explored with respect to their bearing on final cost reserve recommendations and decisions.

The nonstatistical eSBM, though simple in appearance, is arguably a form of risk analysis. The process of defining protect scenarios is a risk activity—one that is focused on identifying technical, management, and cost challenges of concern to the program. Scenario definition encourages a discourse on risk events that otherwise might not be held, thereby allowing risks to become visible, traceable, and estimative to the program's management, stakeholders, and decision-makers.

It is important that an eSBM analysis be continually refined, monitored, and updated as a program matures across its development and operational life cycle. This is true for all analyses that the methods in this book support. Finally, the work of building scenarios for eSBM should be closely connected to and be a regular part of the program's continuous risk management process.

12.3 Statistical eSBM

This section presents eSBM in its statistical form. The statistical eSBM is a closed-form method of moments approach, requiring only a few equations and assumptions. A key feature of the statistical eSBM is its integration of program cost histories into its algorithms.

There are many reasons to implement the statistical eSBM. These include (1) enabling cost estimate confidence measures to be derived, (2) providing a way for management to examine changes in confidence measures as a function of the amount of cost reserve budgeted or planned for the program, and (3) an ability to measure where the protect scenario cost falls on the probability distribution of the program's total cost.

Figure 12.2 illustrates the process flow of the statistical eSBM. To apply the statistical eSBM, three inputs shown on the left in Figure 12.2 are required. They are the PE, the probability that PE will not be exceeded, and the coefficient of variation (CV). The PE is the same as previously defined in the nonstatistical eSBM. The probability that PE will not be exceeded is the value α, such that

$$P(Cost_{WBS} \leq PE) = \alpha \tag{12.1}$$

In Equation 12.1, $Cost_{WBS}$ is the true cost of the program and PE is the program's point estimate cost. The probability α is a judged value, but one that can be guided by historical experience. For example, historical experience

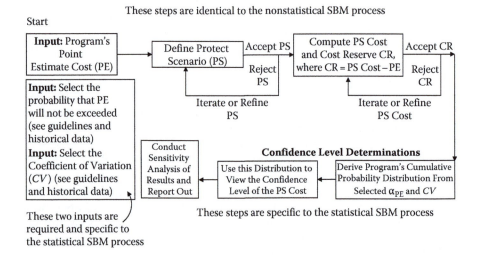

FIGURE 12.2
The statistical eSBM process.

indicates that α typically falls in the interval $0.10 \leq \alpha \leq 0.50$ in the early life cycle phases of a program. This is further discussed later in this section.

The *CV* is the ratio of a probability distribution's standard deviation to its mean. This ratio is given by Equation 12.2. The *CV* is a way to examine the variability of any probability distribution at plus or minus one standard deviation around its mean.

$$CV = \frac{\sigma}{\mu} \qquad (12.2)$$

With values assessed for α and *CV*, the cumulative probability distribution of $Cost_{WBS}$ can then be derived. This distribution is used to view the confidence level associated with the protect scenario cost PS, as well as confidence levels associated with any other cost outcome along this distribution.

The final step in Figure 12.2 is a sensitivity analysis. Here, we can examine the kinds of sensitivities previously described in the nonstatistical eSBM implementation, as well as uncertainties in values chosen for α and *CV*. This allows a broad assessment of confidence level variability, which includes determining a range of possible program cost outcomes for any specified confidence level.

Figure 12.3 illustrates an output from the statistical eSBM process. In this case, a normal probability distribution is shown with point estimate cost PE equal to $100M, α set to 0.25, and *CV* set to 0.50. The range $75M to $226M is plus or minus one standard deviation σ around the mean of $151M.

In the statistical eSBM, the uncertainty in a program's total cost is assumed to follow a normal or a lognormal probability distribution. The reason for choosing them is the frequency with which these forms approximate the

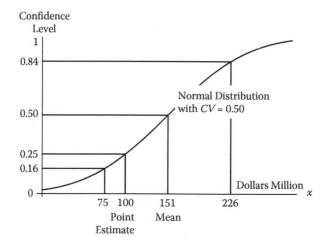

FIGURE 12.3
An output from the statistical eSBM.

probability distribution of a program's total cost, when it is the sum of WBS element costs. The following further discusses their application.

Statistical eSBM: Normal Distribution of $Cost_{WBS}$: The following equations derive from the observation that a program's total cost, denoted by $Cost_{WBS}$, is often normally distributed and the point (PE, α) falls along this distribution. Given PE, α, and CV the mean and standard deviation of $Cost_{WBS}$ are found by the following:

$$\mu = PE - z\frac{(CV)PE}{1 + z(CV)} \tag{12.3}$$

$$\sigma = \frac{(CV)PE}{1 + z(CV)} \tag{12.4}$$

where CV is the coefficient of variation, PE is the program's point estimate cost, and z is the value such that $P(Z \leq z) = \alpha$, where Z is the standard normal random variable. Values for z are available in lookup tables for the standard normal (see Table A.1) or from the Excel function $z = \text{NORM.S.INV(Percentile)}$; for example, $z = 0.525 = \text{NORM.S.INV}(0.70)$. With the values computed from Equations 12.3 and 12.4, the normal distribution function of $Cost_{WBS}$ is fully specified, along with the probability that $Cost_{WBS}$ may take any particular outcome, such as the value of the protect scenario cost PS.

Statistical eSBM: Lognormal Distribution of $Cost_{WBS}$: The following equations derive from the observation that a program's total cost, denoted by $Cost_{WBS}$, is often lognormally distributed and the point (PE, α) falls along this distribution. Given PE, α, and CV the mean and standard deviation of $Cost_{WBS}$ are found by the following:

$$\mu = e^{a + (1/2)b^2} \tag{12.5}$$

$$\sigma = \sqrt{e^{2a + b^2}(e^{b^2} - 1)} = \mu\sqrt{(e^{b^2} - 1)} \tag{12.6}$$

where

$$a = \ln PE - z\sqrt{\ln(1 + (CV)^2)} \tag{12.7}$$

$$b = \sqrt{\ln(1 + (CV)^2)} \tag{12.8}$$

With the values computed from Equations 12.5 and 12.6, the lognormal distribution function of $Cost_{WBS}$ is fully specified, along with the probability that

Cost$_{WBS}$ may take any particular outcome, such as the value of the protect scenario cost PS.

Example 12.1 *Suppose the distribution function of a program's total cost is normal. Suppose the program's point estimate cost is $100M and this was assessed to fall at the 25th percentile. Suppose the type and life cycle phase of the program is such that 30% variability in cost around the mean has been historically seen. Suppose the protect scenario was defined and determined to cost $145M.*

    a. *Compute the mean and standard deviation of Cost*$_{WBS}$.

    b. *Plot the distribution function of Cost*$_{WBS}$.

    c. *Determine the confidence level of the protect scenario cost and its associated cost reserve.*

    d. *Determine the program cost outcome at the 80th percentile confidence level, denoted by* $x_{0.80}$.

Solution

    a. From Equations 12.3 and 12.4

$$\mu = PE - z\frac{(CV)PE}{1 + z(CV)} = 100 - z\frac{(0.30)(100)}{1 + z(0.30)}$$

$$\sigma = \frac{(CV)PE}{1 + z(CV)} = \frac{(0.30)(100)}{1 + z(0.30)}$$

We need z to complete these computations. Since the distribution function of *Cost*$_{WBS}$ was given to be normal, it follows that $P(Cost_{WBS} \le PE) = \alpha = P(Z \le z)$, where Z is a standard normal random variable. Values for z are available in Excel and are computed as follows. Given $\alpha = 0.25$ in this example, then enter this formula into Excel: NORM.S.INV(0.25); that is,

$$z = \text{NORM.S.INV}(\alpha) = \text{NORM.S.INV}(0.25) = -0.6745$$

Therefore,

$$\mu = PE - z\frac{(CV)PE}{1 + z(CV)} = 100 - (-0.6745)\frac{(0.30)(100)}{1 + (-0.6745)(0.30)}$$

$$= 125.4 \text{ (\$M)}$$

$$\sigma = \frac{(CV)PE}{1 + z(CV)} = \frac{(0.30)(100)}{1 + (-0.6745)(0.30)} = 37.6 \text{ (\$M)}$$

    b. A plot of the probability distribution function of *Cost*$_{WBS}$ is shown in Figure 12.4. This is a normal distribution with mean $125.4M and standard deviation $37.6M, as determined from part (a).

    c. To determine the confidence level of the protect scenario, find α_{PS} such that

$$P(Cost_{WBS} \le PS = 145) = \alpha_{PS}$$

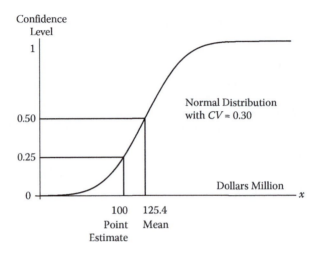

FIGURE 12.4
Normal distribution function of $Cost_{WBS}$.

Finding α_{PS} is equivalent to solving the expression $\mu + z_{PS}\sigma = PS$ for z_{PS}. From this,

$$z_{PS} = \frac{PS - \mu}{\sigma} = \frac{PS}{\sigma} - \frac{1}{CV}$$

Since $PS = 145$, $\mu = 125.4$, and $\sigma = 37.6$ it follows that

$$z_{PS} = \frac{PS - \mu}{\sigma} = \frac{PS}{\sigma} - \frac{1}{CV} = \frac{145}{37.6} - \frac{1}{(0.30)} = 0.523$$

Thus, we want α such that $P(Z \le z_{PS} = 0.523) = \alpha$. Values for α are available in Excel as follows. With $z_{PS} = 0.523$, enter into Excel: NORM.S.DIST(0.523, TRUE); that is,

$$\alpha = \text{NORM.S.DIST}(z_{PS}, \text{TRUE}) = \text{NORM.S.DIST}(0.523, \text{TRUE})$$

$$= 0.70$$

Therefore, the \$145M protect scenario cost falls at the 70th percentile of the distribution. This implies a cost reserve CR equal to \$45M.

d. To determine the 80th percentile confidence level, we need to find $z_{0.80}$ such that

$$P(Z \le z_{0.80}) = 0.80$$

Given $\alpha = 0.80$ in this example, enter into Excel: NORM.S.INV(0.80); that is,

$$z_\alpha = \text{NORM.S.INV}(\alpha) = z_{0.80} = \text{NORM.S.INV}(0.80) = 0.8416$$

Substituting $\mu = 125.4$ and $\sigma = 37.6$ (determined in part (a)) yields the following:

$$\mu + z_{0.80}\,\sigma = 125.4 + 0.8416\,(37.6) = x_{0.80} = 157$$

Therefore, the cost associated with the 80th percentile confidence level is \$157M. Figure 12.5 presents a summary of the results in this example.

In Example 12.1, if a range of possible values for α and CV were used then a range of possible program costs can be generated at any percentile along the distribution. For instance, suppose historical cost data for a particular program indicates its CV varies in the interval $0.20 \le CV \le 0.50$. Given the conditions in Example 12.1, variability in CV affects the mean and standard deviation of program cost. This is illustrated in Table 12.1, given a program's point estimate cost is equal to \$100M and its $\alpha = 0.25$.

Table 12.1 shows a range of possible cost outcomes for the 50th and 80th percentiles. Selecting a particular outcome can be guided by the CV considered most representative of the program's uncertainty at its specific life cycle phase. This is guided by the scenario or scenarios developed at the start of the eSBM process. Figure 12.6 graphically illustrates the results in Table 12.1.

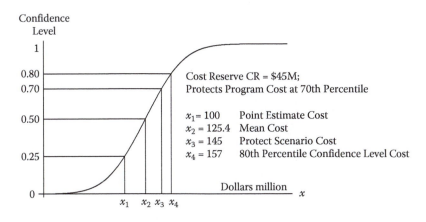

FIGURE 12.5
Normal distribution function of $Cost_{WBS}$ for various confidence intervals.

TABLE 12.1

Ranges of Cost Outcomes in Confidence Levels (Normal Distribution)

Coefficient of Variation (CV)	Standard Deviation ($M)	Confidence Level ($M) 50th Percentile	Confidence Level ($M) 80th Percentile
0.20	23.1	115	125
0.30	37.6	125	157
0.40	54.8	127	183
0.50	75.4	151	214

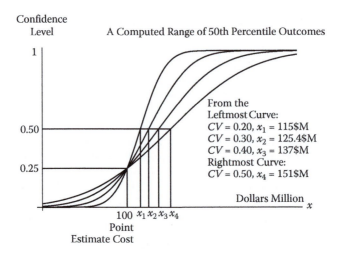

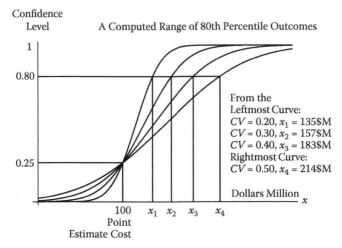

FIGURE 12.6
Range of cost outcomes and their confidence levels (normal distribution).

Example 12.2 *Suppose the distribution function of a program's total cost is lognormal. Suppose the program's point estimate cost is $100M and this was assessed to fall at the 25th percentile. Suppose the type and life cycle phase of the program is such that 30% variability in cost around the mean has been historically seen. Suppose the protect scenario was defined and determined to cost $145M.*

a. Compute the mean and standard deviation of Cost$_{WBS}$.

b. Determine the confidence level of the protect scenario cost and its associated cost reserve.

Solution

a. From Equations 12.7 and 12.8

$$a = \ln PE - z\sqrt{\ln(1+(CV)^2)} = \ln(100) - (-0.6745)\sqrt{\ln(1+(0.30)^2)}$$

$$= 4.80317$$

$$b = \sqrt{\ln(1+(CV)^2)} = \sqrt{\ln(1+(0.30)^2)} = 0.29356$$

From Equations 12.5 and 12.6 the values for a and b are translated into their equivalent statistics μ and σ as follows:

$$\mu = e^{a+(1/2)b^2} = e^{4.80317+(1/2)(0.29356)^2} \approx 127.3 \text{ ($M)}$$

$$\sigma = \sqrt{e^{2a+b^2}(e^{b^2}-1)} = \mu\sqrt{(e^{b^2}-1)}$$

$$= 127.3\sqrt{(e^{(0.29356)^2}-1)} \approx 38.2 \text{ ($M)}$$

b. To determine the confidence level of the protect scenario we need to find α_{PS} such that

$$P(Cost \le PS = 145) = \alpha_{PS}$$

Finding α_{PS} is equivalent to solving $a + z_{PS}(b) = \ln PS$ for z_{PS}. From this,

$$z_{PS} = \frac{\ln PS - a}{b}$$

Since $PS = 145$, $a = 4.80317$, and $b = 0.29356$, it follows that

$$z_{PS} = \frac{\ln PS - a}{b} = \frac{\ln 145 - 4.80317}{0.29356} = 0.59123$$

Thus, we want α such that $P(Z \le z_{PS} = 0.59123) = \alpha$. Values for α are available in Excel as follows. With $z_{PS} = 0.59123$, enter into Excel: NORM.S.DIST(0.59123, TRUE); that is,

$$\alpha = \text{NORM.S.DIST} (z_{PS}, \text{TRUE})$$

$$= \text{NORM.S.DIST} (0.59123, \text{TRUE}) = 0.723$$

$$\Rightarrow P(Z \le z_{PS} = 0.59123) \approx 0.723$$

Therefore, the $145M protect scenario cost falls at the 72nd percentile of the distribution. This implies a CR equal to $45M.

The preceding illustrated the statistical eSBM with values given for α and CV. The following shows how the statistical eSBM can be used to derive values for α and CV, given assessments of α_1 and α_2, where

$$\alpha_1 = P(PE \le Cost_{WBS} \le PS) \text{ and } \alpha_2 = P(Cost_{WBS} \ge PS)$$

as shown in Figure 12.7.

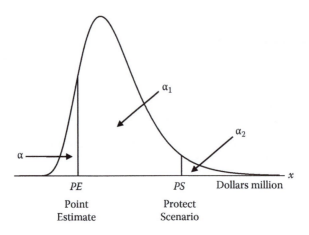

FIGURE 12.7
The eSBM probability α.

Example 12.3 *Suppose the distribution function of a program's total cost is lognormal with $PE = \$100M$ and $PS = \$155M$. In Figure 12.7, if $\alpha_1 = 0.70$ and $\alpha_2 = 0.05$ then derive the eSBM parameters α and CV. Use these results to compute the program's cost at the 80th percentile confidence level.*

Solution Let $Cost_{WBS}$ denote the program's total cost. From the information in Figure 12.7 and with $\alpha_1 = 0.70$ and $\alpha_2 = 0.05$ it follows that

$$\alpha = P(Cost \le PE) = 1 - (\alpha_1 + \alpha_2) = 1 - (0.70 + 0.05) = 0.25$$

It also follows that

$$\alpha_{PS} = P(Cost_{WBS} \le PS) = 1 - \alpha_2 = 1 - 0.05 = 0.95$$

Since the probability distribution of $Cost_{WBS}$ is given to be lognormal, from Chapter 4 it follows that

$$P(Cost_{WBS} \le PE) = P\left(Z \le z = \frac{\ln PE - a}{b}\right) = \alpha$$

$$P(Cost_{WBS} \le PS) = P\left(Z \le z_{PS} = \frac{\ln PS - a}{b}\right) = \alpha_{PS}$$

This implies

$$a + z(b) = \ln PE$$
$$a + z_{PS}(b) = \ln PS$$

Since Z is a standard normal random variable, from Chapter 4 it follows that

$$P(Z \le z) = \alpha = 0.25 \quad \text{when } z = -0.6745$$

and

$$P(Z \le z_{PS}) = \alpha_{PS} = 0.95 \quad \text{when } z = 1.645$$

Given that $PE = \$100M$ and $PS = \$155M$ it follows that

$$a + (-0.6745)(b) = \ln 100$$
$$a + (1.645)(b) = \ln 155$$

Solving these equations yields $a = 4.73262$ and $b = 0.188956$. From Equations 12.5 and 12.6, it follows that $\mu = \$115.64M$ and $\sigma = \$22.05M$. From this, it follows that

$$CV = \frac{\sigma}{\mu} = \frac{22.05}{115.64} = 0.19$$

Thus, given the distribution function of a program's total cost is lognormal with $PE = \$100M$ and $PS = \$155M$, and $\alpha_1 = 0.70$ and $\alpha_2 = 0.05$, then

$$\alpha = P(Cost \leq PE) = 1 - (\alpha_1 + \alpha_2) = 0.25 \text{ and } CV = \frac{\sigma}{\mu} = 0.19$$

To find the program's cost at the 80th percentile confidence level, from the solution to Example 12.1 part (d) recall that $P(Z \leq z_{0.80} = 0.8416) = 0.80$. Since the distribution function of total program cost was given to be lognormal we have

$$a + (0.8416)(b) = \ln x_{0.80}$$

In this case

$$4.73262 + (0.8416)(0.188956) = 4.89165 = \ln x_{0.80}$$

Thus, the program cost associated with the 80th percentile confidence level is

$$e^{4.89165} = x_{0.80} = \$133.2M$$

Figure 12.8 summarizes the results in this example. It shows how eSBM can be used when only the two probabilities $\alpha_1 = 0.70$ and $\alpha_2 = 0.05$

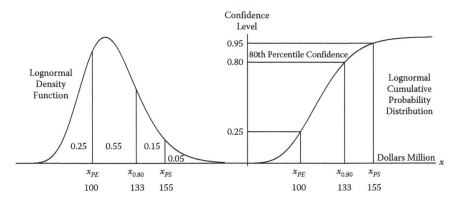

FIGURE 12.8
Resultant lognormal density and distribution functions for Example 12.3.

are given. The following presents how program cost growth histories can be used as a guide for choosing values of the eSBM parameters α and CV.

12.4 Historical Data for eSBM

> In God we trust; all others bring data (Performance of the Defense Acquisition System 2012).
>
> **Attributed to W. Edwards Deming**

As mentioned earlier, a key feature of the statistical eSBM is its integration of program cost growth histories into its algorithms. In recent years, studies by RAND (Arena et al. 2006, Bolten et al. 2008, Younossi et al. 2007) and the U.S. Naval Center for Cost Analysis (NCCA)* have collected historical data on program cost growth. Their evaluations of these data produced findings that can guide the choice of eSBM parameters α and CV. Studies such as these are periodically updated with new data and results. The reader is directed to the cited references and authors for new information that may be available, subsequent to the publication of this second edition.

The RAND and NCCA studies focused on the Department of Defense (DoD) programs; however, root causes for their cost growth are common to many types of today's engineering system developments. Causes for cost growth, found in these studies, include the immaturity of technologies and technical baselines, requirements volatility, system scale and complexity, program definition and execution challenges, and cost-schedule estimation errors.

The intended use of these studies is to guide valuing the eSBM parameters α and CV for the program under consideration. Thus, historical data offers a reference point for analogy comparisons. Sensitivity of the eSBM analysis to the choice or choices of α and CV are always recommended. Another way to look at values for α and CV is from the perspective "if my program experiences values such as these for α and CV, then the amount of cost risk could be x." Addressing this question can be a motivator for focusing risk management actions early in the program, with the aim to ameliorate potential events driving an unacceptable cost risk before going too far into the program's execution.

---

* The collection and analysis of historical program data for use in cost risk analysis was under the auspices of Wendy Kunc, Deputy Assistant Secretary for Cost and Economics, Office of the Assistant Secretary of the Navy (Financial Management and Comptroller) and executive director of the Naval Center for Cost Analysis (NCCA). The analysts who led the statistical analysis of these data were Dr. Brian Flynn, Peter Braxton, and Richard Lee of Technomics, Inc. The reader is directed to the NCCA official website for further information about the NCCA historical cost growth study.

12.4.1 RAND Historical Cost Growth Studies

The RAND Corporation conducted three major cost growth studies in 2006, in 2007, and in 2008 (Arena et al. 2006, Bolten et al. 2008, Younossi et al. 2007). These studies researched and assessed dozens of U.S. DoD acquisition programs to measure and identify root causes for cost growth. In these reports, RAND defined cost growth as the ratio between the most recent Selected Acquisition Report (SAR)* estimate, or the estimate reported in the program's final SAR, and the cost estimate baseline reported in a prior SAR issued at the time of a given milestone (Younossi et al. 2007).

The cost growth studies referenced in this section use terminology specific to the life cycle of DoD programs. Known as milestones, they represent major decision points with respect to whether a program has demonstrated the maturity to advance to its next and subsequent phases. In DoD acquisitions, Milestone A (MS A) is the decision that approves a program to enter the Technology Maturation and Risk Reduction phase. Milestone B (MS B) authorizes a program to enter the Engineering and Manufacturing Development (EMD) phase and for the DoD components to award contracts for EMD. Milestone B commits the required investment resources to the program. Milestone C (MS C) is the point when a program is reviewed for entrance into the procurement phase (production and deployment or for limited deployment). Finally, milestones are typically tied to major contract actions. For example, MS B usually corresponds to the point where DoD authorizes the program to engage in an EMD contract.

Table 12.2 presents a summary of historical cost growth factors (CGFs) by MS B funding category (Arena et al. 2006). The last column in Table 12.2 presents the implied coefficients of variation (*CVs*) for the reported means and standard deviations in columns 3 and 5, respectively.

In Table 12.2, the *CV* data derives from historical observations on program cost variability by Milestone B funding category. These data can be used to form the justification basis for choosing a *CV* value in eSBM. For example, the data in Table 12.2 suggests that programs may experience (on average) a *CV* of 0.50 for development costs for the MS B point estimate and a *CV* of 0.26 for the overall program cost. The other values for *CV* in Table 12.2 can be similarly interpreted.

The data collected in the RAND study indicated a lognormal shape to the total cost growth factor probability distribution. Similar findings were observed in the subsequent 2007 and 2008 RAND studies. This is illustrated

---

* SARs are documents prepared by DoD for the U.S. Congress. They cover all major defense acquisition programs. They are submitted at least annually and are required by Public Law 10 USC 2432. The SARs establish a baseline cost estimate at the time of a program's MS B. Changes to that estimate (or "variances") are made and documented as time passes to explain increases or decreases in current and future budgets. For a more detailed discussion of SARs, see Arena et al. (2006) (footnote excerpted from Bolten et al. [2008]).

TABLE 12.2

RAND Study: MS B Historical Cost Growth Factors (CGFs)

Funding Categories	Number of Observations	Mean	Median	Standard Deviation	Coefficient of Variation (Implied from Data)
Total[a]	46	1.46	1.44	0.38	0.26
Development	46	1.58	1.34	0.79	0.50
Procurement	44	1.44	1.40	0.42	0.29
Military Construction	10	1.33	1.11	0.82	0.62

Source: Arena, M. V. et al. 2006. Historical cost growth of completed weapon system programs, Project Air Force, TR-343-AF, ©2006. The RAND Corporation, Santa Monica, CA.

[a] Total includes development, procurement (adjusted for quantity changes), and military construction (as applicable).

in Figure 12.9. If similar lognormal assumptions are made for the development, procurement, and military construction CGFs in Table 12.2, then Table 12.3 reports the derived 25th, 50th, and the 80th percentile confidence levels for these MS B categories.

Table 12.4 presents a summary of historical CGF by MS C funding category (Arena et al. 2006). The last column in Table 12.4 presents the implied *CVs* for the reported means and standard deviations in columns 3 and 5, respectively.

Although the data in Tables 12.3 and 12.4 offer insights into the eSBM parameter *CV*, they also provide a way to consider values for the eSBM parameter α. Recall that α is the probability that the point estimate cost PE will not be exceeded—it represents the baseline against which change in estimated cost is measured. In these data, the point estimate cost is analogous to

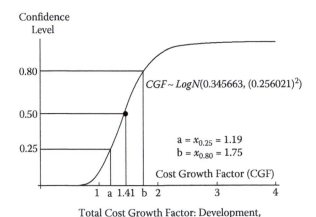

Total Cost Growth Factor: Development, Procurement, and Military Construction

FIGURE 12.9

Lognormal distribution: MS B Total CGF in Table 12.2.

TABLE 12.3

2006 RAND Study: Implied MS B Cost Growth Factor Percentiles

Funding Categories	CGF: 25th Percentile Confidence Level	CGF: 50th Percentile Confidence Level	CGF: 80th Percentile Confidence Level
Total[a] (Figure 12.9)	1.19	1.41	1.75
Development	1.03	1.41	2.10
Procurement	1.14	1.38	1.76
Military Construction	Below 1.0	1.12	1.83

[a] Total includes development, procurement (adjusted for quantity changes), and military construction (as applicable).

TABLE 12.4

RAND Study: MS C Historical Cost Growth Factors (CGFs)

Funding Categories	Number of Observations	Mean	Median	Standard Deviation	Coefficient of Variation (Implied from Data)
Total[a]	68	1.16	1.12	0.26	0.22
Development	65	1.30	1.10	0.64	0.49
Procurement	68	1.19	1.17	0.33	0.28
Military Construction	26	0.81	0.77	0.51	0.63

Source: Arena, M. V. et al. 2006. Historical cost growth of completed weapon system programs, Project Air Force, TR-343-AF, ©2006. The RAND Corporation, Santa Monica, CA.

[a] Total includes development, procurement (adjusted for quantity changes), and military construction (as applicable).

a $CGF = 1$ in Figure 12.9. From this, the percentile of the lognormal distribution associated with $CGF = 1$ offers a representative measure of α. Tables 12.5 and 12.6 provide the results of this analysis.

Mentioned earlier, the eSBM parameter α is a judged value, but one that can be guided by historical experience in ways similar to CV. The results in Tables 12.5 and 12.6 present historical evidence that α typically falls in the interval $0.10 \leq \alpha \leq 0.50$. Moreover, this evidence suggests this interval is true

TABLE 12.5

RAND Study: MS B Historical Data Implied CVs and α's

Funding Categories	Number of Observations	Mean	Standard Deviation	CV, α
Total	46	1.46	0.38	0.26, 0.09
Development	46	1.58	0.79	0.50, 0.23
Procurement	44	1.44	0.42	0.29, 0.13
Military Construction	10	1.33	0.82	0.62, 0.41

TABLE 12.6

RAND Study: MS C Historical Data Implied CVs and α's

Funding Categories	Number of Observations	Mean	Standard Deviation	CV, α
Total	68	1.16	0.26	0.22, 0.29
Development	65	1.30	0.64	0.49, 0.37
Procurement	68	1.19	0.33	0.28, 0.31
Military Construction	26	0.81	0.51	0.63, 0.02

even as a program enters its Milestone C phase. Findings such as these offer, for the first time, experiential insight into the α parameter that previously has only been anecdotally understood.

12.4.2 Naval Center for Cost Analysis: Historical Cost Growth Studies

As mentioned earlier, the U.S. NCCA has been collecting historical cost growth data on the Department of the Navy (DON) programs. Similar to the referenced RAND studies, the NCCA analyses produced findings that can guide the choice of eSBM parameters α and CV.

The NCCA data and analysis findings are based on 100 SARs* that contain raw data on cost outcomes of historical DON major defense acquisition programs (MDAPs). As numerous cost growth studies have indicated, the SARs, while not perfect, are nevertheless a good, convenient, comprehensive, official source of data on cost, schedule, and technical performance of MDAPs. More importantly, they are tied to milestones and they present total program acquisition costs across multiple appropriations categories and life cycle phases.

Of the 100 programs in the NCCA study, 50 were MS B estimates of total program acquisition cost (development, production, and, less frequently, military construction). Platform types included aircraft, helicopters, missiles, ships and submarines, and a few other systems. From the SAR summary sheets, these data elements were captured: base year, baseline type, platform type, baseline and current cost and quantity estimates, changes to date, date of last SAR, and costs in base-year and then-year dollars. The results from the analysis of MS B data are summarized in Table 12.7.

Similar to the RAND studies, the point estimate cost in the NCCA study is analogous to a $CGF = 1$. Furthermore, the NCCA study similarly found the probability distribution of MS B cost growth is approximately lognormal. From this, the percentile of the lognormal distribution associated with $CGF = 1$ offers a representative measure of the eSBM parameter α.

---

* The NCCA has since collected and analyzed cost growth data on over 300 programs. The reader is directed to the NCCA official website for further information about these findings.

TABLE 12.7

NCCA Study: MS B Historical Cost Growth Factors and *CVs*

Statistics	Without Quantity Adjustment		Quantity-Adjusted	
	Base-Year $	Then-Year $	Base-Year $	Then-Year $
Mean	1.48	1.84	1.23	1.36
Standard Deviation	0.94	1.60	0.44	0.69
CV	0.63	0.87	0.36	0.51

In Figure 12.10, this is shown by $x=1$. From Figure 12.10, and the data in the last column of Table 12.7, it can be shown that $x=1$ falls at the 34th percentile confidence level. This means $\alpha=0.34$ for the NCCA MS B program histories in Table 12.7, with respect to quantity-adjusted then-year dollars.

Of the 100 programs in the NCCA study, 43 were MS C estimates of total program acquisition costs. An analysis similar to that conducted for the MS B data was done for the MS C information. Results from the analysis of MS C data are summarized in Table 12.8.

The values in Table 12.8 show an across-the-board drop in the MS C cost growth statistics when compared to the MS B values. Reasons for this include the effects of near-settled development costs in these data, but also from increased program knowledge, maturity, and stability of programs at MS C.

Similar to the RAND studies, the point estimate cost in the NCCA study is analogous to a $CGF=1$. Furthermore, the NCCA study similarly found the probability distribution of MS C cost growth is approximately lognormal. From this, the percentile of the lognormal distribution associated with $CGF=1$ offers a representative measure of the eSBM parameter α.

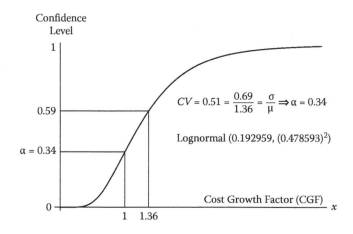

FIGURE 12.10

NCCA study: eSBM parameter α for MS B historical programs.

TABLE 12.8

NCCA Study: MS C Historical Cost Growth Factors and *CVs*

Statistics	Without Quantity Adjustment		Quantity-Adjusted	
	Base-Year $	Then-Year $	Base-Year $	Then-Year $
Mean	1.11	1.08	1.11	1.10
Standard Deviation	0.50	0.58	0.21	0.28
CV	0.45	0.53	0.19	0.26

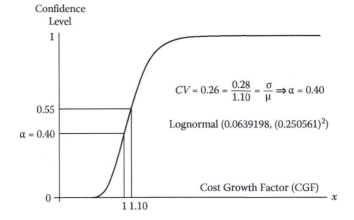

FIGURE 12.11
NCCA study: eSBM parameter α for MS C historical programs.

In Figure 12.11, this is shown by $x = 1$. From Figure 12.11, and the data in the last column of Table 12.8, it can be shown that $x = 1$ falls at the 40th percentile confidence level. This means $\alpha = 0.40$ for the NCCA MS C program histories in Table 12.8, with respect to quantity-adjusted then-year dollars.

The historical data in this section is presented to illustrate how it can be used to guide the choice of eSBM parameters α and *CV*. The RAND and NCCA studies exemplify the value such data brings in building defensible, traceable, and realistic cost uncertainty analyses. The reader is directed to these and other cited studies (Performance of the Defense Acquisition System 2012) as new information is available subsequent to the publication of this second edition.

12.5 Summary

The eSBM was created as an alternative to the use of advanced statistical methods for generating measures of cost risk (Garvey 2008, Garvey et al. 2012).

A central feature of eSBM is its emphasis on defining and costing the impacts of scenarios as the basis for deriving a range of possible program costs and assessing cost estimate confidence. As discussed in Section 12.1, scenarios are written narratives about potential risk events that, if they occur, increase program cost beyond what was estimated or planned. The aim is to protect the program from the realization of these scenarios to avoid the potential cost risk they pose.

Defining scenarios that identify risk events a program may face is an ideal practice. It builds the rationale and arguments to justify contingencies that may be needed to protect program cost from the realization of these events. The clear expression of scenarios, and the description of the risk events they contain, is a critical part of the analysis process that is often lacking in practice. Scenarios can lead to cost reserve decisions made with an understanding of which risk events, if they occur, these reserves can cover.

Another key feature of eSBM is its simplicity, requiring only a few equations that can be easily programmed in a spreadsheet. Moreover, its emphasis on directly integrating program cost histories into its algorithms is an ideal practice in cost uncertainty analysis. In general, eSBM

- Provides an analytic argument for deriving the amount of cost reserve needed to guard against well-defined "scenarios."
- Brings the discussion of "scenarios" and their credibility to the decision-makers; this is a more meaningful topic to focus on, instead of statistical abstractions that advanced simulation approaches can sometimes create.
- Does not require the use of statistical methods to develop a valid measure of cost risk reserve—the nonstatistical eSBM form.
- Allows percentiles (confidence measures) to be designed into the approach with a minimum set of statistical assumptions—the statistical eSBM form.
- Allows percentiles and the mean, median, mode, and variance to be calculated algebraically in near–real time within a simple spreadsheet environment; this is because eSBM in its statistical form is a closed form method of moments approach.
- Avoids the requirement to develop probability distribution functions for all the uncertain variables in a WBS. In eSBM, the only decision needed is whether the total program cost uncertainty is best represented by a normal or a lognormal probability distribution.
- Captures correlation indirectly in the analysis by the magnitude of the coefficient of variation applied in the statistical eSBM; this means there is no need to induce correlations in an eSBM analysis. Furthermore, if past program histories are incorporated into the

eSBM algorithms, then the correlations implied by those programs are implicitly captured in the analysis.

- Supports a wide range of parameter sensitivity analyses. This facilitates traceability and focuses attention on key risk events identified in the written scenarios that have the potential to drive cost higher than expected.

In summary, eSBM emphasizes a careful and deliberative approach to cost uncertainty analysis—one that also encourages the incorporation of program cost growth histories into its algorithms. eSBM requires the development of scenarios that represent the program's "risk story" rather than debate what type of WBS cost element distribution or correlation value to choose. Time is best spent building the scenarios (the case arguments) for why and how a confluence of risk events that form a scenario may increase a program's cost. This is where the discussions and analyses are best focused.

Exercises

12.1 Derive Equations 12.3 and 12.4 for the case where a program's cost follows a normal distribution.

12.2 Derive Equations 12.7 and 12.8 for the case where a program's cost follows a lognormal distribution.

12.3 Suppose the distribution function of a program's total cost is normal. Suppose the program's point estimate cost is $200M and this was assessed to fall at the 30th percentile. Suppose the type and life cycle phase of the program is such that 25% variability in cost around the mean has been historically seen. Suppose the protect scenario was defined and determined to cost $280M.

   a. Compute the mean and standard deviation of $Cost_{WBS}$.

   b. Determine the confidence level of the protect scenario cost and its associated cost reserve.

12.4 In the statistical form of eSBM, α was defined as the probability that the true program cost will be less than or equal to its point estimate cost PE; that is, from Equation 12.1.

$$P(Cost_{WBS} \leq PE) = \alpha$$

Instead, suppose the program team has a better understanding of the probability that the true program cost will be less than or equal to its

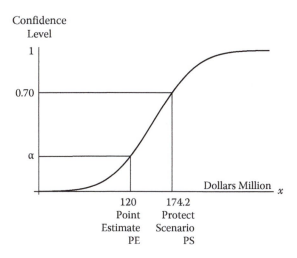

FIGURE 12.12
Program cost probability distribution for Exercise 12.4b.

protect scenario cost PS, denoted by α_{PS}; that is,

$$P(Cost_{WBS} \le PS) = \alpha_{PS}$$

a. Modify eSBM Equations 12.3 and 12.4 in terms of α_{PS} instead of α.

b. Suppose a program team developed their protect scenario and estimated its cost to be $174.2M. Suppose the point estimate cost for this program is $120M. Suppose the program team assessed $\alpha_{PS} = 0.70$, $CV = 0.30$, and that the program's cost follows a normal distribution. From this, derive the confidence level α for the program's point estimate cost. Refer to Figure 12.12 for this exercise.

12.5 Verify the α values in Tables 12.5 and 12.6, if the mean and standard deviation of each funding category are parameters of a lognormal distribution.

References

Arena, M. V., R. S. Leonard, S. E. Murray, and O. Younossi. 2006. Historical cost growth of completed weapon system programs, Project Air Force, TR-343-AF, ©2006. The RAND Corporation, Santa Monica, CA.

Bolten, J. G., R. S. Leonard, M. V. Arena, O. Younossi, and J. M. Sollinger. 2008. Sources of weapon system cost growth: An analysis of 35 major defense acquisition programs, ©2008. The RAND Corporation, Santa Monica, CA.

Garvey, P. R. 2008 (Spring). A scenario-based method for cost risk analysis. *Journal of Cost Analysis and Parametrics*, 1, 65—76.

Garvey, P. R., B. J. Flynn, P. Braxton, and R. Lee. December 2012. Enhanced scenario-based method for cost risk analysis: Theory, application, and implementation. *Journal of Cost Analysis and Parametrics*, 5(2), 98–142.

Performance of the Defense Acquisition System. 2012. Annual Report. Washington, DC: Office of the Under Secretary of Defense, Acquisition, Technology and Logistics, 2012; 3000 Defense Pentagon, Washington, DC, www.acq.osd.mil.

Younossi, O., M. V. Arena, R. S. Leonard, C. R. R. Roll, Jr., A. Jain and J. M. Sollinger. 2007. Is weapon system cost growth increasing? A quantitative assessment of completed and ongoing programs, ©2007. The RAND Corporation, Santa Monica, CA.

13

Cost Uncertainty Analysis Practice Points

The following provides recommended practices and considerations when performing cost uncertainty analyses. They reflect the authors' insights and experiences in developing, refining, and applying many of the techniques presented in this book.

13.1 Treating Cost as a Random Variable

The cost of a future system can be significantly affected by uncertainty. The existence of uncertainty implies the existence of a range of possible costs. How can a decision-maker be shown the chance a particular cost in the range of possible costs will be realized? The probability distribution is a recommended approach for providing this insight. Probability distributions result when variables (e.g., weight, power-output, staff-level) used to derive a system's cost randomly assume values across ranges of possible values. For instance, the cost of a satellite might be derived on the basis of a range of possible weight values, with each value randomly occurring. This approach treats cost as a random variable. It is recognition that values for these variables (such as weight) are not typically known with sufficient precision to perfectly predict cost, at a time when such predictions are needed. This point is further expressed by S.A. Book.*

> The mathematical vehicle for working with a range of possible costs is the probability distribution, with cost itself viewed as a "random variable." Such terminology does not imply, of course, that costs are "random" (though well they may be!) but rather that they are composed of a large number of very small pieces, whose individual contributions to the whole we do not have the ability to investigate in a degree of detail sufficient to calculate the total cost precisely. It is much more efficient for us to recognize that virtually all components of cost are simply "uncertain" and to find some way to assign probabilities to various possible ranges of costs. An analogue is the situation in coin tossing where, in theory, if we knew

---

* Book, S. A. 1997. Cost risk analysis—A tutorial. In *Risk Management Symposium Proceedings*. Los Angeles, CA: The Aerospace Corporation.

all the physics involved and solved all the differential equations, we could predict with certainty whether a coin would fall "heads" or "tails." However, the combination of influences acting on the coin is too complicated to understand in sufficient detail to calculate the physical parameters of the coin's motion. So we do the next best thing: we bet that the uncertainties will probably average out in such a way that the coin will fall "heads" half the time and "tails" the other half. It is much more efficient to consider the deterministic physical process of coin tossing to be a "random" statistical process and to assign probabilities of 0.50 to each of the two possible outcomes, heads or tails.

13.2 Risk versus Uncertainty

In this book, we make a distinction between the terms risk and uncertainty. Risk is the chance of loss or injury. In a situation that includes favorable and unfavorable events, risk is the probability an unfavorable event occurs. Uncertainty is the indefiniteness about the outcome of a situation. We analyze uncertainty for the purpose of measuring risk. In systems engineering, the analysis might focus on measuring the risk of failing to achieve performance objectives, overrunning the budgeted cost, or delivering the system too late to meet user needs.

13.3 Subjective Probability Assessments

Probability theory is a well-established formalism for quantifying uncertainty. Introduced in Chapter 2, its application to real-world systems engineering problems often involves the use of subjective probabilities. Subjective probabilities are those assigned to events on the basis of personal judgment. They are measures of a person's degree-of-belief that an event will occur. Subjective probabilities are associated with one-time, non-repeatable events—those whose probabilities cannot be objectively determined from a sample space of outcomes developed by repeated trials, or experimentation. Subjective probabilities must be consistent with the axioms of probability (refer to Chapter 2). For instance, if an engineer assigns a probability of 0.70 to the event "the number of gates for the new processor chip will not exceed 12,000," then it must follow the chip will exceed 12,000 gates with probability 0.30. Subjective probabilities are conditional on the state of the person's knowledge, which changes with time.

To be credible, subjective probabilities should only be assigned to events by subject matter experts—persons with significant experience with events similar to the one under consideration. Instead of assigning a single subjective probability to an event, subject experts often find it easier to describe a function that depicts a distribution of probabilities. Such a distribution is sometimes called a subjective probability distribution. Subjective probability distributions are governed by the same mathematical properties of probability distributions associated with discrete or continuous random variables (described in Chapter 3). Subjective probability distributions are most common in cost uncertainty analysis, particularly on the input-side of the process (refer to Figure 1.3 and the case discussions in Chapter 6). Because of their nature, subjective probability distributions can be thought of as "belief functions." They describe a subject expert's belief in the distribution of probabilities for an event under consideration. Probability theory provides the mathematical formalism with which we operate (add, subtract, multiply, and divide) on these belief functions.

13.4 Subjectivity in Systems Engineering and Analysis Problems

Systems engineering and analysis practices often necessitate the use of historical experience and expert judgments. These aspects must be recognized and properly addressed when designing and applying the formal methods in this book. What then is our analytic perspective in dealing with the ever-present reality of incomplete information, knowledge, and judgments in systems engineering and analysis problems? Consider the view by R.L. Keeney, a renowned theoretician, practitioner, and Ramsey medalist* in field of risk and decision analysis.

> The final issue concerns the charge that value (utility) models are not scientific or objective. With that, I certainly agree in the narrow sense. Indeed values are subjective, but they are undeniably a part of decision situations. Not modeling them does not make them go away. It is simply a question of whether these values are included implicitly and perhaps unknowingly in a decision process or whether there is an attempt to make them explicit and consistent and logical. In

---

* The Frank P. Ramsey Medal is the highest award of the Decision Analysis Society of the Institute for Operations Research and Management Science (INFORMS). It is named in honor of Frank Plumpton Ramsey, a Cambridge University mathematician who pioneered decision theory in the twentieth century. His 1926 essay "Truth and Probability" anticipated many of the developments in mathematical decision theory later made by renowned scholars John von Neumann, Oskar Morgenstern, and Leonard J. Savage.

a broader sense, the systematic development of a model of values is definitely scientific and objective. It lays out the assumptions on which the model is based, the logic supporting these assumptions, and the basis for data (that is, specific value judgments). This makes it possible to appraise the implications of different value judgments. All of this is very much in the spirit of scientific analysis. It seems more reasonable—even more scientific—to approach important decisions with the relevant values explicit and clarified rather than implicit and vague.

Keeney, R.L. (1992)

13.5 Correlation

As shown throughout this book, correlation can have a significant effect on the measure of a program's cost risk. Correlations can exist across pairs of work breakdown structure (WBS) element costs. It can also exist between variables that define a WBS element's cost. The importance of correlation as a critical consideration in cost uncertainty analysis cannot be understated. Ignoring correlation is equivalent to setting its value to zero between *all* cost element pairs, which can underestimate the program's true measure of cost risk (refer to Figure 9.2).

Statistical theory offers a number of ways to measure correlation. Two common measures are Pearson's product-moment correlation and Spearman's rank correlation. Subtleties concerning these measures must be understood to avoid errors in a cost uncertainty analysis. Pearson's product-moment correlation strictly measures the linearity between two random variables. Spearman's rank correlation measures whether a change in the value of one random variable is associated with a linear or nonlinear change in the value of another. Thus, Pearson and Spearman measures of correlation can be very different. This is illustrated in Figure 5.10 and discussed extensively in Chapters 8 and 9. Mathematically, the variance of a sum of random variables is a function of Pearson's product-moment correlation, not Spearman's rank correlation. Thus, Pearson correlation is the only technically correct measure of correlation to use when computing the variance of a sum of WBS element costs.

Pearson product-moment correlations are automatically captured in the results from Monte Carlo simulations where functional relationships between WBS element costs are present. Thus, if cost associations across and within the WBS are specified by functional relationships, then there is no need to create a correlation matrix. In fact, doing so can lead to a double counting correlation that can lead to overestimating a program's total cost risk (refer to Figures 9.3 and 9.4). Pearson correlations implied by logically defined functional relationships, which determine a WBS element's cost, are more intuitive for reviewers of the analysis than debating the merits of subjectively assigned correlations between some or all WBS cost element pairs.

Do not introduce rank correlation into cost uncertainty analysis, whether using Monte Carlo simulation or method of moments approaches.* If rank correlations are introduced into an analysis that is already capturing Pearson product-moment correlations (e.g., from the presence of functional relationships) then this leads to (1) double counting the effects of correlation on a program's total cost risk and (2) an analysis with mixed types of correlation measures leading to results whose interpretation is unknown.

Care must be taken if it is necessary to subjectively assign Pearson correlations. Pearson correlations can be restricted to a subinterval of $-1 \leq \rho \leq 1$. For example, the correlation coefficient of the bivariate lognormal distribution is bounded by the interval $-e^{-1} < \rho < 1$. Thus, the Pearson correlation between any two random variables cannot be arbitrarily assigned a value. Caution is needed to avoid potentially assigning impermissible values. If analysts need to assign values to correlations not automatically captured in a Monte Carlo simulation or a method of moments approach, then do so in accordance with the guidelines in Chapter 9. It is better to incorporate a judged level of correlation between WBS element costs known to co-vary than ignoring their associations altogether. The latter comes with the implicit assignment of zero to the correlation between such pairs of WBS element costs. Finally, sensitivity analyses should always be conducted around assigned correlation values to assess the reasonableness of their effects on a program's overall measure of cost risk.

13.6 Capturing Cost-Schedule Uncertainties

Decision-makers require understanding how uncertainties between a system's cost and schedule interact. A decision-maker might bet on a "high-risk" schedule in the hopes of keeping the system's cost within requirements. On the other hand, the decision-maker may be willing to assume "more cost" for a schedule with a small chance of being exceeded. This is a common trade-off

---

* The Lurie–Goldberg algorithm is an alternative to using rank correlations in a Monte Carlo simulation. Published in the late 1990s, the algorithm provides a method of generating Pearson-correlated random numbers.

faced by decision-makers on systems engineering projects. The family of distributions in Chapter 7 provides an analytical basis for computing this trade-off, using joint and conditional cost-schedule probabilities. This family is a set of mathematical models that might be hypothesized for capturing the joint interactions between cost and schedule.

A parameter required by these models is the correlation between cost and schedule.* Direct computation is one approach for determining this parameter, as illustrated in Case Discussion 6.2. However, in some instances this might not be analytically possible or practical. Another approach is to obtain an estimate of the correlation from sample values generated by Monte Carlo simulation. This is a reasonable method that can be done regardless of the complexity of the cost-schedule estimation relationships. Subjective assessments might be used. However, care must again be taken to specify an admissible correlation. Furthermore, there may already exist an implied correlation by virtue of how the cost-schedule estimation relationships are mathematically defined (refer to Case Discussion 6.2). Subjectively specifying a correlation when one is already present (only its magnitude is unknown) is double counting correlation.

13.7 Distribution Function of a System's Total Cost

Cost analysts should study the mathematical relationships they define in a system's work breakdown structure, to see whether analytical approximations to the distribution function of (a system's total cost) can be argued. Analytical approximations can reveal much information about the "cost-behavior" in a system's WBS. Section 6.2.2 presented five cases when the normal distribution approximates the distribution function of a system's total cost. There are many reasons for this approximation. Primary among them is a summation of WBS cost element costs.

Seen in the Chapter 6 case discussions, it is typical to have a mixture of independent and correlated cost element costs within a system's WBS. Because of the central limit theorem (Theorem 5.10), the greater the number of independent cost element costs the more it is that the distribution function of $Cost_{WBS}$ is approximately normal. The central limit theorem is very powerful. It does not take many independent cost element costs for the distribution function of $Cost_{WBS}$ to move toward normality. Such a move is

---

* Because these models treat cost and schedule as correlated random variables, it is important to recognize that *they do not capture causal impacts* that schedule compression or extension has on cost.

evidenced when (1) a sufficient number of independent cost element costs are summed and (2) when no cost element's cost distribution has a much larger standard deviation than the standard deviations of the other cost element cost distributions. If conditions in the WBS result in cost being positively skewed, then the lognormal often approximates the distribution function of program cost.

Monte Carlo simulation is another approach for developing an empirical approximation to the distribution function of program cost. The Monte Carlo method, discussed in Section 6.3, is often needed when a system's WBS contains cost estimating relationships too complex for strict analytical studies. In Monte Carlo simulations, a question frequently asked is "How many trials are necessary to have confidence in the output of the simulation?" As a guideline, 10,000 trials (samples) should be sufficient to meet the precision requirements for most Monte Carlo simulations; particularly those for cost uncertainty analyses.

This practice point provides guidance for approximating the probability distribution of a program's total cost $Cost_{WBS}$. The method of moments produces an analytically derived measure of the mean and variance of $Cost_{WBS}$. Monte Carlo simulation produces an empirically derived basis for these two measures, as well as an empirically derived probability distribution of $Cost_{WBS}$. To produce the probability distribution of $Cost_{WBS}$ using the method of moments, the form or shape of this distribution must be assumed. Best practice observations, published evidence, and statistical tests indicate the probability distribution of $Cost_{WBS}$ is often well approximated by normal or lognormal forms. Seen throughout this book, the normal and lognormal distributions well approximate the empirically derived probability distribution of $Cost_{WBS}$. There are many technical and empirically observed reasons for this. Primary among them is that a program's total cost is a summation of WBS element costs, including a summation of costs derived from non-linear cost estimation relationships. Within the WBS, it is typical to have a mixture of independent and correlated element costs. The greater the number of independent WBS element costs, the more it is that the probability distribution of $Cost_{WBS}$ is approximately normal. Why is this? Mentioned earlier, it is essentially the phenomenon explained by the central limit theorem.

The central limit theorem is very powerful in that it does not take many independent WBS element costs for the probability distribution of $Cost_{WBS}$ to approach normality. This central tendency is evidenced when (1) a sufficient number of independent WBS element costs are summed and (2) no WBS element's probability distribution has a much larger standard deviation than the standard deviations of the other WBS element distributions. When conditions in the WBS result in $Cost_{WBS}$ being positively skewed (i.e., a nonnormal distribution), then the lognormal often well approximates the distribution function of $Cost_{WBS}$. There is an extensive theoretical and practical discussion on this topic in Young (1995), Garvey (2000), and in Section 5.3.

▬▬▬▬▬▬▬

13.8 Benefits of Cost Uncertainty Analysis

Cost uncertainty analysis provides decision-makers many benefits and important insights, discussed next.

Establishing a Cost and Schedule Risk Baseline: Baseline probability distributions of a system's cost and schedule can be developed for a given system configuration, acquisition strategy, and cost-schedule estimation approach. This baseline provides decision-makers visibility into potentially high-payoff areas for risk reduction initiatives. Baseline distributions assist in determining a system's cost and schedule that simultaneously have a specified probability of not being exceeded (Chapter 7). They can also provide decision-makers an assessment of the likelihood of achieving a budgeted (or proposed) cost and schedule, or cost for a given feasible schedule.

Determining Cost Reserve: Cost uncertainty analysis provides a basis for determining cost reserve as a function of the uncertainties specific to a system. The analysis provides the direct link between the amount of cost reserve to recommend and the probability that a system's cost will not exceed a prescribed (or desired) magnitude (refer to Figure 1.6). An analysis should be conducted to verify the recommended cost reserve covers fortuitous events (e.g., unplanned code growth, unplanned schedule delays) deemed possible by the system's engineering team. Finally, it is sometimes necessary to allocate cost reserve dollars into the cost elements of a system's WBS. The reader is directed to the Book Young algorithm* as an approach for making this allocation.

Conducting Risk Reduction Trade-Off Analyses: Cost uncertainty analyses can be conducted to study the payoff of implementing risk reduction initiatives (e.g., rapid prototyping) on lessening a system's cost and schedule risks. Furthermore, families of probability distribution functions can be generated to compare the cost and cost risk impacts of alternative system requirements, schedule uncertainties, and competing system configurations or acquisition strategies.

Documenting the Cost Uncertainty Analysis: The validity and meaningfulness of a cost uncertainty analysis relies on the engineering team's experience, judgment, and knowledge of the system's uncertainties. Formulating and documenting a supporting rationale that summarizes the team's collective insights into these uncertainties is the critical part of the process. Without a well-documented rationale, the credibility of the analysis can be easily questioned. The details of the analysis methodology are important and should also be documented. The methodology must be technically sound

---

* Book, S. A. 1997. Cost risk analysis—A tutorial. In *Risk Management Symposium Proceedings*. Los Angeles, CA: The Aerospace Corporation.

and offer value-added problem structure, analyses, and insights otherwise not visible. Decisions that successfully eliminate uncertainty, or reduce it to acceptable levels, are ultimately driven by human judgment. This at best is aided by, not directed by, the methods presented in this book.

Management Perspectives: Cost risk analysis inputs should have a traceable and defensible basis in terms of their origin, pedigree, and soundness. Inputs derived from or based on evidence, historical data, or subject opinion should be documented and summarized in ways that support independent reviews.

The process of defining risk scenarios or narratives is a good practice. It builds the rationale and case arguments to justify the reserve needed to protect program cost from the realization of unwanted events. This is lacking in Monte Carlo simulation if designed as arbitrary randomizations of possible program costs, a practice which can lead to reserve recommendations absent clear program context for what these funds are to protect.

Analyze the consequences of identified risk scenarios on cost. Use these findings as a basis for identifying the amount of cost reserve needed to protect the budget from unexpected cost increases. Read from the cost probability distribution the percentile associated with the recommended cost reserve, to determine its level of confidence instead of arbitrarily budgeting to a predetermined confidence level. Time is best spent building the analysis and case arguments to justify how a confluence of identified risk events, that form one or more risk scenarios, may drive the cost of a program to a particular percentile. This is the perspective from which to make risk-informed budget and cost risk reserve decisions.

As a management practice, encourage and emphasize a careful and deliberative approach to cost risk analysis, regardless of the analysis method employed. Require the development of realistic excursions from a system's technical baseline or its cost analysis requirements description document that represents its risk scenarios.

14

Collected Works of Dr. Stephen A. Book

This chapter lists the major technical works of Dr. Stephen A. Book that advanced cost risk analysis theory and practice. The chapter is organized into works that were formally published in professional journals and those that were delivered as briefings to various conferences and technical gatherings.

14.1 Textbooks

1. Book, S. A. 1995. *Essentials of Statistics.* New York: Mc Graw Hill.
2. Book, S. A. and M. J. Epstein. 1982. *Statistical Analysis: Resolving Decision Problems in Business and Management.* Glenview, IL: Scott, Foresman and Co.
3. Book, S. A. 1977. *Statistics: Basic Techniques for Solving Applied Problems.* New York: McGraw Hill.

14.2 Journal Publications

Contact the professional organization associated with the identified journal to inquire about the availability of these papers.

1. Book, S. 2007. Quantifying the relationship between schedule and cost. *The Measurable News,* Winter 2007–2008, 11–15.
2. Book, S. and P. Young. 2006. The trouble with R^2. *Journal of Parametrics,* 25(1), 87–114.
3. Book, S. 2006. Unbiased percentage-error CERs with smaller standard errors. *The Journal of Cost Analysis & Management,* 8(1), 55–71.
4. Book, S. 2006. The mathematics of deriving factor CERs from cost data. *Parametric World,* 2006, pp. 16–22.
5. Book, S. 2006 (Spring). 'Earned schedule' and its possible unreliability as an indicator. *The Measurable News,* Project Management Institute College of Performance Management, Spring 2006, 24–30 (Correction Note: Fall 2006, 22–24).
6. Book, S. 2003. Issues associated with basing decisions on schedule variance in an earned value management system. *National Estimator,* Society of Cost Estimating and Analysis, Fall 2003, 11–15.

7. Book, S. 1999. Problems of correlation in the probabilistic approach to cost analysis. In *Proceedings of the Fifth Annual U.S. Army Conference on Applied Statistics*, October 19–21, U.S. Army Research Laboratory, Adelphi, MD, Report No. ARL-SR-110, July 2001, pp. 77–86.

8. Book, S. and Lao, N. (1999, November). Minimum-percentage-error regression under zero-bias constraints. In *Proceedings of the Fourth Annual U.S. Army Conference on Applied Statistics*, October 21–23, 1998, U.S. Army Research Laboratory, Adelphi, MD, Report No. ARL-SR-84, pp. 47–56.

9. Book, S. and Young, P. 1997. General-error regression for deriving cost-estimating relationships. *The Journal of Cost Analysis*, 14(2), 1–28. Later presented at *33rd DODCAS*, Williamsburg, VA, February 1–4, 2000.

10. Book, S. 1996. The learning rate's overpowering impact on cost estimates and how to diminish it (with E.L. Burgess). *Journal of Parametrics*, 16, 33–57.

11. Book, S. 1982. Least-absolute-deviations position finding. *Naval Research Logistics Quarterly*, 29, 235–246.

14.3 Conference Presentations and Proceedings

Contact the professional organization associated with the identified conference to inquire about the availability of these presentations.

1. Book, S. (June 26–29, 2012). Significant reasons to eschew log-log OLS regression when deriving estimating relationships. In *2012 SCEA/ISPA Joint Annual Conference and Training Workshop*, Orlando, FL.

2. Book, S. (June 8–11, 2010). Schedule risk analysis: Why it is important and how to do it. In *Integration Training Track, 2010 SCEA-ISPA Joint Annual Conference & Training Workshop*, San Diego, CA. Also presented to six previous *SCEA-ISPA Annual Conferences*.

3. Book, S. (June 8–11, 2010). Multiplicative-error regression. In *Practitioner Training Track, 2010 ISPA/SCEA Joint Annual Conference & Training Workshop*, San Diego, CA. This work was also presented to six previous *ISPASCEA Annual Conferences and the European Aerospace Working Group on Cost Engineering (EACE)*, Cranfield University, Milton Keynes/Bedford, U.K., October 19–21, 2004.

4. Book, S., M. Broder, and D. Feldman. (June 2–4, 2009). Statistical Foundations of adaptive cost-estimating relationships. In *ISPA/SCEA Joint Annual Conference & Training Workshop*, St Louis, MO.

5. Book, S. (April 26–28, 2009). Quantifying the relationship between cost and schedule. In *NASA Cost Symposium*, Cape Canaveral, FL. Also presented to *First Annual Department of Energy Cost Analysis and Training Symposium*, Santa Clara, CA, May 19–20, 2010 and *77th MORS Symposium*, Fort Leavenworth, KS, June 16–18, 2009.

6. Book, S. 2009 (Summer). Combining probabilistic estimates to reduce uncertainty. *Journal of Cost Analysis and Parametrics*, 47–54. Previously presented to *NASA*

Program Management Conference, Daytona Beach, FL, February 24–25, 2009 and *NASA Cost Analysis Symposium*, Portland, OR, August 26–28, 2008.

7. Book, S. 2009. Estimating using subsystem vs. box-level CERs: In-progress report (with N.J. Menton). In *42nd Department of Defense Cost Analysis Symposium (DODCAS)*, Williamsburg, VA, February 2009.

8. Book, S. and M. Broder. (June 24–27, 2008). Adaptive cost-estimating relationships. In *2008 SCEA-ISPA Joint Annual Conference & Training Workshop*, Industry Hills, CA.

9. Book, S. 2008. Cost estimating initiatives at NASA (with T.J. Coonce). In *41st DODCAS*, Williamsburg, VA, February 18–22, 2008.

10. Book, S. 2008. Cost risk as a discriminator in trade studies. In *Department of the Navy Cost Analysis Symposium*, Quantico, VA, September 4, 2008. Also presented to *76th MORS Symposium*, New London, CT, June 10–12, 2008; *SCEA-ISPA Joint International Conference & Workshop*, New Orleans, LA, June 12–16, 2007; and *NASA Project Management Conference*, Galveston, TX, February 7–8, 2007.

11. Book, S. 2008. The meaning and use of S-curves in cost estimating. In *NASA Project Management Conference*, Daytona Beach, FL, February 26–27, 2008 and *NASA Cost Symposium*, Denver, CO, July 17–19, 2007.

12. Book, S. February 14–17, 2006. Prediction bounds for general-error-regression CERs. In *39th Department of Defense Cost Analysis Symposium (DODCAS)*, Williamsburg, VA.

13. Book, S. 2006. Unbiased percentage-error CERs with smaller standard errors. *The Journal of Cost Analysis & Management*, 8(1), 55–71.

14. Book, S. 2006. Allocating risk dollars back to WBS elements. In *ISPA/SCEA Joint Conference and Training Workshop*, Seattle, WA. Also presented at *Department of the Navy Cost Analysis Symposium*, Quantico, VA, October 11, 2007; *40th DODCAS*, Williamsburg, VA, February 13–16, 2007; *Joint Meeting of SSCAG/EACE/SCAF*, London, U.K., September 19–21, 2006; and *NASA Cost Symposium*, Cleveland, OH, June 20–22, 2006.

15. Book, S. 2006. IRLS/MUPE CERs are not MPE-ZPB CERs. In *ISPA International Conference*, Bellevue, WA, May 23–26, 2006.

16. Book, S. 2006. Prediction bounds for general-error-regression CERs. In *SCEA National Conference*, Tysons Corner, VA, June 13–16, 2006. Also presented to *EACE, Abbey Wood*, Bristol, U.K., April 26–27, 2006; *39th DODCAS*, Williamsburg, VA, February 14–17, 2006; *72nd MORS Symposium*, Monterey, CA, June 22–24, 2004; and *ISPA International Conference*, Frascati, Italy, May 10–12, 2004.

17. Book, S. 2005. A theory of modeling correlations for use in cost-risk analysis. In *NASA Project Management Conference*, Galveston, TX, March 21–22, 2006. Also presented to *73rd MORS Symposium*, West Point, NY, June 21–23, 2005 and *SCEA-ISPA Joint International Conference & Educational Workshop*, Broomfield, CO, June 14–17, 2005.

18. Book, S. 2005. Risks in costing software vs. high confidence in estimates. In *SCEA-ISPA Joint International Conference & Educational Workshop*, Broomfield, CO, June 14–17, 2005.

19. Book, S. 2005. Universal risk issues in source selection. In *38th DOD-CAS*, Williamsburg, VA, February 15–18, 2005.

20. Book, S. 2005. Performance of the Interquartile Range (IQR) as a marker for the Cost Readiness Level (CRL) quality metric. In *NASA Cost Analysis Symposium*, New Orleans, LA, April 12–14, 2005.

21. Book, S. (October 19–21, 2004). Multiplicative-error regression. In *European Aerospace Working Group on Cost Engineering (EACE)*, Cranfield University, Milton Keynes/Bedford, U.K. (also presented to *2010 ISPA/SCEA Joint Annual Conference & Training Workshop (Practitioner Training Track)*, San Diego, CA, October 19–21, 2004; and six previous *ISPA/SCEA Annual Conferences*).

22. Book, S. 2004. How to make your point estimate look like a cost-risk analysis (so it can be used for decision-making). In *SCEA 2004 National Conference*, Manhattan Beach, CA, June 15–18, 2004; *SCEA 2004 National Conference*, Manhattan Beach, CA, June 15–18, 2004. Also presented to *EACE*, Immenstadt, Germany, November 4–6, 2003.

23. Book, S., and E. Burgess. 2003. A way out of the learning-rate morass: Quantity as an Independent Variable (QAIV). In *36th Annual DoD Cost Analysis Symposium (Advanced Training Track)*, Williamsburg, VA.

24. Book, S. 2002. Issues in specifying a triangular cost distribution. In *ISPA International Conference*, San Diego, CA, May 21–24, 2002.

25. Book, S. 2001. Risk assessment and probability. In *European Aerospace Working Group on Cost Engineering*, Noordwijk, the Netherlands, December 6, 2001.

26. Book, S. 2001. Effects of inflation on earned-value-based EACs (with J.E. Gayek). In *Joint ISPA/SCEA International Conference*, Tysons Corner, VA, June 12–15, 2001.

27. Book, S. 2001. Estimating probable system cost. *Crosslink: The Aerospace Corporation's* magazine of advances in aerospace technology, Winter 2000/2001, pp. 12–21. Also presented to *EACE*, Frascati, Italy, May 2–4, 2001.

28. Book, S. 2000. Do not sum earned-value-based WBS-element estimates-at-completion. In *SCEA National Conference & Training Workshop*, Manhattan Beach, CA, June 13–16, 2000.

29. Book, S. 1999. What we can and cannot learn from earned value. In *EACE*, Frascati, Italy, November 17–18, 1999. Also presented at *SCEA National Conference*, San Antonio, TX, June 8–11, 1999.

30. Book, S. 1999. Costs of reusable launch vehicles: Should we pay up front to build in high reliability or pay later to buy more vehicles? In *AIAA Space Technology & Applications International Forum (STAIF-99)*, Albuquerque, NM, January 31–February 2, 1999.

31. Book, S. 1999. Why correlation matters in cost estimating. In *32nd DODCAS*, Williamsburg, VA, February 2–5, 1999.

32. Book, S. 1998. Cost risk as a figure of merit. In *31st DODCAS*, Williamsburg, VA, February 3–6, 1998.

33. Book, S. A. 1997. Cost risk analysis—A tutorial. In *Risk Management Symposium Proceedings*, Los Angeles, CA, The Aerospace Corporation.

34. Book, S. 1996. Fictions we live by. In *29th DODCAS*, Leesburg, VA, February 21–23, 1996.

35. Book, S. 1994. Do not sum most likely costs. In *1994 NASA Cost Estimating Symposium*, Houston, TX, November 7–10, 1994.

36. Book, S. 1991. System error analysis based on one-at-a-time perturbations (with M.R. Chernick). *Mathematical and Computer Modelling* [sic——U.K. Publication], 15, 77–84.

37. Book, S. 1990. Deriving cost-estimating relationships using weighted least-squares regression. In *IAA/ISPA/AIAA Space Systems Cost Methodologies and Applications Symposium*, San Diego, CA, p. 19.

38. Book, S. A. and P. H. Young. 1990. Monte Carlo generation of total-cost distributions when WBS-element costs are correlated. In *24th Annual Department of Defense Cost Analysis Symposium*, Leesburg, VA, September 6–7, 1990.

39. Book, S. A. and P. H. Young. 1990. Optimality considerations related to the USCM-6 'Ping factor'. In *ICA/NES National Conference*, Los Angeles, CA, p. 40.

40. Book, S., W. Brady, and P. Mazaika. 1980. The nonuniform GPS constellation. In *IEEE 1980 Position Location and Navigation Symposium Record*, Atlantic City, NJ, pp. 1–8.

Appendix A: Statistical Tables and Related Integrals

A.1 Percentiles of the Standard Normal Distribution

Table A.1 presents values of the cumulative distribution function of the standard normal distribution. These values are denoted by $F_Z(z)$, which is given by

$$F_Z(z) = P(Z \leq z) = \int_{-\infty}^{z} \frac{1}{\sqrt{2\pi}} e^{-y^2/2} \, dy \qquad (A.1)$$

Example: (a) What is $P(Z \leq 0.33)$? (b) What is $P(Z \leq -0.33)$?

a. From Equation A.1 $F_Z(0.33) = P(Z \leq 0.33)$; from Table A.1 $F_Z(0.33) = 0.6293$

b. Since $Z \sim N(0,1)$ we have $P(Z \leq -z) = P(Z > z) = 1 - P(Z \leq z)$; therefore, in this example, $F_Z(-0.33) = P(Z \leq -0.33) = P(Z > 0.33) = 1 - P(Z \leq 0.33) = 1 - 0.6293 = 0.3707$

A.2 Kolmogorov–Smirnov Goodness-of-Fit Test

Table A.2 is used for the Kolmogorov–Smirnov goodness-of-fit test. The values in Table A.2 apply *only when all the parameters of the hypothesized distribution are known,* that is, none of the distribution's parameters are estimated (or derived) from the sample data. The reader is directed to Law and Kelton (1991) and Stephens (1974) for an expanded discussion of Table A.2. In Table A.2, D is the Kolmogorov–Smirnov test statistic defined as

$$D = \max_{x} \left| F_X(x) - \hat{F}_X(x) \right|$$

This statistic measures the largest vertical distance between the hypothesized cumulative distribution function $F_X(x)$ and the empirical (observed) cumulative distribution function $\hat{F}_X(x)$, developed from the sample data.

TABLE A.1

Percentiles of the Standard Normal Distribution (the 3-Digit Columns are z, the 8-Digit Columns are $F_Z(z)$)

0.00	0.5000000	0.21	0.5831661	0.42	0.6627572	0.63	0.7356528
0.01	0.5039894	0.22	0.5870644	0.43	0.6664021	0.64	0.7389138
0.02	0.5079784	0.23	0.5909541	0.44	0.6700314	0.65	0.7421540
0.03	0.5119665	0.24	0.5948348	0.45	0.6736448	0.66	0.7453732
0.04	0.5159535	0.25	0.5987063	0.46	0.6772419	0.67	0.7485712
0.05	0.5199389	0.26	0.6025681	0.47	0.6808225	0.68	0.7517478
0.06	0.5239223	0.27	0.6064198	0.48	0.6843863	0.69	0.7549030
0.07	0.5279032	0.28	0.6102612	0.49	0.6879331	0.70	0.7580364
0.08	0.5318814	0.29	0.6140918	0.50	0.6914625	0.71	0.7611480
0.09	0.5358565	0.30	0.6179114	0.51	0.6949743	0.72	0.7642376
0.10	0.5398279	0.31	0.6217195	0.52	0.6984682	0.73	0.7673050
0.11	0.5437954	0.32	0.6255158	0.53	0.7019441	0.74	0.7703501
0.12	0.5477585	0.33	0.6293000	0.54	0.7054015	0.75	0.7733727
0.13	0.5517168	0.34	0.6330717	0.55	0.7088403	0.76	0.7763728
0.14	0.5556700	0.35	0.6368306	0.56	0.7122603	0.77	0.7793501
0.15	0.5596177	0.36	0.6405764	0.57	0.7156612	0.78	0.7823046
0.16	0.5635595	0.37	0.6443087	0.58	0.7190427	0.79	0.7852362
0.17	0.5674949	0.38	0.6480272	0.59	0.7224047	0.80	0.7881447
0.18	0.5714237	0.39	0.6517317	0.60	0.7257469	0.81	0.7910300
0.19	0.5753454	0.40	0.6554217	0.61	0.7290692	0.82	0.7938920
0.20	0.5792597	0.41	0.6590970	0.62	0.7323712	0.83	0.7967307
0.84	0.7995459	1.05	0.8531409	1.26	0.8961653	1.47	0.9292191
0.85	0.8023375	1.06	0.8554277	1.27	0.8979576	1.48	0.9305633
0.86	0.8051055	1.07	0.8576903	1.28	0.8997274	1.49	0.9318879
0.87	0.8078498	1.08	0.8599289	1.29	0.9014746	1.50	0.9331928
0.88	0.8105704	1.09	0.8621434	1.30	0.9031995	1.51	0.9344783
0.89	0.8132671	1.10	0.8643339	1.31	0.9049020	1.52	0.9357445
0.90	0.8159399	1.11	0.8665004	1.32	0.9065824	1.53	0.9369916
0.91	0.8185888	1.12	0.8686431	1.33	0.9082408	1.54	0.9382198
0.92	0.8212136	1.13	0.8707618	1.34	0.9098773	1.55	0.9394292
0.93	0.8238145	1.14	0.8728568	1.35	0.9114919	1.56	0.9406200
0.94	0.8263912	1.15	0.8749280	1.36	0.9130850	1.57	0.9417924
0.95	0.8289439	1.16	0.8769755	1.37	0.9146565	1.58	0.9429466
0.96	0.8314724	1.17	0.8789995	1.38	0.9162066	1.59	0.9440826
0.97	0.8339768	1.18	0.8809998	1.39	0.9177355	1.60	0.9452007
0.98	0.8364569	1.19	0.8829767	1.40	0.9192433	1.61	0.9463011
0.99	0.8389129	1.20	0.8849303	1.41	0.9207301	1.62	0.9473839
1.00	0.8413447	1.21	0.8868605	1.42	0.9221961	1.63	0.9484493
1.01	0.8437523	1.22	0.8887675	1.43	0.9236414	1.64	0.9494974
1.02	0.8461358	1.23	0.8906514	1.44	0.9250663	1.65	0.9505285

(Continued)

TABLE A.1 (*Continued*)

Percentiles of the Standard Normal Distribution (the 3-Digit Columns are z, the 8-Digit Columns are $F_Z(z)$)

1.03	0.8484950	1.24	0.8925122	1.45	0.9264707	1.66	0.9515428
1.04	0.8508300	1.25	0.8943502	1.46	0.9278549	1.67	0.9525403
1.68	0.9535214	1.89	0.9706211	2.10	0.9821356	2.31	0.9895559
1.69	0.9544861	1.90	0.9712835	2.11	0.9825709	2.32	0.9898296
1.70	0.9554346	1.91	0.9719335	2.12	0.9829970	2.33	0.9900969
1.71	0.9563671	1.92	0.9725711	2.13	0.9834143	2.40	0.9918025
1.72	0.9572838	1.93	0.9731967	2.14	0.9838227	2.50	0.9937903
1.73	0.9581849	1.94	0.9738102	2.15	0.9842224	2.60	0.9953388
1.74	0.9590705	1.95	0.9744120	2.16	0.9846137	2.70	0.9965330
1.75	0.9599409	1.96	0.9750022	2.17	0.9849966	2.80	0.9974448
1.76	0.9607961	1.97	0.9755809	2.18	0.9853713	2.90	0.9981341
1.77	0.9616365	1.98	0.9761483	2.19	0.9857379	3.00	0.9986500
1.78	0.9624621	1.99	0.9767046	2.20	0.9860966	3.10	0.9990323
1.79	0.9632731	2.00	0.9772499	2.21	0.9864475	3.20	0.9993128
1.80	0.9640697	2.01	0.9777845	2.22	0.9867907	3.30	0.9995165
1.81	0.9648522	2.02	0.9783084	2.23	0.9871263	3.40	0.9996630
1.82	0.9656206	2.03	0.9788218	2.24	0.9874546	3.50	0.9997673
1.83	0.9663751	2.04	0.9793249	2.25	0.9877756	3.60	0.9998409
1.84	0.9671159	2.05	0.9798179	2.26	0.9880894	3.70	0.9998922
1.85	0.9678433	2.06	0.9803008	2.27	0.9883962	3.80	0.9999276
1.86	0.9685573	2.07	0.9807739	2.28	0.9886962	3.90	0.9999519
1.87	0.9692582	2.08	0.9812373	2.29	0.9889894	4.00	0.9999683
1.88	0.9699460	2.09	0.9816912	2.30	0.9892759	5.00	0.9999997

TABLE A.2

Modified Critical Values for the Kolmogorov–Smirnov Test Statistic

α	$1 - \alpha$	$c_{1-\alpha}$
0.010	0.990	1.628
0.025	0.975	1.480
0.050	0.950	1.358
0.100	0.900	1.224
0.150	0.850	1.138

Notes: Applicable when the parameters of the hypothesized distribution $F_X(x)$ are known and not estimated from the sample data. Let n denote the number of samples. If $\left(\sqrt{n} + 0.12 + \frac{0.11}{\sqrt{n}} \right) D > c_{1-\alpha}$ reject the claim that the observed values come from the hypothesized distribution; otherwise accept it.

A.3 Integrals Related to the Normal Probability Density Function

The following integrals are often useful in proofs and computations involving the normal probability density function (PDF). In each integral, a is a real number and b is a positive real number. The first integral is the integral of the normal PDF. The second integral is the mean of a normally distributed random variable. The third integral is the second moment of a normally distributed random variable, with mean a and variance b^2.

$$\int_{-\infty}^{\infty} \frac{1}{\sqrt{2\pi}b} e^{-\frac{(x-a)^2}{2b^2}} dx = 1 \tag{A.2}$$

$$\int_{-\infty}^{\infty} x \frac{1}{\sqrt{2\pi}b} e^{-\frac{(x-a)^2}{2b^2}} dx = a \tag{A.3}$$

$$\int_{-\infty}^{\infty} x^2 \frac{1}{\sqrt{2\pi}b} e^{-\frac{(x-a)^2}{2b^2}} dx = a^2 + b^2 \tag{A.4}$$

A.4 Sums of Independent Uniform Random Variables

Suppose the random variable U is defined as the sum of n uniformly distributed *independent* random variables, that is

$$U = U_1 + U_2 + U_3 + \cdots + U_n$$

where $U_i \sim Unif(0,1)$ for $i = 1, 2, 3, \ldots, n$. Let $f_U(u)$ denote the PDF of U. From Theorem 5.12, a general expression for $f_U(u)$ can be developed. A convenient form of this expression is given here:

$$f_U(u) = \frac{1}{(n-1)!} \left[u^{n-1} - \binom{n}{1} (u-1)^{n-1} + \binom{n}{2} (u-2)^{n-1} - \cdots \right]$$

In this expression, $0 < u < n$ and the summation is continued as long as the arguments u, $(u-1)$, $(u-2)$, $\ldots$ are positive (Cramer 1966). From the central limit theorem, as n increases the distribution function of U will approach a normal distribution with mean $\frac{n}{2}$ and variance $\frac{n}{12}$. This is illustrated in Figure A.1.

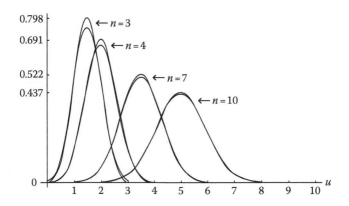

FIGURE A.1
Probability density functions for sums of uniform independent random variables.

Figure A.1 shows pairs of PDFs plotted for $n = 3, 4, 7,$ and 10. The left-most pair show plots of $f_{Normal}(u)$ and $f_U(u)$, respectively, for $n = 3$; specifically,

$$f_{Normal}(u) = \frac{1}{\sqrt{2\pi}} \frac{1}{\sqrt{\frac{3}{12}}} e^{-\frac{1}{2}\left[\frac{\left(u-\frac{3}{2}\right)^2}{\frac{3}{12}}\right]}$$

$$f_U(u) = \begin{cases} \frac{1}{2}u^2, & 0 < u < 1 \\ \frac{1}{2}(u^2 - 3(u-1)^2), & 1 < u < 2 \\ \frac{1}{2}(u^2 - 3(u-1)^2 + 3(u-2)^2), & 2 < u < 3 \end{cases}$$

The second pair of PDFs (from the left) show plots of $f_{Normal}(u)$ and $f_U(u)$, respectively, for $n = 4$; specifically,

$$f_{Normal}(u) = \frac{1}{\sqrt{2\pi}} \frac{1}{\sqrt{\frac{4}{12}}} e^{-\frac{1}{2}\left[\frac{\left(u-\frac{4}{2}\right)^2}{\frac{4}{12}}\right]}$$

$$f_U(u) = \begin{cases} \frac{1}{6}u^3, & 0 < u < 1 \\ \frac{1}{6}(u^3 - 4(u-1)^3), & 1 < u < 2 \\ \frac{1}{6}(u^3 - 4(u-1)^3 + 6(u-2)^3), & 2 < u < 3 \\ \frac{1}{6}(u^3 - 4(u-1)^3 + 6(u-2)^3 - 4(u-3)^3), & 3 < u < 4 \end{cases}$$

A similar convention holds for the two remaining pairs of PDFs plotted in Figure A.1. The values shown along the vertical axis, in Figure A.1, correspond to values for $f_{Normal}(u)$.

Table A.3 compares the cumulative probabilities derived from each PDF pair in Figure A.1. In Table A.3, the columns labeled $F_U(u)$ and $F_{Normal}(u)$ are defined as follows:

$$F_U(u) = \int_0^u f_U(t)\, dt$$

and

$$F_{Normal}(u) = \int_{-\infty}^u f_{Normal}(t)\, dt$$

where

$$f_{Normal}(t) = \frac{1}{\sqrt{2\pi}} \frac{1}{\sqrt{\frac{n}{12}}} e^{-\frac{1}{2}\left[\frac{\left(t-\frac{n}{2}\right)^2}{\frac{n}{12}}\right]}$$

for $n = 3, 4, 7$, and 10.

TABLE A.3

Sums of Independent Uniform Random Variables—Cumulative Probability

		$U = U_1 + U_2 + U_3 + \cdots + U_n$			
		$U_i \sim Unif(0,1)$ $i = 1,2,3,\cdots,n$			
$n = 3$	$F_U(u)$	$F_{Normal}(u)$	$n = 4$	$F_U(u)$	$F_{Normal}(u)$
$0 < u < 1$	0.16666667	0.158655	$0 < u < 1$	0.041666667	0.0416323
$0 < u < 2$	0.83333334	0.841345	$0 < u < 2$	0.499999997	0.5
$0 < u < 3$	1	0.99865	$0 < u < 3$	0.958333327	0.958368
			$0 < u < 4$	1	0.999734
$n = 7$	$F_U(u)$	$F_{Normal}(u)$	$n = 10$	$F_U(u)$	$F_{Normal}(u)$
$0 < u < 1$	0.0001984127	0.000531557	$0 < u < 1$	0.00000027557	0.00000588567
$0 < u < 2$	0.0240079367	0.0247673	$0 < u < 2$	0.00027943121	0.0005075
$0 < u < 3$	0.2603174567	0.256345	$0 < u < 3$	0.01346285321	0.0142299
$0 < u < 4$	0.7396825367	0.743655	$0 < u < 4$	0.13890156321	0.136661
$0 < u < 5$	0.9759920567	0.975233	$0 < u < 5$	0.49999999321	0.5
$0 < u < 6$	0.9998015807	0.999468	$0 < u < 6$	0.86109842321	0.863339
$0 < u < 7$	1	0.999998	$0 < u < 7$	0.98653713321	0.98577
			$0 < u < 8$	0.99972055521	0.999492
			$0 < u < 9$	0.99999971085	0.999994
			$0 < u < 10$	1	0.999999978398

References

Cramer, H. 1966. *Mathematical Methods of Statistics*, pp. 245. Princeton, NJ: Princeton University Press.

Law, A. M. and W. D. Kelton. 1991. *Simulation Modeling and Analysis*, 2nd edn. New York: McGraw-Hill, Inc.

Stephens, M. A. 1974. EDF statistics for goodness of fit and some comparisons. *Journal of the American Statistical Association*, 69, 730–737.

Appendix B: Bivariate Normal-Lognormal Distribution

Let $Y_1 = X_1$ and $Y_2 = \ln X_2$, where X_1 and X_2 are random variables defined on $-\infty < x_1 < \infty$ and $0 < x_2 < \infty$. If Y_1 and Y_2 each have a normal distribution, then

$$E(Y_1) = \mu_{Y_1} = \mu_{X_1} = \mu_1 \quad Var(Y_1) = \sigma_{Y_1}^2 = \sigma_{X_1}^2 = \sigma_1^2$$

$$E(Y_2) = \mu_{Y_2} = \mu_2 = \frac{1}{2}\ln\left[\frac{(\mu_{X_2})^4}{(\mu_{X_2})^2 + \sigma_{X_2}^2}\right]$$

$$Var(Y_2) = \sigma_{Y_2}^2 = \sigma_2^2 = \ln\left[\frac{(\mu_{X_2})^2 + \sigma_{X_2}^2}{(\mu_{X_2})^2}\right]$$

The pair of random variables

$$(X_1, X_2) \sim Bivariate\ NLogN\left((\mu_1, \mu_2), \left(\sigma_1^2, \sigma_2^2, \rho_{1,2}\right)\right)$$

has a bivariate normal-lognormal distribution if

$$f_{X_1,X_2}(x_1, x_2) = \frac{1}{(2\pi)\sigma_1\sigma_2\sqrt{1 - \rho_{1,2}^2}\, x_2} e^{-\frac{1}{2}w}$$

where

$$-1 < \rho_{1,2} = \rho_{Y_1,Y_2} = \rho_{X_1,\ln X_2} < 1$$

and

$$w = \frac{1}{1 - \rho_{1,2}^2}\left\{\left(\frac{x_1 - \mu_1}{\sigma_1}\right)^2 - 2\rho_{1,2}\left(\frac{x_1 - \mu_1}{\sigma_1}\right)\left(\frac{\ln x_2 - \mu_2}{\sigma_2}\right) + \left(\frac{\ln x_2 - \mu_2}{\sigma_2}\right)^2\right\}$$

Theorem B.1   If $(X_1, X_2) \sim Bivariate\ NLogN((\mu_1, \mu_2), (\sigma_1^2, \sigma_2^2, \rho_{1,2}))$ then

$$\rho_{1,2} = \rho_{X_1,X_2}\frac{\left(e^{\sigma_2^2} - 1\right)^{1/2}}{\sigma_2}$$

Proof. *By definition*

$$\rho_{X_1,X_2} = \frac{Cov(X_1, X_2)}{\sigma_{X_1}\sigma_{X_2}} = \frac{\sigma_{X_1X_2}}{\sigma_{X_1}\sigma_{X_2}}$$

where

$$\sigma_{X_1X_2} = \int_0^\infty \int_{-\infty}^\infty (x_1 - \mu_1)(x_2 - \mu_2) f_{X_1,X_2}(x_1, x_2) dx_1 dx_2$$

and $\sigma_{X_1} = \sigma_1$. Since X_2 is lognormal

$$\sigma_{X_2} = \left(e^{2\mu_2 + \sigma_2^2}\left(e^{\sigma_2^2} - 1\right)\right)^{1/2} = E(X_2)(e^{\sigma_2^2} - 1)^{1/2}$$

Thus, $\rho_{X_1,X_2} = \dfrac{\sigma_{X_1X_2}}{\sigma_{X_1}\sigma_{X_2}} = \dfrac{\sigma_{X_1X_2}}{\sigma_1 E(X_2)\left(e^{\sigma_2^2} - 1\right)^{1/2}}$

To compute $\sigma_{X_1X_2}$, let $t_1 = \dfrac{x_1 - \mu_1}{\sigma_1}$ and $t_2 = \dfrac{\ln x_2 - \mu_2}{\sigma_2}$; therefore,

$$\sigma_{X_1X_2} = \frac{1}{2\pi\sqrt{1 - \rho_{1,2}^2}} \int_{-\infty}^\infty \int_{-\infty}^\infty (\sigma_1 t_1)(e^{\mu_2 + \sigma_2 t_2} - \mu_2) e^{-\frac{1}{2(1-\rho_{1,2}^2)}(t_1^2 - 2\rho_{1,2}t_1 t_2 + t_2^2)} dt_1 dt_2$$

$$= \frac{1}{2\pi\sqrt{1 - \rho_{1,2}^2}} \int_{-\infty}^\infty (\sigma_1 t_1)\left[I_1 - \mu_2 I_2\right] dt_1$$

where

$$I_1 = \int_{-\infty}^\infty e^{\mu_2 + \sigma_2 t_2} e^{-\frac{1}{2(1-\rho_{1,2}^2)}(t_1^2 - 2\rho_{1,2}t_1 t_2 + t_2^2)} dt_2$$

and

$$I_2 = \int_{-\infty}^\infty e^{-\frac{1}{2(1-\rho_{1,2}^2)}(t_1^2 - 2\rho_{1,2}t_1 t_2 + t_2^2)} dt_2$$

To determine I_1, note the integrand can be written as

$$I_1 = e^{\mu_2} \int_{-\infty}^\infty e^{-\frac{1}{2(1-\rho_{1,2}^2)}\left(t_1^2 - 2\left[\rho_{1,2}t_1 + (1-\rho_{1,2}^2)\sigma_2\right]t_2 + t_2^2\right)} dt_2$$

Letting

$$A = A(t_1) = \rho_{1,2}t_1 + (1 - \rho_{1,2}^2)\sigma_2$$

and noting that

$$t_2^2 - 2At_2 = (t_2 - A)^2 - A^2$$

we can write

$$I_1 = e^{\mu_2}e^{-\frac{1}{2(1-\rho_{1,2}^2)}t_1^2}e^{\frac{1}{2(1-\rho_{1,2}^2)}A^2}\int_{-\infty}^{\infty}e^{-\frac{1}{2(1-\rho_{1,2}^2)}(t_2-A)^2}dt_2$$

$$I_1 = e^{\mu_2}e^{-\frac{1}{2(1-\rho_{1,2}^2)}t_1^2}e^{\frac{1}{2(1-\rho_{1,2}^2)}A^2}\sqrt{2\pi}\sqrt{(1-\rho_{1,2}^2)}$$

To determine I_2, note the integrand can be written as

$$I_2 = e^{-\frac{1}{2(1-\rho_{1,2}^2)}t_1^2}\int_{-\infty}^{\infty}e^{-\frac{1}{2(1-\rho_{1,2}^2)}(t_2^2-2\rho_{1,2}t_1t_2)}dt_2$$

Letting $B = B(t_1) = \rho_{1,2}t_1$ and noting that $t_2^2 - 2Bt_2 = (t_2 - B)^2 - B^2$ we have

$$I_2 = e^{-\frac{1}{2(1-\rho_{1,2}^2)}t_1^2}e^{\frac{1}{2(1-\rho_{1,2}^2)}B^2}\int_{-\infty}^{\infty}e^{-\frac{1}{2(1-\rho_{1,2}^2)}(t_2-B)^2}dt_2$$

$$I_2 = e^{-\frac{1}{2(1-\rho_{1,2}^2)}t_1^2}e^{\frac{1}{2(1-\rho_{1,2}^2)}B^2}\sqrt{2\pi}\sqrt{(1-\rho_{1,2}^2)}$$

Thus,

$$I_1 - \mu_2 I_2 = e^{\frac{-t_1^2}{2(1-\rho_{1,2}^2)}}\sqrt{2\pi}\sqrt{(1-\rho_{1,2}^2)}\left[e^{\mu_2}e^{\frac{A^2}{2(1-\rho_{1,2}^2)}} - \mu_2 e^{\frac{B^2}{2(1-\rho_{1,2}^2)}}\right]$$

and

$$\sigma_{X_1 X_2} = \frac{1}{\sqrt{2\pi}} \int_{-\infty}^{\infty} (\sigma_1 t_1) e^{\frac{-t_1^2}{2(1-\rho_{1,2}^2)}} \left[e^{\mu_2} e^{\frac{A^2}{2(1-\rho_{1,2}^2)}} - \mu_2 e^{\frac{B^2}{2(1-\rho_{1,2}^2)}} \right] dt_1$$

$$\sigma_{X_1 X_2} = \frac{1}{\sqrt{2\pi}} \left[e^{\mu_2} \sigma_1 \int_{-\infty}^{\infty} t_1 e^{\frac{-(t_1^2 - A^2)}{2(1-\rho_{1,2}^2)}} dt_1 - \mu_2 \sigma_1 \int_{-\infty}^{\infty} t_1 e^{\frac{-(t_1^2 - B^2)}{2(1-\rho_{1,2}^2)}} dt_1 \right]$$

$$\sigma_{X_1 X_2} = \frac{1}{\sqrt{2\pi}} \left[e^{\mu_2} \sigma_1 \int_{-\infty}^{\infty} t_1 e^{-\frac{1}{2}(t_1 - \rho_{1,2}\sigma_2)^2 + \frac{1}{2}\sigma_2^2} dt_1 - \mu_2 \sigma_1 \int_{-\infty}^{\infty} t_1 e^{-t_1^2/2} dt_1 \right]$$

$$\sigma_{X_1 X_2} = \frac{1}{\sqrt{2\pi}} \left[e^{\mu_2} \sigma_1 e^{\frac{1}{2}\sigma_2^2} \int_{-\infty}^{\infty} t_1 e^{-\frac{1}{2}(t_1 - \rho_{1,2}\sigma_2)^2} dt_1 - \mu_2 \sigma_1 \cdot 0 \right]$$

$$\sigma_{X_1 X_2} = \frac{1}{\sqrt{2\pi}} \left[e^{\mu_2 + \frac{\sigma_2^2}{2}} \sigma_1 \rho_{1,2} \sigma_2 \sqrt{2\pi} \right] = E(X_2) \rho_{1,2} \sigma_1 \sigma_2$$

Hence,

$$\rho_{X_1, X_2} = \frac{\sigma_{X_1 X_2}}{\sigma_{X_1} \sigma_{X_2}} = \frac{E(X_2) \rho_{1,2} \sigma_1 \sigma_2}{\sigma_1 \left[e^{2\mu_2 + \sigma_2^2}(e^{\sigma_2^2} - 1) \right]^{1/2}} = \frac{E(X_2) \rho_{1,2} \sigma_1 \sigma_2}{\sigma_1 \left[E(X_2)(e^{\sigma_2^2} - 1)^{1/2} \right]}$$

Thus,

$$\rho_{1,2} = \rho_{X_1, X_2} \frac{(e^{\sigma_2^2} - 1)^{1/2}}{\sigma_2} \tag{B.1}$$

Theorem B.2　　*If $(X_1, X_2) \sim$ Bivariate $NLogN\left((\mu_1, \mu_2), (\sigma_1^2, \sigma_2^2, \rho_{1,2})\right)$, then*

$$f_1(x_1) = \frac{1}{\sqrt{2\pi}\,\sigma_1} e^{-\frac{1}{2}[(x_1 - \mu_1)^2/\sigma_1^2]}$$

and

$$f_2(x_2) = \frac{1}{\sqrt{2\pi}\,\sigma_2 x_2} e^{-\frac{1}{2}[(\ln x_2 - \mu_2)^2/\sigma_2^2]}$$

Proof. By definition

$$f_1(x_1) = \int_0^\infty f_{X_1,X_2}(x_1, x_2)\, dx_2$$

$$f_2(x_2) = \int_{-\infty}^\infty f_{X_1,X_2}(x_1, x_2)\, dx_1$$

The density function $f_{X_1,X_2}(x_1, x_2)$ can be factored as

$$f_{X_1,X_2}(x_1, x_2) = \left\{ \frac{1}{\sqrt{2\pi}\sigma_1} e^{-(x_1-\mu_1)^2/2\sigma_1^2} \right\} Q(x_1, x_2) \tag{B.2}$$

where

$$Q(x_1, x_2) = \left\{ \frac{1}{\sqrt{2\pi}(\sigma_2\sqrt{1-\rho_{1,2}^2})x_2} e^{-(\ln x_2 - b)^2/2\sigma_2^2(1-\rho_{1,2}^2)} \right\}$$

and

$$b = \mu_2 + \frac{\sigma_2}{\sigma_1}\rho_{1,2}(x_1 - \mu_1)$$

Therefore,

$$f_1(x_1) = \int_0^\infty \left\{ \frac{1}{\sqrt{2\pi}\sigma_1} e^{-(x_1-\mu_1)^2/2\sigma_1^2} \right\} Q(x_1, x_2)\, dx_2$$

$$= \left\{ \frac{1}{\sqrt{2\pi}\sigma_1} e^{-(x_1-\mu_1)^2/2\sigma_1^2} \right\} \int_0^\infty Q(x_1, x_2)\, dx_2$$

$$= \frac{1}{\sqrt{2\pi}\,\sigma_1} e^{-\frac{1}{2}[(x_1-\mu_1)^2/\sigma_1^2]}$$

since the integrand is the density function of a $LogN\left(b, \sigma_2^2\left(1 - \rho_{1,2}^2\right)\right)$ random variable. To compute $f_2(x_2)$, the density function $f_{X_1,X_2}(x_1, x_2)$ is factored as

$$f_{X_1,X_2}(x_1, x_2) = Q^*(x_1, x_2) \left\{ \frac{1}{\sqrt{2\pi}\sigma_2} \cdot \frac{1}{x_2} e^{-(\ln x_2 - \mu_2)^2/2\sigma_2^2} \right\} \tag{B.3}$$

where

$$Q^*(x_1, x_2) = \left\{ \frac{1}{\sqrt{2\pi}(\sigma_1\sqrt{1 - \rho_{1,2}^2})} e^{-(x_1 - b^*)^2/2\sigma_1^2(1 - \rho_{1,2}^2)} \right\}$$

and

$$b^* = \mu_1 + \frac{\sigma_1}{\sigma_2}\rho_{1,2}(\ln x_2 - \mu_2)$$

Therefore,

$$f_2(x_2) = \int_{-\infty}^{\infty} \left\{ \frac{1}{\sqrt{2\pi}\sigma_2} \cdot \frac{1}{x_2} e^{-(\ln x_2 - \mu_2)^2/2\sigma_2^2} \right\} Q^*(x_1, x_2)\, dx_1$$

$$= \left\{ \frac{1}{\sqrt{2\pi}\sigma_2} \cdot \frac{1}{x_2} e^{-(\ln x_2 - \mu_2)^2/2\sigma_2^2} \right\} \int_{-\infty}^{\infty} Q^*(x_1, x_2)\, dx_1$$

$$= \frac{1}{\sqrt{2\pi}\,\sigma_2\, x_2} e^{-\frac{1}{2}[(\ln x_2 - \mu_2)^2/\sigma_2^2]}$$

since the integrand is the density function of a $N\left(b^*, \sigma_1^2\left(1 - \rho_{1,2}^2\right)\right)$ *random variable.*

Theorem B.3   *If* $(X_1, X_2) \sim$ *Bivariate* $NLogN((\mu_1, \mu_2), (\sigma_1^2, \sigma_2^2, \rho_{1,2}))$, *then*

$$X_1\,|x_2 \sim N\left(\mu_1 + \frac{\sigma_1}{\sigma_2}\rho_{1,2}(\ln x_2 - \mu_2)\,,\; \sigma_1^2\left(1 - \rho_{1,2}^2\right)\right)$$

$$X_2\,|x_1 \sim LogN\left(\mu_2 + \frac{\sigma_2}{\sigma_1}\rho_{1,2}(x_1 - \mu_1)\,,\; \sigma_2^2\left(1 - \rho_{1,2}^2\right)\right)$$

Proof. By definition,

$$f_{X_1|x_2}(x_1) = \frac{f_{X_1,X_2}(x_1, x_2)}{f_2(x_2)} = \frac{\left\{ \frac{1}{\sqrt{2\pi}\,\sigma_2 x_2} e^{-\frac{1}{2}[(\ln x_2 - \mu_2)^2/\sigma_2^2]} \right\} Q^*(x_1, x_2)}{\frac{1}{\sqrt{2\pi}\,\sigma_2 x_2} e^{-\frac{1}{2}[(\ln x_2 - \mu_2)^2/\sigma_2^2]}}$$

$$f_{X_1|x_2}(x_1) = Q^*(x_1, x_2)$$

Thus, from Equation B.3

$$X_1 \mid x_2 \sim N\left(b^*, \sigma_1^2\left(1 - \rho_{1,2}^2\right)\right)$$

where

$$b^* = \mu_1 + \frac{\sigma_1}{\sigma_2}\rho_{1,2}(\ln x_2 - \mu_2)$$

Similarly,

$$f_{X_2|x_1}(x_2) = \frac{f_{X_1,X_2}(x_1,x_2)}{f_1(x_1)} = \frac{\left\{\frac{1}{\sqrt{2\pi}\,\sigma_1}e^{-\frac{1}{2}[(x_1-\mu_1)^2/\sigma_1^2]}\right\}Q(x_1,x_2)}{\frac{1}{\sqrt{2\pi}\,\sigma_1}e^{-\frac{1}{2}[(x_1-\mu_1)^2/\sigma_1^2]}}$$

$$f_{X_2|x_1}(x_2) = Q(x_1,x_2)$$

Thus, from Equation B.2

$$X_2 \mid x_1 \sim LogN\left(b, \sigma_2^2\left(1 - \rho_{1,2}^2\right)\right)$$

where

$$b = \mu_2 + \frac{\sigma_2}{\sigma_1}\rho_{1,2}(x_1 - \mu_1)$$

Theorem B.4 *If $(X_1, X_2) \sim$ Bivariate $NLogN\left((\mu_1, \mu_2), (\sigma_1^2, \sigma_2^2, \rho_{1,2})\right)$, then*

$$E(X_2 \mid x_1) = e^{\mu_2 + \frac{\sigma_2}{\sigma_1}\rho_{1,2}(x_1-\mu_1) + \frac{1}{2}\sigma_2^2(1-\rho_{1,2}^2)}$$

$$Var(X_2 \mid x_1) = e^{2(\mu_2 + \frac{\sigma_2}{\sigma_1}\rho_{1,2}(x_1-\mu_1))}e^z(e^z - 1)$$

$$E(X_1 \mid x_2) = \mu_1 + \frac{\sigma_1}{\sigma_2}\rho_{1,2}(\ln x_2 - \mu_2)$$

$$Var(X_1 \mid x_2) = \sigma_1^2\left(1 - \rho_{1,2}^2\right)$$

where $z = \sigma_2^2\left(1 - \rho_{1,2}^2\right)$.

Proof. Theorem B.3 proved that

$$X_2 \mid x_1 \sim LogN\left(\mu_2 + \frac{\sigma_2}{\sigma_1}\rho_{1,2}(x_1 - \mu_1), \sigma_2^2\left(1 - \rho_{1,2}^2\right)\right)$$

Therefore,

$$E(X_2 \mid x_1) = e^{\mu_2 + \frac{\sigma_2}{\sigma_1}\rho_{1,2}(x_1 - \mu_1) + \frac{1}{2}\sigma_2^2(1 - \rho_{1,2}^2)}$$

$$Var(X_2 \mid x_1) = e^{2(\mu_2 + \frac{\sigma_2}{\sigma_1}\rho_{1,2}(x_1 - \mu_1))} e^z (e^z - 1)$$

where $z = \sigma_2^2 \left(1 - \rho_{1,2}^2\right)$. *Theorem B.3 also proved that*

$$X_1 \mid x_2 \sim N\left(\mu_1 + \frac{\sigma_1}{\sigma_2}\rho_{1,2}(\ln x_2 - \mu_2)\,,\ \sigma_1^2\left(1 - \rho_{1,2}^2\right)\right)$$

Therefore, it follows immediately from the properties of the normal distribution that

$$E(X_1 \mid x_2) = \mu_1 + \frac{\sigma_1}{\sigma_2}\rho_{1,2}(\ln x_2 - \mu_2)$$

$$Var(X_1 \mid x_2) = \sigma_1^2\left(1 - \rho_{1,2}^2\right)$$

Theorem B.5   *If* $(X_1, X_2) \sim$ *Bivariate NLogN* $\left((\mu_1, \mu_2), (\sigma_1^2, \sigma_2^2, \rho_{1,2})\right)$, *then*

$$Median(X_2 \mid x_1) = e^{\mu_2 + \frac{\sigma_2}{\sigma_1}\rho_{1,2}(x_1 - \mu_1)}$$

$$Mode(X_2 \mid x_1) = e^{\mu_2 + \frac{\sigma_2}{\sigma_1}\rho_{1,2}(x_1 - \mu_1) - \sigma_2^2(1 - \rho_{1,2}^2)}$$

$$Median(X_1 \mid x_2) = E(X_1 \mid x_2)$$

$$Mode(X_1 \mid x_2) = E(X_1 \mid x_2)$$

Proof.   Since $X_2 \mid x_1$ *is lognormally distributed,*

$$Median(X_2 \mid x_1) = e^{\mu_2 + \frac{\sigma_2}{\sigma_1}\rho_{1,2}(x_1 - \mu_1)}$$

and

$$Mode(X_2 \mid x_1) = e^{\mu_2 + \frac{\sigma_2}{\sigma_1}\rho_{1,2}(x_1 - \mu_1) - \sigma_2^2(1 - \rho_{1,2}^2)}$$

Since $X_1 \mid x_2$ *is normally distributed, it follows immediately that*

$$Median(X_1 \mid x_2) = E(X_1 \mid x_2)$$

$$Mode(X_1 \mid x_2) = E(X_1 \mid x_2)$$

Property B.1   *If $(X_1, X_2) \sim$ Bivariate $NLogN\big((\mu_1, \mu_2), (\sigma_1^2, \sigma_2^2, \rho_{1,2})\big)$, then*

$$E(X_1 \,|Median(X_2 \,|\mu_1)) = \mu_1$$

Proof.   From Theorem B.4, it was established that

$$E(X_1 \,|x_2) = \mu_1 + \frac{\sigma_1}{\sigma_2}\rho_{1,2}\,(\ln x_2 - \mu_2)$$

From Theorem B.5,

$$Median(X_2 \,|x_1 = \mu_1) = e^{\mu_2}$$

It follows that

$$E(X_1 \,|Median(X_2 \,|\mu_1)) = E(X_1 \,\big|e^{\mu_2}) = \mu_1 + \frac{\sigma_1}{\sigma_2}\rho_{1,2}(\ln e^{\mu_2} - \mu_2)$$

$$= \mu_1 + \frac{\sigma_1}{\sigma_2}\rho_{1,2}(\mu_2 - \mu_2)$$

$$= \mu_1$$

Appendix C: Bivariate Lognormal Distribution

Let $Y_1 = \ln X_1$ and $Y_2 = \ln X_2$, where X_1 and X_2 are random variables defined on $0 < x_1 < \infty$ and $0 < x_2 < \infty$. If Y_1 and Y_2 each have a normal distribution, then

$$E(Y_1) = \mu_{Y_1} = \mu_1 = \frac{1}{2}\ln\left[\frac{(\mu_{X_1})^4}{(\mu_{X_1})^2 + \sigma_{X_1}^2}\right]$$

$$Var(Y_1) = \sigma_{Y_1}^2 = \sigma_1^2 = \ln\left[\frac{(\mu_{X_1})^2 + \sigma_{X_1}^2}{(\mu_{X_1})^2}\right]$$

$$E(Y_2) = \mu_{Y_2} = \mu_2 = \frac{1}{2}\ln\left[\frac{(\mu_{X_2})^4}{(\mu_{X_2})^2 + \sigma_{X_2}^2}\right]$$

$$Var(Y_2) = \sigma_{Y_2}^2 = \sigma_2^2 = \ln\left[\frac{(\mu_{X_2})^2 + \sigma_{X_2}^2}{(\mu_{X_2})^2}\right]$$

The pair of random variables

$$(X_1, X_2) \sim Bivariate\ LogN\left((\mu_1, \mu_2), \left(\sigma_1^2, \sigma_2^2, \rho_{1,2}\right)\right)$$

has a bivariate lognormal distribution if

$$f_{X_1, X_2}(x_1, x_2) = \frac{1}{(2\pi)\sigma_1\sigma_2\sqrt{1 - \rho_{1,2}^2}\, x_1 x_2}e^{-\frac{1}{2}w}$$

where

$$-1 < \rho_{1,2} = \rho_{Y_1, Y_2} = \rho_{\ln X_1, \ln X_2} < 1$$

and

$$w = \frac{1}{1 - \rho_{1,2}^2}\left\{\left(\frac{\ln x_1 - \mu_1}{\sigma_1}\right)^2 - 2\rho_{1,2}\left(\frac{\ln x_1 - \mu_1}{\sigma_1}\right)\left(\frac{\ln x_2 - \mu_2}{\sigma_2}\right) + \left(\frac{\ln x_2 - \mu_2}{\sigma_2}\right)^2\right\}$$

Theorem C.1   *If* $(X_1, X_2) \sim$ *Bivariate* $LogN((\mu_1, \mu_2), (\sigma_1^2, \sigma_2^2, \rho_{1,2}))$ *then*

$$\rho_{X_1, X_2} = \frac{e^{\rho_{1,2}\sigma_1\sigma_2} - 1}{\sqrt{e^{\sigma_1^2} - 1}\sqrt{e^{\sigma_2^2} - 1}}$$

Proof. By definition,

$$\rho_{X_1, X_2} = \frac{Cov(X_1, X_2)}{\sigma_{X_1}\sigma_{X_2}} = \frac{E(X_1 X_2) - E(X_1)E(X_2)}{\sigma_{X_1}\sigma_{X_2}} \tag{C.1}$$

Since $Y_1 = \ln X_1$ *and* $Y_2 = \ln X_2$,

$$E(X_1 X_2) = E(e^{Y_1} e^{Y_2}) = E(e^{Y_1 + Y_2})$$

Since $Y_i \sim N\left(\mu_i, \sigma_i^2\right)$ *(for* $i = 1, 2$*), the expectation* $E(e^{Y_1 + Y_2})$ *is a special evaluation of the moment generating function (Ross 1994) of a bivariate normal, which is*

$$M(t_1, t_2) = E(e^{t_1 Y_1 + t_2 Y_2}) = \int_{-\infty}^{\infty} \int_{-\infty}^{\infty} e^{t_1 y_1 + t_2 y_2} f(y_1, y_2) dy_1 dy_2$$

$$= e^{(\mu_1 t_1 + \mu_2 t_2) + \frac{1}{2}\left(\sigma_1^2 t_1^2 + 2\rho_{Y_1, Y_2}\sigma_1\sigma_2 t_1 t_2 + \sigma_2^2 t_2^2\right)}$$

for some real t_1 *and* t_2*. Therefore,*

$$E(X_1 X_2) = E(e^{Y_1} e^{Y_2}) = E(e^{Y_1 + Y_2}) = e^{(\mu_1 + \mu_2) + \frac{1}{2}\left(\sigma_1^2 + \sigma_2^2 + 2\rho_{Y_1, Y_2}\sigma_1\sigma_2\right)}$$

To determine the remaining terms in Equation C.1, for $r \geq 0$ *the moments of* X_1 *and* X_2 *are*

$$E(X_i^r) = e^{r\mu_i + \frac{1}{2}r^2\sigma_i^2} \tag{C.2}$$

Thus,

$$E(X_1) = e^{\mu_1 + \frac{1}{2}\sigma_1^2}$$

$$E(X_2) = e^{\mu_2 + \frac{1}{2}\sigma_2^2}$$

and

$$\sigma_{X_1}^2 = Var(X_1) = E(X_1^2) - (E(X_1))^2 = e^{2\mu_1 + 2\sigma_1^2} - \left(e^{\mu_1 + \frac{1}{2}\sigma_1^2}\right)^2$$

$$= e^{2\mu_1 + 2\sigma_1^2} - e^{2\mu_1 + \sigma_1^2}$$

$$\sigma_{X_2}^2 = Var(X_2) = E(X_2^2) - (E(X_2))^2 = e^{2\mu_2 + 2\sigma_2^2} - \left(e^{\mu_2 + \frac{1}{2}\sigma_2^2}\right)^2$$

$$= e^{2\mu_2 + 2\sigma_2^2} - e^{2\mu_2 + \sigma_2^2}$$

Substituting into Equation C.1,

$$\rho_{X_1,X_2} = \frac{E(X_1 X_2) - E(X_1)E(X_2)}{\sigma_{X_1}\sigma_{X_2}}$$

$$\rho_{X_1,X_2} = \frac{e^{(\mu_1 + \mu_2) + \frac{1}{2}\left(\sigma_1^2 + 2\rho_{Y_1,Y_2}\sigma_1\sigma_2 + \sigma_2^2\right)} - \left(e^{\mu_1 + \frac{1}{2}\sigma_1^2}\right)\left(e^{\mu_2 + \frac{1}{2}\sigma_2^2}\right)}{\sqrt{e^{2\mu_1 + 2\sigma_1^2} - e^{2\mu_1 + \sigma_1^2}}\sqrt{e^{2\mu_2 + 2\sigma_2^2} - e^{2\mu_2 + \sigma_2^2}}}$$

This can be factored as

$$\rho_{X_1,X_2} = \frac{e^{(\mu_1 + \mu_2) + \frac{1}{2}\left(\sigma_1^2 + \sigma_2^2\right)}(e^{\rho_{1,2}\sigma_1\sigma_2} - 1)}{e^{(\mu_1 + \mu_2) + \frac{1}{2}(\sigma_1^2 + \sigma_2^2)}\sqrt{e^{\sigma_1^2} - 1}\sqrt{e^{\sigma_2^2} - 1}}$$

Thus,

$$\rho_{X_1,X_2} = \frac{e^{\rho_{1,2}\sigma_1\sigma_2} - 1}{\sqrt{e^{\sigma_1^2} - 1}\sqrt{e^{\sigma_2^2} - 1}} \tag{C.3}$$

Theorem C.2 *If $(X_1, X_2) \sim$ Bivariate $LogN\left((\mu_1, \mu_2), (\sigma_1^2, \sigma_2^2, \rho_{1,2})\right)$, then*

$$f_1(x_1) = \frac{1}{\sqrt{2\pi}\,\sigma_1 x_1}e^{-\frac{1}{2}\left[(\ln x_1 - \mu_1)^2/\sigma_1^2\right]}$$

and

$$f_2(x_2) = \frac{1}{\sqrt{2\pi}\,\sigma_2 x_2}e^{-\frac{1}{2}\left[(\ln x_2 - \mu_2)^2/\sigma_2^2\right]}$$

Proof. *By definition,*

$$f_1(x_1) = \int_0^\infty f_{X_1,X_2}(x_1,x_2)\,dx_2$$

$$f_2(x_2) = \int_0^\infty f_{X_1,X_2}(x_1,x_2)\,dx_1$$

The density function $f_{X_1,X_2}(x_1,x_2)$ can be factored as

$$f_{X_1,X_2}(x_1,x_2) = \left\{\frac{1}{\sqrt{2\pi}\sigma_1}\cdot\frac{1}{x_1}e^{-(\ln x_1-\mu_1)^2/2\sigma_1^2}\right\}Q(x_1,x_2) \qquad\text{(C.4)}$$

where

$$Q(x_1,x_2) = \left\{\frac{1}{\sqrt{2\pi}(\sigma_2\sqrt{1-\rho_{1,2}^2})x_2}e^{-(\ln x_2-b)^2/2\sigma_2^2\left(1-\rho_{1,2}^2\right)}\right\}$$

and

$$b = \mu_2 + \frac{\sigma_2}{\sigma_1}\rho_{1,2}(\ln x_1 - \mu_1)$$

Therefore,

$$f_1(x_1) = \int_0^\infty \left\{\frac{1}{\sqrt{2\pi}\sigma_1}\cdot\frac{1}{x_1}e^{-(\ln x_1-\mu_1)^2/2\sigma_1^2}\right\}Q(x_1,x_2)\,dx_2$$

$$= \left\{\frac{1}{\sqrt{2\pi}\sigma_1}\cdot\frac{1}{x_1}e^{-(\ln x_1-\mu_1)^2/2\sigma_1^2}\right\}\int_0^\infty Q(x_1,x_2)\,dx_2$$

$$= \frac{1}{\sqrt{2\pi}\,\sigma_1 x_1}e^{-\frac{1}{2}[(\ln x_1-\mu_1)^2/\sigma_1^2]}$$

since the integrand is the density function of a $LogN\left(b,\sigma_2^2\left(1-\rho_{1,2}^2\right)\right)$ random variable. To compute $f_2(x_2)$, the density function $f_{X_1,X_2}(x_1,x_2)$ is factored as

$$f_{X_1,X_2}(x_1,x_2) = Q^*(x_1,x_2)\left\{\frac{1}{\sqrt{2\pi}\sigma_2}\cdot\frac{1}{x_2}e^{-(\ln x_2-\mu_2)^2/2\sigma_2^2}\right\} \qquad\text{(C.5)}$$

where

$$Q^*(x_1, x_2) = \left\{ \frac{1}{\sqrt{2\pi}(\sigma_1\sqrt{1-\rho_{1,2}^2})x_1} e^{-(\ln x_1 - b^*)^2/2\sigma_1^2(1-\rho_{1,2}^2)} \right\}$$

and

$$b^* = \mu_1 + \frac{\sigma_1}{\sigma_2}\rho_{1,2}(\ln x_2 - \mu_2)$$

Therefore,

$$f_2(x_2) = \int_0^\infty \left\{ \frac{1}{\sqrt{2\pi}\sigma_2} \cdot \frac{1}{x_2} e^{-(\ln x_2 - \mu_2)^2/2\sigma_2^2} \right\} Q^*(x_1, x_2)\, dx_1$$

$$= \left\{ \frac{1}{\sqrt{2\pi}\sigma_2} \cdot \frac{1}{x_2} e^{-(\ln x_2 - \mu_2)^2/2\sigma_2^2} \right\} \int_0^\infty Q^*(x_1, x_2)\, dx_1$$

$$= \frac{1}{\sqrt{2\pi}\sigma_2} \cdot \frac{1}{x_2} e^{-(\ln x_2 - \mu_2)^2/2\sigma_2^2}$$

since the integrand is the density function of a $LogN\left(b^*, \sigma_1^2\left(1-\rho_{1,2}^2\right)\right)$ *random variable.*

Theorem C.3 *If* $(X_1, X_2) \sim$ *Bivariate* $LogN((\mu_1, \mu_2), (\sigma_1^2, \sigma_2^2, \rho_{1,2}))$, *then*

$$X_1 \mid x_2 \sim LogN\left(\mu_1 + \frac{\sigma_1}{\sigma_2}\rho_{1,2}(\ln x_2 - \mu_2),\ \sigma_1^2\left(1-\rho_{1,2}^2\right)\right)$$

$$X_2 \mid x_1 \sim LogN\left(\mu_2 + \frac{\sigma_2}{\sigma_1}\rho_{1,2}(\ln x_1 - \mu_1),\ \sigma_2^2\left(1-\rho_{1,2}^2\right)\right)$$

Proof. By definition,

$$f_{X_1|x_2}(x_1) = \frac{f_{X_1,X_2}(x_1, x_2)}{f_2(x_2)} = \frac{\left\{ \frac{1}{\sqrt{2\pi}\,\sigma_2 x_2} e^{-\frac{1}{2}\left[(\ln x_2 - \mu_2)^2/\sigma_2^2\right]} \right\} Q^*(x_1, x_2)}{\frac{1}{\sqrt{2\pi}\,\sigma_2 x_2} e^{-\frac{1}{2}\left[(\ln x_2 - \mu_2)^2/\sigma_2^2\right]}}$$

$$f_{X_1|x_2}(x_1) = Q^*(x_1, x_2)$$

Thus, from Equation C.5,

$$X_1 \mid x_2 \sim LogN\left(b^*, \sigma_1^2\left(1 - \rho_{1,2}^2\right)\right)$$

where

$$b^* = \mu_1 + \frac{\sigma_1}{\sigma_2}\rho_{1,2}(\ln x_2 - \mu_2)$$

Similarly,

$$f_{X_2|x_1}(x_2) = \frac{f_{X_1,X_2}(x_1,x_2)}{f_1(x_1)} = \frac{\left\{\frac{1}{\sqrt{2\pi}\,\sigma_1 x_1}e^{-\frac{1}{2}\left[(\ln x_1-\mu_1)^2/\sigma_1^2\right]}\right\}Q(x_1,x_2)}{\frac{1}{\sqrt{2\pi}\,\sigma_1 x_1}e^{-\frac{1}{2}\left[(\ln x_1-\mu_1)^2/\sigma_1^2\right]}}$$

$$f_{X_2|x_1}(x_2) = Q(x_1,x_2)$$

Thus, from Equation C.4,

$$X_2 \mid x_1 \sim LogN\left(b, \sigma_2^2\left(1 - \rho_{1,2}^2\right)\right)$$

where

$$b = \mu_2 + \frac{\sigma_2}{\sigma_1}\rho_{1,2}(\ln x_1 - \mu_1)$$

Theorem C.4   *If* $(X_1, X_2) \sim$ *Bivariate* $LogN((\mu_1, \mu_2), (\sigma_1^2, \sigma_2^2, \rho_{1,2}))$, *then*

$$E(X_2 \mid x_1) = x_1^{\frac{\sigma_2}{\sigma_1}\rho_{1,2}}e^{\mu_2 - \frac{\sigma_2}{\sigma_1}\rho_{1,2}\mu_1 + \frac{1}{2}\sigma_2^2\left(1-\rho_{1,2}^2\right)}$$

$$Var(X_2 \mid x_1) = x_1^{2\frac{\sigma_2}{\sigma_1}\rho_{1,2}}e^{2(\mu_2 - \frac{\sigma_2}{\sigma_1}\rho_{1,2}\mu_1)}e^z(e^z - 1)$$

$$E(X_1 \mid x_2) = x_2^{\frac{\sigma_1}{\sigma_2}\rho_{1,2}}e^{\mu_1 - \frac{\sigma_1}{\sigma_2}\rho_{1,2}\mu_2 + \frac{1}{2}\sigma_1^2\left(1-\rho_{1,2}^2\right)}$$

$$Var(X_1 \mid x_2) = x_2^{2\frac{\sigma_1}{\sigma_2}\rho_{1,2}}e^{2(\mu_1 - \frac{\sigma_1}{\sigma_2}\rho_{1,2}\mu_2)}e^{z^*}(e^{z^*} - 1)$$

where

$$z = \sigma_2^2\left(1 - \rho_{1,2}^2\right) \quad and \quad z^* = \sigma_1^2\left(1 - \rho_{1,2}^2\right)$$

Proof. *Theorem C.3 proved that*

$$X_2 \mid x_1 \sim LogN\left(\mu_2 + \frac{\sigma_2}{\sigma_1}\rho_{1,2}(\ln x_1 - \mu_1), \ \sigma_2^2\left(1 - \rho_{1,2}^2\right)\right)$$

Therefore,

$$E(X_2 \mid x_1) = e^{\mu_2 + \frac{\sigma_2}{\sigma_1}\rho_{1,2}(\ln x_1 - \mu_1) + \frac{1}{2}\sigma_2^2\left(1 - \rho_{1,2}^2\right)}$$

$$= x_1^{\frac{\sigma_2}{\sigma_1}\rho_{1,2}} e^{\mu_2 - \frac{\sigma_2}{\sigma_1}\rho_{1,2}\mu_1 + \frac{1}{2}\sigma_2^2\left(1 - \rho_{1,2}^2\right)}$$

and

$$Var(X_2 \mid x_1) = e^{2\left(\mu_2 + \frac{\sigma_2}{\sigma_1}\rho_{1,2}(\ln x_1 - \mu_1)\right)} e^{\sigma_2^2\left(1 - \rho_{1,2}^2\right)}\left(e^{\sigma_2^2\left(1 - \rho_{1,2}^2\right)} - 1\right)$$

$$= x_1^{2\frac{\sigma_2}{\sigma_1}\rho_{1,2}} e^{2\left(\mu_2 - \frac{\sigma_2}{\sigma_1}\rho_{1,2}\mu_1\right)} e^z (e^z - 1)$$

Theorem C.3 also proved that

$$X_1 \mid x_2 \sim LogN\left(\mu_1 + \frac{\sigma_1}{\sigma_2}\rho_{1,2}(\ln x_2 - \mu_2), \ \sigma_1^2\left(1 - \rho_{1,2}^2\right)\right)$$

Therefore,

$$E(X_1 \mid x_2) = e^{\mu_1 + \frac{\sigma_1}{\sigma_2}\rho_{1,2}(\ln x_2 - \mu_2) + \frac{1}{2}\sigma_1^2\left(1 - \rho_{1,2}^2\right)}$$

$$= x_2^{\frac{\sigma_1}{\sigma_2}\rho_{1,2}} e^{\mu_1 - \frac{\sigma_1}{\sigma_2}\rho_{1,2}\mu_2 + \frac{1}{2}\sigma_1^2\left(1 - \rho_{1,2}^2\right)}$$

and

$$Var(X_1 \mid x_2) = e^{2\left(\mu_1 + \frac{\sigma_1}{\sigma_2}\rho_{1,2}(\ln x_2 - \mu_2)\right)} e^{\sigma_1^2\left(1 - \rho_{1,2}^2\right)}\left(e^{\sigma_1^2\left(1 - \rho_{1,2}^2\right)} - 1\right)$$

$$= x_2^{2\frac{\sigma_1}{\sigma_2}\rho_{1,2}} e^{2\left(\mu_1 - \frac{\sigma_1}{\sigma_2}\rho_{1,2}\mu_2\right)} e^{z^*} (e^{z^*} - 1)$$

Theorem C.5 *If* $(X_1, X_2) \sim$ *Bivariate* $LogN\big((\mu_1, \mu_2), (\sigma_1^2, \sigma_2^2, \rho_{1,2})\big)$, *then*

$$Median(X_2 \mid x_1) = x_1^{\frac{\sigma_2}{\sigma_1}\rho_{1,2}} e^{\mu_2 - \frac{\sigma_2}{\sigma_1}\rho_{1,2}\mu_1}$$

$$Mode(X_2 \mid x_1) = x_1^{\frac{\sigma_2}{\sigma_1}\rho_{1,2}} e^{\mu_2 - \frac{\sigma_2}{\sigma_1}\rho_{1,2}\mu_1 - \sigma_2^2\left(1-\rho_{1,2}^2\right)}$$

$$Median(X_1 \mid x_2) = x_2^{\frac{\sigma_1}{\sigma_2}\rho_{1,2}} e^{\mu_1 - \frac{\sigma_1}{\sigma_2}\rho_{1,2}\mu_2}$$

$$Mode(X_1 \mid x_2) = x_2^{\frac{\sigma_1}{\sigma_2}\rho_{1,2}} e^{\mu_1 - \frac{\sigma_1}{\sigma_2}\rho_{1,2}\mu_2 - \sigma_1^2\left(1-\rho_{1,2}^2\right)}$$

Proof. *From Theorem C.3, it follows that*

$$Median(X_2 \mid x_1) = e^{\mu_2 + \frac{\sigma_2}{\sigma_1}\rho_{1,2}(\ln x_1 - \mu_1)} = x_1^{\frac{\sigma_2}{\sigma_1}\rho_{1,2}} e^{\mu_2 - \frac{\sigma_2}{\sigma_1}\rho_{1,2}\mu_1}$$

$$Mode(X_2 \mid x_1) = e^{\mu_2 + \frac{\sigma_2}{\sigma_1}\rho_{1,2}(\ln x_1 - \mu_1) - \sigma_2^2\left(1-\rho_{1,2}^2\right)}$$

$$= x_1^{\frac{\sigma_2}{\sigma_1}\rho_{1,2}} e^{\mu_2 - \frac{\sigma_2}{\sigma_1}\rho_{1,2}\mu_1 - \sigma_2^2\left(1-\rho_{1,2}^2\right)}$$

$$Median(X_1 \mid x_2) = e^{\mu_1 + \frac{\sigma_1}{\sigma_2}\rho_{1,2}(\ln x_2 - \mu_2)} = x_2^{\frac{\sigma_1}{\sigma_2}\rho_{1,2}} e^{\mu_1 - \frac{\sigma_1}{\sigma_2}\rho_{1,2}\mu_2}$$

$$Mode(X_1 \mid x_2) = e^{\mu_1 + \frac{\sigma_1}{\sigma_2}\rho_{1,2}(\ln x_2 - \mu_2) - \sigma_1^2\left(1-\rho_{1,2}^2\right)}$$

$$= x_2^{\frac{\sigma_1}{\sigma_2}\rho_{1,2}} e^{\mu_1 - \frac{\sigma_1}{\sigma_2}\rho_{1,2}\mu_2 - \sigma_1^2\left(1-\rho_{1,2}^2\right)}$$

Property C.1 *If* $(X_1, X_2) \sim$ *Bivariate* $LogN\big((\mu_1, \mu_2), (\sigma_1^2, \sigma_2^2, \rho_{1,2})\big)$, *then the conditional coefficients of dispersion are*

$$D_{F_{X_1 \mid x_2}} = \frac{[Var(X_1 \mid x_2)]^{1/2}}{E(X_1 \mid x_2)} = \sqrt{(e^{z^*} - 1)}$$

$$D_{F_{X_2 \mid x_1}} = \frac{[Var(X_2 \mid x_1)]^{1/2}}{E(X_2 \mid x_1)} = \sqrt{(e^{z} - 1)}$$

where $F_{X_1|x_2}$ and $F_{X_2|x_1}$ are the cumulative distributions of $f_{X_1|x_2}$ and $f_{X_2|x_1}$, where $z = \sigma_2^2 \left(1 - \rho_{1,2}^2 \right)$ and $z^* = \sigma_1^2 \left(1 - \rho_{1,2}^2 \right)$. This property is stated without proof. It is a direct algebraic consequence of Theorem C.4.

Reference

Ross, S. 1994. *A First Course in Probability*, 4th edn. New York: Macmillan College Publishing Company.

Appendix D: Method of Moments WBS Example

Suppose the cost of an electronic system is represented by the work breakdown structure (WBS) in Table D.1. This WBS consists of element costs $X_1, X_2, X_3, \ldots, X_{10}$. Suppose

$$Cost_{WBS} = X_1 + X_2 + X_3 + \cdots + X_{10} \tag{D.1}$$

and the random variables $X_1, W, X_5, X_7, X_8, X_9$ in Table D.1 are independent—they are uncorrelated. Use the method of moments to determine the following:

a. $E(Cost_{WBS})$ and $Var(Cost_{WBS})$
b. A probability distribution function that approximates the distribution of $Cost_{WBS}$
c. The value of $Cost_{WBS}$ that has a 95% chance of not being exceeded

Solution

a. Given Equation D.1 and the relationships in Table D.1, Equation D.1 can be written as

$$Cost_{WBS} = X_1 + \frac{1}{2}X_1 + \left(\frac{1}{4}X_1 + \frac{1}{8}X_2 + W\right) + \frac{1}{10}X_1 + X_5$$

$$+ \frac{1}{10}X_1 + X_7 + X_8 + X_9 + \frac{1}{4}X_1 \tag{D.2}$$

Combining and simplifying this expression yields the following:

$$Cost_{WBS} = \frac{181}{80}X_1 + W + X_5 + X_7 + X_8 + X_9 \tag{D.3}$$

$$E(Cost_{WBS}) = \frac{181}{80}E(X_1) + E(W) + E(X_5)$$

$$+ E(X_7) + E(X_8) + E(X_9) \tag{D.4}$$

TABLE D.1

An Electronic System Work Breakdown Structure

Cost Element Name	Cost Element Cost X_i ($M)	Distribution of X_i or the Applicable Functional Relationship
Prime Mission Product (PMP)	X_1	$N(12.5, 6.6)$
System Engineering and Program Management (SEPM)	X_2	$X_2 = \frac{1}{2}X_1$
System Test Evaluation (STE)	X_3	$X_3 = \frac{1}{4}X_1 + \frac{1}{8}X_2 + W,$ where $W \sim Unif(0.6, 1.0)$
Data and Technical Orders	X_4	$X_4 = \frac{1}{10}X_1$
Site Survey and Activation	X_5	$Trng(5.1, 6.6, 12.1)$
Initial Spares	X_6	$X_6 = \frac{1}{10}X_1$
System Warranty	X_7	$Unif(0.9, 1.3)$
Early Prototype Phase	X_8	$Trng(1.0, 1.5, 2.4)$
Operations Support	X_9	$Trng(0.9, 1.2, 1.6)$
System Training	X_{10}	$X_{10} = \frac{1}{4}X_1$

$$Var(Cost_{WBS}) = \left(\frac{181}{80}\right)^2 Var(X_1) + Var(W) + Var(X_5)$$
$$+ Var(X_7) + Var(X_8) + Var(X_9) \tag{D.5}$$

The variance of $Cost_{WBS}$ given by Equation D.5 captures all Pearson product-moment correlations present in the WBS from the *cost estimating functional relationships* defined in Table D.1. Equation D.5 also reflects that $X_1, W, X_5, X_7, X_8,$ and X_9 were given to be independent random variables. To compute the mean and variance of $Cost_{WBS}$ given by Equations D.4 and D.5, respectively, the means and variances of $X_1, W, X_5, X_7, X_8,$ and X_9 are needed. Table D.2 presents these statistics. They are determined by standard formulas available in Chapter 4.

TABLE D.2

Cost Statistics for $X_1, W, X_5, X_7, X_8,$ and X_9

Cost Element Cost X_i($M)	$E(X_i)$($M)	$Var(X_i)$($M)^2
X_1	12.500	6.6
W	0.800	0.16/12
X_5	7.933	40.75/18
X_7	1.100	0.16/12
X_8	1.633	1.51/18
X_9	1.233	0.37/18

Substituting the values in Table D.2 into Equations D.4 and D.5 produces the mean and variance of the electronic system's total WBS cost, given by Equations D.6 and D.7, respectively.

$$E(Cost_{WBS}) = 40.98 \ (\$M) \tag{D.6}$$

$$Var(Cost_{WBS}) = 36.18 \ (\$M)^2 \tag{D.7}$$

$$\sigma_{Cost_{WBS}} = \sqrt{Var(Cost_{WBS})} = 6.015 \ (\$M) \tag{D.8}$$

b. To approximate the distribution function of $Cost_{WBS}$, observe the following. First, the random variables $X_1, W, X_5, X_7, X_8,$ and X_9 are independent. Hence, the central limit theorem will draw the shape of the distribution of $Cost_{WBS}$ toward a normal distribution. Second, the random variables $X_2, X_3, X_4, X_6,$ and X_{10} are highly correlated to X_1, which is normally distributed. With this, it can be shown that $\rho_{X_v,X_1} = 1 \ (v = 2, 4, 6, 10)$ and $\rho_{X_3,X_1} = 0.9898$. Thus, it is reasonable to conclude (for this example) the probability distribution function of $Cost_{WBS}$ is approximately normal—with mean and variance given by Equations D.6 and D.7, respectively. Figure D.1 presents a graph of the probability distribution function of $Cost_{WBS}$.

c. From Figure D.1, it can be seen that $Cost_{WBS} = 50.87 \ (\$M)$ has a 95% chance of not being exceeded. Thus, a cost equal to 50.87 (\$M) has only a 5% chance of being exceeded. Equivalently, 50.87 (\$M) is the 0.95-fractile of $Cost_{WBS}$; that is, $x_{0.95} = 50.87 \ (\$M)$. If the electronic system is considering budgeting at the 95th percentile, then a cost reserve of nearly 10 (\$M) is needed above its cost at the 50th percentile for this high level of cost estimate confidence.

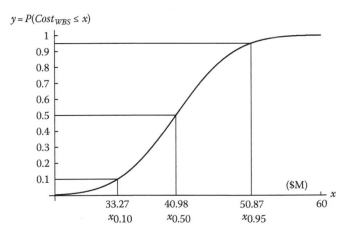

FIGURE D.1
Normal probability distribution for $Cost_{WBS}$.

Appendix E: Unraveling the S-Curve

One of the least addressed topics in cost risk analysis is conveying its findings to decision-makers. The cost community's focus has been on the front end of the process (methods and practices) rather than on the back end, specifically, what the analysis reveals and how best to convey its findings. This section discusses one aspect of this topic by addressing the question "I've generated an S-curve—what does it reveal about my program's cost risk and how should I present these findings to decision-makers?"

E.1 What is the S-Curve?

The S-curve is an informal term for the probability distribution of the cost of a program. Figure E.1 illustrates two ways to present a probability distribution. One way is the probability density function (PDF), as shown on the left of Figure E.1. The other way is the cumulative distribution function (CDF), as shown on the right of Figure E.1. The CDF is informally called the S-curve.

In Figure E.1, the range of possible cost outcomes for a program is given by the interval $a \leq x \leq b$. These distributions reveal the confidence that the actual cost of a program will not exceed any cost in the range of possible outcomes. In Figure E.1, if the probability that the actual cost of the program will be less than or equal to x is 25%, then in a PDF this probability is an area under the curve. In a CDF, this probability is the value 0.25 along the vertical axis, as shown on the right in Figure E.1.

The PDF is the most common form of a probability distribution used to characterize the cost uncertainties of elements that comprise a program's work breakdown structure (WBS). This is shown in Figure E.2 by the elements on the left, which is the input side of a cost uncertainty analysis. The right side of Figure E.2 shows the outputs of a cost uncertainty analysis, where the CDF, or S-curve, is the most common form used to express percentile levels of confidence that the actual cost of a program will be less than or equal to a value x.

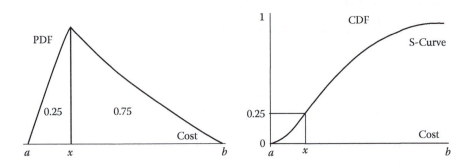

FIGURE E.1
Ways to view a program's cost probability distribution.

E.2 Unraveling the S-Curve

The S-curve provides decision-makers a basis for examining trade-offs between a program's point estimate cost* and its confidence level. For example, in Figure E.2, there is a 25% chance the actual program cost will be less than or equal to x_1 dollars, a 50% chance the actual program cost will be less than or equal to x_2 dollars, and an 80% chance the actual program cost will be less than or equal to x_3 dollars.

The variance of the S-curve, in Figure E.2, affects the amount of risk dollars needed to budget a program at a given confidence level, relative to (say) the program's point estimate cost. For example, in Figure E.2, if a program is budgeted to the 50th percentile cost x_2, then relative to x_1 there are h_1 risk dollars contained in x_2. If a program is budgeted to the 80th percentile cost x_3, then relative to x_1 there are $(h_1 + h_2)$ risk dollars contained in x_3. Thus, the risk dollars between x_1 and cost outcomes greater than x_1 is from the accumulation of cost risk from individual WBS elements that comprise a program's total cost. How can the S-curve be unraveled to reveal which WBS elements drive the amount of risk dollars needed for a given confidence? The following algorithm is used to address this question.

---

* In this book, the point estimate (PE) is taken to be the cost that does not include allowances for cost uncertainty. The PE cost is the sum of the WBS element costs summed across a program's WBS without adjustments for uncertainty. The PE cost is often developed from a program's cost analysis requirements description (CARD).

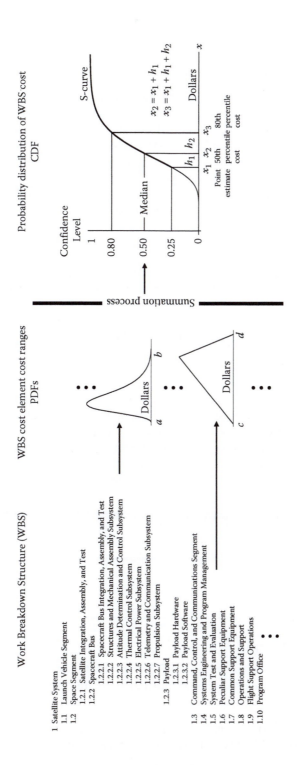

FIGURE E.2
Probability distributions in WBS cost risk analysis.

E.3 Book's Cost Risk Allocation Algorithm*

Book's algorithm is designed to allocate the risk dollars of a program into the individual WBS elements as a function of the cost risk of each element. The total risk dollars of a program is the difference between its point estimate cost and its cost at a confidence level greater than that of the point estimate, such as the dollars given by h_1 or $(h_1 + h_2)$ in Figure E.2. WBS elements allocated the largest fraction of risk dollars are the cost risk driving elements of the program and signal potential priorities for risk management actions. The following are the key terms and equations of Book's algorithm. A numerical example of the algorithm is then presented.

From Figure E.2, suppose a program is budgeted to the 80th percentile cost x_3. Then, relative to x_1 there are $(h_1 + h_2)$ risk dollars contained in x_3. To allocate $(h_1 + h_2)$ to the individual WBS elements, define the *need* of element k as the difference between its 80th percentile cost and its point estimate cost; that is,

$$Need_k = x_{k,0.80} - PE_k \tag{E.1}$$

where $x_{k,0.80}$ is the 80th percentile cost of WBS element k and PE_k is its point estimate cost.

Equation E.1 is set equal to zero if $x_{k,0.80} \leq PE_k$ plus any correlation effects due the impacts of the needs of other WBS elements with which element k is correlated. Equation E.1 is the above-average portion of σ_k measuring only the possible shortfall in dollars for WBS element k, if any of the identified risks associated with this element are realized. The fraction of risk dollars to be allocated to WBS element k, including correlation effects, is given by

$$Alloc_k = \frac{\text{Correlated Need of WBS Element } k}{\text{Total Need Base}} = \frac{\left(\sum_{j=1}^{n} \rho_{jk} Need_j Need_k\right)}{\sum_{k=1}^{n} \sum_{j=1}^{n} \rho_{jk} Need_j Need_k} \tag{E.2}$$

E.4 Numerical Example

Consider the WBS on the left in Figure E.3. Suppose the cost probability distribution of each WBS element is lognormal with the mean and standard

---

* This algorithm was created by S. A. Book, PhD. It is published in the 2008 *NASA Cost Estimating Handbook*, NASA Cost Analysis Division, http://www.nasa.gov/offices/ipce/CA.html.

WBS Element	Point Estimate ($M)	Mean ($M)	Sigma ($M)
1. Satellite	200	230	69
2. Launch	80	90	9
3. Ground	100	70	14
4. Data Distribution	300	350	140
Program Cost	680	740	

Monte Carlo Simulation Results (First 20 of 5000 Trials)					
Trial Number	1. Satellite	2. Launch	3. Ground	4. Data Distribution	Total ($M)
1	$ 205.07	$ 84.49	$ 59.97	$ 432.92	$ 782.44
2	$ 227.80	$ 86.78	$ 69.61	$ 482.22	$ 866.40
3	$ 159.15	$ 94.66	$ 64.89	$ 497.95	$ 816.66
4	$ 261.52	$ 76.97	$ 60.33	$ 151.53	$ 550.36
5	$ 302.11	$ 99.12	$ 56.54	$ 242.11	$ 699.88
6	$ 225.99	$ 101.40	$ 68.41	$ 368.03	$ 763.83
7	$ 335.86	$ 99.47	$ 82.75	$ 845.43	$ 1363.51
8	$ 279.25	$ 98.92	$ 70.53	$ 443.87	$ 892.57
9	$ 160.00	$ 91.61	$ 82.53	$ 392.05	$ 726.18
10	$ 250.66	$ 87.61	$ 74.07	$ 280.87	$ 693.21
11	$ 197.80	$ 82.91	$ 60.02	$ 123.35	$ 464.08
12	$ 203.37	$ 90.22	$ 68.88	$ 360.09	$ 722.55
13	$ 400.09	$ 91.35	$ 68.83	$ 560.93	$ 1121.21
14	$ 307.00	$ 100.34	$ 77.11	$ 254.32	$ 738.77
15	$ 310.87	$ 88.79	$ 67.98	$ 328.28	$ 795.91
16	$ 260.09	$ 91.97	$ 47.55	$ 224.12	$ 623.74
17	$ 149.99	$ 72.46	$ 82.96	$ 306.97	$ 612.38
18	$ 251.13	$ 78.63	$ 58.47	$ 215.72	$ 603.95
19	$ 259.10	$ 100.79	$ 96.57	$ 491.69	$ 948.16
20	$ 259.51	$ 90.62	$ 91.14	$ 396.04	$ 837.30

FIGURE E.3

A program WBS and Monte Carlo simulation (partial results).

deviation (sigma) given in Figure E.3. Suppose a Monte Carlo simulation was run on the WBS that generated the first 20 of 5000 trials on the right in Figure E.3.

The next step in the algorithm is to compute Pearson correlation coefficients between each pair of WBS element costs. This can be performed by (1) exporting the WBS element costs from all trials of the Monte Carlo simulation (the first 20 of 5000 trials are shown in Figure E.3) then (2) computing Pearson correlation coefficients for each column pair of WBS element costs from the exported values. The Excel function CORREL(*array1*, *array2*) can be used to perform this computation, where the columns of the WBS element costs shown in Figure E.3 are the entries for *array1* and *array2*. The results of this operation are provided in Figure E.4.*

Suppose the program represented by the WBS in Figure E.3 is required to be budgeted at the 80th percentile confidence level. Given this, Figure E.5 presents the costs of the program and its WBS elements at this confidence level. The Excel function PERCENTILE.INC(*array*, 0.80) can be used to compute these costs. For example, if *array* is set equal to the satellite column of cost data from the Monte Carlo simulation in Figure E.3, then this

Correlation Matrix	1. Satellite	2. Launch	3. Ground	4. Data Distribution
1. Satellite	1	0.227186	0.2526355	0.3208879
2. Launch	0.227186	1	0.20326533	0.19046209
3. Ground	0.2526355	0.20326533	1	0.29252485
4. Data Distribution	0.3208879	0.19046209	0.29252485	1

FIGURE E.4
Pearson correlation coefficients computed from the simulation data in Figure E.3.

Costs ($M)	Point Estimate	80th Percentile	*Need$_k$*
1. Satellite	200	283.56	83.56
2. Launch	80	97.15	17.15
3. Ground	100	81.21	0.00
4. Data Distribution	300	447.84	147.84
Program Cost	680	878.48	198.48

FIGURE E.5
Costs at 80th percentile.

---

* The number of significant digits in Figure E.4 is for traceability in this example; in practice, 3 or 4 significant digits is best.

Excel function produces 283.56 ($M) as the 80th percentile cost for the satellite WBS element. Equation E.1 is then applied to compute $Need_k$ as shown in Figure E.5.

From Figure E.5, the fraction of risk dollars to be allocated to WBS element k, including correlation effects, can now be computed from Equation E.2. For example, for the satellite WBS element ($k = 1$) we have

$$Alloc_1 = \frac{\text{Correlated Need of WBS Element 1}}{\text{Total Need Base}} = \frac{\left(\sum_{j=1}^{4} \rho_{j1} Need_j Need_1\right)}{\sum_{k=1}^{4} \sum_{j=1}^{4} \rho_{jk} Need_j Need_k}$$

where

$$\sum_{j=1}^{4} \rho_{j1} Need_j Need_1 = \rho_{11} Need_1 Need_1 + \rho_{21} Need_2 Need_1$$
$$+ \rho_{31} Need_3 Need_1 + \rho_{41} Need_4 Need_1$$
$$= (1)(83.56)(83.56) + (0.227186)(17.15)(83.56)$$
$$+ (0.2526355)(0)(83.56) + (0.3208879)(147.84)(83.56)$$
$$= 11272$$

and

$$\sum_{k=1}^{4} \sum_{j=1}^{4} \rho_{jk} Need_j Need_k = [83.56, 17.15, 0, 147.48]$$

$$\times \begin{pmatrix} 1 & 0.227186 & 0.2526355 & 0.3208879 \\ 0.227186 & 1 & 0.20326533 & 0.19046209 \\ 0.2526355 & 0.20326533 & 1 & 0.29252485 \\ 0.3208879 & 0.19046209 & 0.29252485 & 1 \end{pmatrix} \begin{bmatrix} 83.56 \\ 17.15 \\ 0 \\ 147.48 \end{bmatrix}$$

$$= 38,678$$

From this, it follows that

$$Alloc_1 = \frac{\left(\sum_{j=1}^{4} \rho_{j1} Need_j Need_1\right)}{\sum_{k=1}^{4} \sum_{j=1}^{4} \rho_{jk} Need_j Need_k} = \frac{11,272}{38,678} = 0.29$$

Therefore, the satellite WBS element requires 29% of the 198.48 ($M) risk dollars—given the cost of the program is to be budgeted at the 80th percentile confidence level. Similar computations made for the other WBS elements reveal the results shown in Figure E.6. From this, the following additional findings can be conveyed to the decision-maker:

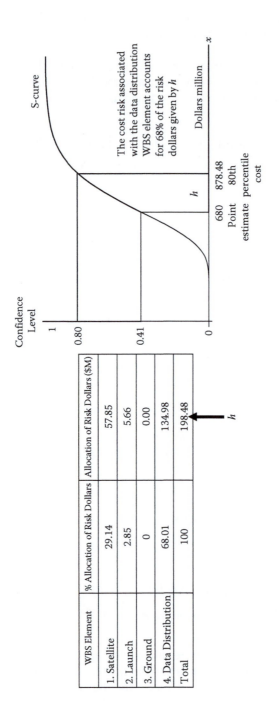

FIGURE E.6
Unraveling the S-curve: allocation of 80th percentile risk dollars.

- The Ground WBS element may be overestimated since it needs no additional risk dollars.
- The largest cost risk driver of the program is the data distribution WBS element. This element consumes 68% of the risk dollars needed for the program to be budgeted at the 80th percentile.

The Data Distribution WBS element is a prime candidate for further management attention to reduce its high cost risk to the program. Its high demand for risk dollars is a consequence of its large cost uncertainty (seen by its sigma value in Figure E.3) and its point estimate cost, relative to the costs of the other WBS elements. The risk dollars might be reduced simply by providing a more complete and less uncertain system definition combined with better cost estimating methods.

This appendix discussed a key aspect of communicating findings from a cost uncertainty analysis to decision-makers, that is, unraveling a program's S-curve to identify the elements of cost that drive the greatest amount of cost risk. In the preceding numerical example, it was shown that one WBS element more than others is driving the program's cost risk. Identifying cost risk drivers, in the way presented, fosters risk-reducing management actions to be taken as early as possible—such that program cost, schedule, and technical goals are more likely to be achieved.

Appendix F: Iteratively Reweighted Least Squares

This appendix presents a brief introduction to iteratively reweighted least squares (IRLS*). IRLS (Bickel and Doksom 1977) is a least squares procedure where the constants of a regression function are determined by iterating through a minimization of a weighted sum of additive squared errors. Though IRLS is sometimes characterized as a multiplicative error method, its iterative process operates from an additive error perspective (Book 2006, Jennrich and Moore 1975).

F.1 IRLS Formulation

The IRLS minimization problem is formulated as follows:

$$f(x_i, \vec{v}_{j+1}) = \text{Min} \sum_{i=1}^{n} \left[\frac{y_i - f(x_i, \vec{v})}{f(x_i, \vec{v}_j)} \right]^2 \tag{F.1}$$

where $f(x_i, \vec{v})$ is a specified regression function and $\vec{v}$ is a vector of the regression function's constants.

In Equation F.1, the subscript i is the number of observations in a dataset, j represents the jth iteration in a sequence of iterations through a vector of constants $\vec{v}_1, \vec{v}_2, \vec{v}_3, \ldots, \vec{v}_j, \vec{v}_{j+1}, \ldots, \vec{v}_k$. The IRLS procedure is stimulated by guessing a starter vector of regression constants denoted by $\vec{v}_0$. The vector $\vec{v}_k$ is defined as the vector of regression constants that have converged to a set of values that are no longer significantly changing with further successive iterations. The decision criterion for determining convergence between $\vec{v}_j$ and $\vec{v}_{j+1}$ is a tolerance level τ established by the model builder or its users.

In Equation F.1, only the vector $\vec{v}$ in the numerator is subject to optimization; the denominator $f(x_i, \vec{v}_j)$ is a constant with respect to the optimization

---

* The method of IRLS is also referred to as the minimum unbiased percentage error (MUPE) technique (Binkley 1994, Hu and Sjovold 1994, Nguyen 1994).

process, having been evaluated from the prior jth iteration. The operative element in Equation F.1 can be rewritten in the following way:

$$\sum_{i=1}^{n} \left[\frac{y_i - f(x_i, \vec{v})}{f(x_i, \vec{v}_j)} \right]^2 = \sum_{i=1}^{n} \frac{1}{f(x_i, \vec{v}_j)^2} [y_i - f(x_i, \vec{v})]^2 \qquad (\text{F.2})$$

Equation F.2 demonstrates the earlier comment that IRLS is really a minimization of a weighted sum of additive squared errors. Hence, IRLS is not truly a multiplicative error approach in its iteration toward a vector of convergent model parameters. This is an important consideration, as explained in Chapter 10, when deciding whether a dataset of observations exhibits additive or multiplicative errors.

Like the general error regression method (GERM), presented in Chapter 10, IRLS is a regression technique that allows flexibility in specifying a regression model's form. To illustrate this, consider the dataset of observations in Table 10.1 and its scatterplot in Figure 10.1. The following applies IRLS to find a set of converged values for a, b, and c such that the nonlinear function $f(x) = a + bx^c$ minimizes the weighted sum of additive squared errors.

Given $f(x) = a + bx^c$, minimizing the weighted sum of additive squared errors means finding regression constants a, b, and c such that

$$\left| f(x_i, \vec{v}_{j+1}) - f(x_i, \vec{v}_j) \right| \leq \tau \qquad (\text{F.3})$$

where $\vec{v}$ is the vector of regression constants a_j, b_j, and c_j determined from the jth iteration in a sequence of iterations through Equation F.1, beginning with a vector $\vec{v}$ of starter values a_0, b_0, and c_0. In Equation F.3, τ is the decision criterion (the tolerance level) to determine whether convergence between $\vec{v}_j$ and $\vec{v}_{j+1}$ is achieved, or equivalently when $f(x_i, \vec{v}_{j+1}) \approx f(x_i, \vec{v}_j)$ for all i at iteration $(j + 1)$. If convergence is achieved at the $(j + 1)$st iteration, then the vector $\vec{v}_K$ contains the regression constants a_{j+1}, b_{j+1}, and c_{j+1}. The parameters of the regression model $f(x) = a + bx^c$ are then given by $a = a_{j+1}$, $b = b_{j+1}$, and $c = c_{j+1}$.

The IRLS procedure is stimulated by a starter vector of regression constants denoted by $\vec{v}_0 = [a_0, b_0, c_0]$. Suppose $a_0 = 100$, $b_0 = 100$, and $c_0 = 0.50$. Table F.1 is the dataset of observations in Table 10.1 and the analysis setup to begin the IRLS procedure.

In Table F.1, the values in columns D and E are computed the same way at this stage of the IRLS procedure as shown in Table 10.6. The column titled "Estimated Cost, Regression Model" contains values from $f_0(x_i) = a_0 + b_0 x_i^{c_0}$. The sum of squared errors (*SSE*) is shown in the lower right corner of

TABLE F.1

Dataset and IRLS Starting Values and Setup

	A	B	C	D	E
1	Number of Program Data Points		$n = 7$	**Regression Model Form**	
2	Number of Regression Constants		$m = 3$	$f(x) = a + bx^c$	
3	Degrees of Freedom		$n - m = 4$		
4	Program Number	Number of Staff, x	Observed Cost, y	Estimated Cost, Regression Model	Error
5	1	7.9	1.380	381.069	0.9964
6	2	8.2	3.395	386.356	0.9912
7	3	9.8	7.201	413.050	0.9826
8	4	11.5	10.900	439.116	0.9752
9	5	16.4	15.434	504.969	0.9694
10	6	19.7	16.074	543.847	0.9704
11	7	23.6	17.274	585.798	0.9705
12	Starter Values			Sum of Squared Errors (SSE)	
13	$a_0 = 100$			$SSE = 6.715141$	
14	$b_0 = 100$				
15	$c_0 = 0.50$				

Table F.1. In IRLS, the *SSE* is the minimization objective* subject to varying regression constants a_j, b_j, and c_j for each iteration j. Table F.2 presents the results from the first three sequential IRLS iterations.

For Iteration 1, the values for a_1, b_1, and c_1 derive from minimizing the *SSE* in Table F.1 while varying a_0, b_0, and c_0. The column titled "Estimated Cost, Regression Model" contains values from $f_1(x_i) = a_1 + b_1 x_i^{c_1}$. The column titled "Error" contains values from the Equation F.4.

$$e_i = \frac{y_i - f_1(x_i)}{f_0(x_i)} = \frac{y_i - \left(a_1 + b_1 x_i^{c_1}\right)}{a_0 + b_0 x_i^{c_0}} \tag{F.4}$$

For Iteration 2, in Table F.2, the values for a_2, b_2, and c_2 derive from minimizing the *SSE* from Iteration 1 while varying a_1, b_1, and c_1. The column titled "Estimated Cost, Regression Model" contains values from $f_2(x_i) = a_2 + b_2 x_i^{c_2}$. The column titled "Error" contains values from the equation

$$e_i = \frac{y_i - f_2(x_i)}{f_1(x_i)} = \frac{y_i - \left(a_2 + b_2 x_i^{c_2}\right)}{a_1 + b_1 x_i^{c_1}} \tag{F.5}$$

---

* Throughout these computations, the minimization is accomplished by the Excel *Solver* add-in using the generalized reduced gradient algorithm (discussed in Chapter 10).

TABLE F.2

IRLS Iterations 1, 2, and 3

Iteration 1		Iteration 2		Iteration 3	
Estimated Cost, Regression Model	Error	Estimated Cost, Regression Model	Error	Estimated Cost, Regression Model	Error
2.26204	0.0023	1.57560	0.0865	1.49256	0.0714
3.16199	−0.0006	2.88508	−0.1613	2.89458	−0.1735
6.99286	−0.0005	7.88303	0.0975	8.11975	0.1165
9.84894	−0.0024	10.99891	0.0100	11.25036	0.0319
14.65376	−0.0015	15.05360	−0.0260	15.10337	−0.0220
16.49383	0.0008	16.20820	0.0081	16.12959	0.0034
17.97920	0.0012	16.97765	−0.0165	16.78502	−0.0288
Solver Values: Iteration 1		**Solver Values: Iteration 2**		**Solver Values: Iteration 3**	
$a_1 =$	24.77334	$a_2 =$	18.60165	$a_3 =$	18.00494
$b_1 =$	−216.26157	$b_2 =$	−1440.41987	$b_3 =$	−2263.23252
$c_1 =$	−1.09464	$c_2 =$	−2.14719	$c_3 =$	−2.38063
Minimize Sum of Squared Errors		**Minimize Sum of Squared Errors**		**Minimize Sum of Squared Errors**	
SSE	0.000016	SSE	0.044109	SSE	0.051111
Model Quality Measures		**Model Quality Measures**		**Model Quality Measures**	
SEE	0.002009	SEE	0.105011	SEE	0.113039
Bias	−0.000108	Bias	−0.000218	Bias	−0.000135
R^2	0.986754	R^2	0.995940	R^2	0.993763

For Iteration 3, in Table F.2 the values for a_3, b_3, and c_3 derive from minimizing the SSE from Iteration 2 while varying a_2, b_2, and c_2. The column titled "Estimated Cost, Regression Model" contains values from $f_3(x_i) = a_3 + b_3 x_i^{c_3}$. The column titled "Error" contains values from the equation

$$e_i = \frac{y_i - f_3(x_i)}{f_2(x_i)} = \frac{y_i - \left(a_3 + b_3 x_i^{c_3}\right)}{a_2 + b_2 x_i^{c_2}} \tag{F.6}$$

Table F.3 presents the results of nine successive iterations through the IRLS procedure for the dataset given in Table F.1. At Iteration 7, it follows that $\left| f(x_i, \vec{v}_7) - f(x_i, \vec{v}_6) \right| = 0$. Thus, the vector $\vec{v}_K$ contains the regression constants [17.94669, −2421.11675, −2.41455]. Therefore, the final regression model found by IRLS for this dataset is

$$f(x) = 17.95 - 2421.12x^{-2.415} \tag{F.7}$$

TABLE F.3

IRLS Iterations to Convergence

Iteration No.	1	2	3
$a =$	24.77334	18.60165	18.00494
$b =$	−216.26157	−1440.41987	−2263.23252
$c =$	−1.09464	−2.14719	−2.38063
Iteration No.	4	5	6
$a =$	17.94151	17.94671	17.94669
$b =$	−2419.38539	−2421.11440	−2421.11675
$c =$	−2.41440	−2.41455	−2.41455
Iteration No.	7	8	9
$a =$	17.94669	17.94669	17.94669
$b =$	−2421.11675	−2421.11675	−2421.11675
$c =$	−2.41455	−2.41455	−2.41455

Figure F.1 shows the increasing improvement in the fit of regression model as its constants are derived through the first six IRLS iterations. The open circles are values predicted by the model. The dark circles are the dataset of observations. Observe there is no significant change in the fit of the regression model given by Equation F.7 to the observed data from Iteration 3 and thereafter.

In summary, IRLS can operate on any regression model form with additive or multiplicative error. It provides an unbiased solution, but its iterative procedure requires a somewhat complex implementation. Using IRLS, regression constants of models of the form $y = bx^c \varepsilon$ can be algebraically derived (an analytic solution). However, this is not true for nonlinear "triad" models such as $y = a + bx^c \varepsilon$ or models of greater complexity such as $y = (a + bx^c d^x)\varepsilon$. In these cases, IRLS requires the use of optimization techniques, as shown herein, to search for a vector containing a converged set of values for these constants. As in the GERM approach (Chapter 10), the use of optimization in IRLS may also produce locally optimal rather than globally optimal results.

Exercise

F.1 Build an Excel spreadsheet model that performs 10 iterations of IRLS through the dataset of observations given in Table F.1. Use the spreadsheet with its *Solver* add-in feature to verify the results in Table F.3, with the starter values $a_0 = 100$, $b_0 = 100$, and $c_0 = 0.50$. *Note: Results may vary somewhat depending on the Solver algorithm available in Excel.*

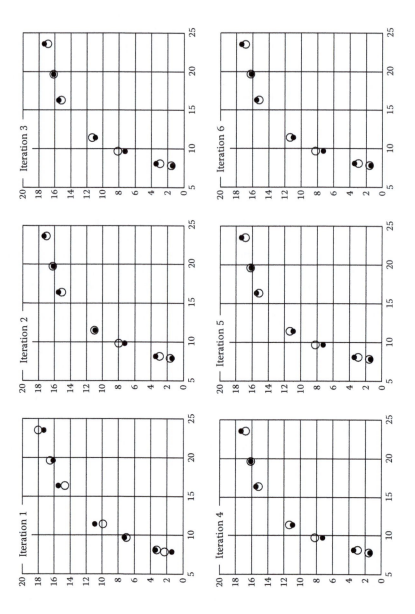

FIGURE F.1
IRLS regression for $f(x) = a + bx^c$ through Iteration 6.

References

Bickel, P. J. and K. A. Doksum. 1977. *Mathematical Statistics: Basic Ideas and Selected Topics*. San Francisco, CA: Holden-Day, Inc., pp. 132–141.

Binkley, J. 1994. *A Statistical Review of Tecolote's Technique for Determining Cost Estimating Relationships*. The Aerospace Corporation, 7pp.

Book, S. A. 2006. IRLS/MUPE CERs are not MPE-ZPB CERs. In *ISPA/SCEA Joint Conference and Training Workshop*, Seattle, WA.

Hu, S. P. and A. R. Sjovold. 1994. *Multiplicative Error Regression Techniques*. Tecolote Research, Inc., 31 pp. (Also available as Appendix C of Reference 41.)

Jennrich, R. I. and R. H. Moore. 1975. Maximum Likelihood Estimation by Means of Nonlinear Least Squares. In *Proceedings Statistical Computing Section*. American Statistical Association, pp. 57–65.

Nguyen, P. N. 1994, August. *Unmanned Space Vehicle Cost Model*, 7th edn. Los Angeles, CA: U.S. Air Force Space and Missile Systems Center (SMC/FMC), xvi + 452pp.

Additional Reading

SAS Institute Inc. (1990). *SAS/Statistical User's Guide*, Vol. 2.

Seber, G. A. and C. J. Wild. 1989. *Nonlinear Regression*. New York: John Wiley & Sons, pp. 37, 46, 86–88.

Appendix G: Sample Lot Cost and Quantity Data

This appendix presents the sample lot cost and quantity data (Table G.1) referred to in Chapter 11.

TABLE G.1

Sample Lot Cost and Quantity Data

Lot ID	Weight, lbs.	Lot Size	AUC	Cumulative Lot Size	Cumulative Program Cost
Lot A1	985	37	2937.1840	37	108,675.81
Lot A2	985	26	2250.4850	63	167,188.42
Lot A3	985	9	2271.1420	72	187,628.70
Lot A4	985	69	1745.2730	141	308,052.53
Lot A5	985	240	686.0282	381	472,699.29
Lot A6	985	180	686.4231	561	596,255.44
Lot A7	985	284	522.0096	845	744,506.18
Lot A8	985	450	448.3847	1295	946,279.28
Lot A9	985	432	410.8124	1727	1,123,750.23
Lot A10	985	430	398.7651	2157	1,295,219.23
Lot A11	985	300	419.5685	2457	1,421,089.79
Lot B1	985	15	1886.1080	15	28,291.62
Lot B2	985	30	2150.4310	45	92,804.55
Lot B3	985	60	1233.5180	105	166,815.65
Lot B4	985	132	1220.1440	237	327,874.69
Lot B5	985	108	943.0884	345	429,728.24
Lot B6	985	265	948.2006	610	681,001.39
Lot B7	985	265	788.8112	875	890,036.37
Lot B8	985	265	785.5494	1140	1,098,206.96
Lot B9	985	149	811.9428	1289	1,219,186.44
Lot B10	985	180	747.8859	1469	1,353,805.91
Lot B11	985	195	686.7638	1664	1,487,724.86
Lot B12	985	420	432.0905	2084	1,669,202.88
Lot C1	510	125	336.6665	125	42,083.31
Lot C2	510	390	355.0541	515	180,554.40

(Continued)

TABLE G.1 (*Continued*)

Sample Lot Cost and Quantity Data

Lot ID	Weight, lbs.	Lot Size	AUC	Cumulative Lot Size	Cumulative Program Cost
Lot C3	510	1490	227.8711	2005	520,082.36
Lot C4	510	1593	195.9469	3598	832,225.73
Lot C5	510	1560	174.4292	5158	1,104,335.34
Lot C6	510	2147	166.8957	7305	1,462,660.31
Lot C7	510	1679	164.7038	8984	1,739,197.93
Lot C8	510	2527	134.8004	11,511	2,079,838.66
Lot C9	510	947	174.4672	12,458	2,245,059.09
Lot D1	190	1200	46.1547	1200	55,385.67
Lot D2	190	2793	30.7965	3993	141,400.30
Lot D3	190	2603	28.2268	6596	214,874.77
Lot D4	190	1682	27.6394	8278	261,364.15
Lot D5	190	2542	27.3006	10,820	330,762.36
Lot D6	190	784	31.5043	11,604	355,461.69
Lot D7	190	1204	28.7777	12,808	390,110.04
Lot E1	190	65	336.8508	65	21,895.30
Lot E2	190	1857	55.2581	1922	124,509.61
Lot E3	190	1999	46.6230	3921	217,708.88
Lot E4	190	1535	49.3162	5456	293,409.29
Lot E5	190	2602	33.4887	8058	380,546.90
Lot E6	190	3224	28.2329	11,282	471,569.77
Lot E7	190	3461	26.8077	14,743	564,351.37
Lot E8	190	2060	26.4079	16,803	618,751.64
Lot E9	190	3667	22.8784	20,470	702,646.75
Lot E10	190	710	27.7968	21,180	722,382.50
Lot E11	190	788	31.0326	21,968	746,836.17
Lot F1	190	920	70.7552	920	65,094.78
Lot F2	190	900	42.3159	1820	103,179.09
Lot F3	190	1100	40.6714	2920	147,917.60
Lot F4	190	2487	27.1378	5407	215,409.26
Lot G1	190	1534	70.6727	1534	108,411.88
Lot G2	190	1020	61.7886	2554	171,436.20
Lot G3	190	2000	38.3292	4554	248,094.60
Lot G4	190	2245	29.6672	6799	314,697.48
Lot G5	190	2014	34.3181	8813	383,814.14
Lot H1	510	65	709.2363	65	46,100.36
Lot H2	510	29	746.0290	94	67,735.20
Lot H3	510	100	840.9082	194	151,826.02
Lot H4	510	225	425.0159	419	247,454.59

(Continued)

TABLE G.1 (*Continued*)

Sample Lot Cost and Quantity Data

Lot ID	Weight, lbs.	Lot Size	AUC	Cumulative Lot Size	Cumulative Program Cost
Lot H5	510	600	216.0518	1019	377,085.65
Lot H6	510	880	204.0192	1899	556,622.58
Lot H7	510	1110	161.1088	3009	735,453.30
Lot H8	510	1398	139.8911	4407	931,021.03
Lot H9	510	900	126.8681	5307	1,045,202.33
Lot H10	510	1144	111.0231	6451	1,172,212.72

Index